Geological Excursions in Powys Central Wales

Edited by

N. H. WOODCOCK *and* **M. G. BASSETT**

UNIVERSITY OF WALES PRESS
NATIONAL MUSEUM OF WALES

Published on behalf of the Geologists' Association, South Wales Group

CARDIFF, August 1993

ISBN 0-7083-1217-9

REFERENCES TO THIS VOLUME

It is recommended that reference to the whole or part of this volume should be made in one of the following forms, as appropriate:

WOODCOCK, N.H. and BASSETT, M.G. (eds), 1993. *Geological excursions in Powys, central Wales*, 366pp., University of Wales Press, National Museum of Wales, Cardiff.

LENG, M.J. and CAVE, R. 1993. The Machynlleth and Llanidloes areas. pp. 129–53. *In* WOODCOCK, N.H. and BASSETT, M.G. (eds), 1993. *Geological excursions in Powys, central Wales*. 366pp., University of Wales Press, National Museum of Wales, Cardiff.

BRITISH LIBRARY CATALOGUING IN PUBLICATION DATA

A catalogue record for this book is available from the British Library.

NOTE

This volume also represents No. 14 in the National Museum of Wales GEOLOGICAL SERIES.

Typset by Alden Multimedia, Northampton
Printed in England at the Bath Press

FOREWORD

THE modern county of Powys covers an area of some 7,000 square kilometres in central Wales, abutting along its eastern border onto the English counties of Shropshire and Hereford and Worcester. In its original form Powys was a kingdom established in early Medieval (or 'Dark Age') times, following the departure from Britain of the Romans. The derivation and meaning of the name are somewhat obscure, but it is believed by some to come via the Roman term *Pagenses*, which was used for the rural area covering much of what is now Shropshire and a large area of adjoining mid Wales (from the Latin *pagensis* < *pagus*, a district or province). One alternative is that Powys might be from the Welsh verb *gorffwys*, meaning to rest or repose (medieval Welsh *gorffywys* or *gorffowys*), in possible allusion to a specific resting place within the original region. After the thirteenth century the name fell into disuse, and in the sixteenth century the legal unification of Wales with England during the latter part of the reign of Henry VIII brought with it administrative changes that included shire government more or less on the English model.

The Local Government Act of April 1974 led to an amalgamation of the former shires of Brecon, Radnor and Montgomery into a resurrected and considerably enlarged Powys as an elongated administrative unit spanning well over half the length of Wales. It is mostly a thinly populated region, extending from Llanfyllin in the north and Machynlleth in the west as far south as Ystradgynlais in the Swansea Valley, on the margin of the South Wales Coalfield. This extended scope of the county, embracing a considerable area unconnected with medieval Powys, is extremely fortuitous in the context of this volume. The old Powys was set largely on Ordovician–Silurian successions dominated by graptolitic mudrocks and 'greywackes', mapped in broad outline in the nineteenth century but until fairly recently considered to be generally monotonous and unprepossessing; this was part of the deep-water heart of the 'Welsh Basin' or 'geosyncline'. In its new form the county incorporates an outstanding variety of bedrock geology ranging from the Precambrian to the Triassic, all of which apart from the latter (and the missing Permian) are featured here in excursion itineraries.

Notwithstanding the largely rural nature of the region, rock exposure is excellent in places, which coupled with the overall fascinating variety makes Powys ideal for teaching geology at both school and university levels, and for amateur study. There is also the added bonus that sites are generally less used than equivalents in north, south, and west Wales. The fact that the scenery is beautiful throughout the county gives yet further pleasure in teaching geology and working here.

Three particular and concurrent factors provided an appropriate impetus and timing for the production of this guidebook. An upsurge of academic research interest in the geology of Powys began in the 1980s, led largely by a vigorous Cambridge school under the direction of Nigel Woodcock, with some emphasis on Lower Palaeozoic sedimentological processes and tectonism; numerous other individuals and groups have also added to a fascinating and rapidly changing picture across a wide spectrum of topics such as Ordovician volcanism, low-grade metamorphism, Devonian floras, Dinantian calcretes, graptolites, and patterns of trilobite evolution. At the same time there was an initiation of a comprehensive mapping programme by officers of the British Geological Survey, which has also led to substantial re-appraisal of earlier views. Not least, there is now a refined stratigraphy and understanding of geological environments and processes as detailed as anywhere else in the UK. And thirdly, these new investigations took place at a time when the South Wales Group of the Geologists' Association was seeking to expand its publication programme, in which field guidance at all levels of interest and expertise has always been an important aspect. It was both natural and inevitable, therefore, that the Group would see an excursion guide to Powys as a companion to the guide to neighbouring Dyfed that had been published in 1982 and which has proved to be highly successful both educationally and financially (*Geological excursions in Dyfed, south-west Wales*, edited by M. G. Bassett, 1982, 327pp., published for the Group by the National Museum of Wales). The editors of this guide therefore combined forces to set the Powys project in motion, with one (N. H. W.) responsible mainly for soliciting contributions and the other (M. G. B.) agreeing to seek funding and a medium for publication. The willing co-operation of the British Geological Survey team was an essential element in ensuring an up-to-date and comprehensive coverage. The results have exceeded expectations, not least in the fact that much of what is published here is not simply a compendium of previously published research but is original and appearing in print for the first time. In that sense the volume forms a unique reference to our current state of knowledge of the geology of a large part of central Wales. Two of the excursions transgress into north-east Dyfed, reflecting recent work relevant to Powys geology and coverage that did not feature in the Dyfed guide.

Most of what has been published previously about Powys geology has appeared in 'academic' papers and memoirs; however, this is not the first attempt to give site-based guidance to a wider audience throughout the county. In 1978 J. H. Davies *et al.* published a *Geology of Powys in Outcrop*, which listed and described ninety-one sites of geological interest arranged on a geographical basis from north to south (fourth version published in 1984 by Meirionydd Field Press). Useful information on specific sites (including ownership and conservation problems) can be gained from that guide, but it differs from this volume in not attempting to provide itineraries that will give an understanding of successions and processes within a particular area. All the authors contributing to chapters published here were encouraged to bear in mind the educational value of their itineraries to wide-ranging audiences, and also the practicalities of gaining an appreciation of the geology of an area within a single day excursion.

Just as authors of guides such as this have a responsibility to direct field parties to localities where no physical or scientific damage can be done, so there is an equal responsibility on the users. Everyone involved in geological field excursions should be aware of the increasing pressure on sites; they share with the authors the responsibility to use the sites properly. Even though, as implied above, Powys is relatively under-used for field teaching at present, there is a full awareness that the publication of this guidebook will fairly certainly increase the pressures. With appropriate care and common sense there should be no misuse of the environment. In particular, **please obtain permission for access** wherever specified, in advance where parties are involved, and ensure that no damage is caused. Full codes of conduct are included in this guide (pages 9–15) and should be followed carefully. Any misuse of a site might lead to a ban on access for other geologists in the future, which would be both sad and harmful for the continuation of field teaching and research.

Acknowledgements. On behalf of the authors and both editors it is a pleasure to thank the joint publishers for their help and financial support in bringing this volume to fruition. Many people in their respective institutes have given assistance to authors in preparation of their contributions, but in particular we thank Ms V. K. Deisler (National Museum of Wales) for her considerable contribution in co-ordinating editorial and bibliographic matters, and Mrs D. G. Evans and Mrs L. Norton (also NMW) for redrawing all the diagrams from submitted drafts into a common style.

M. G. BASSETT
Keeper, Department of Geology, National Museum of Wales
Editor, Geologists' Association, South Wales Group

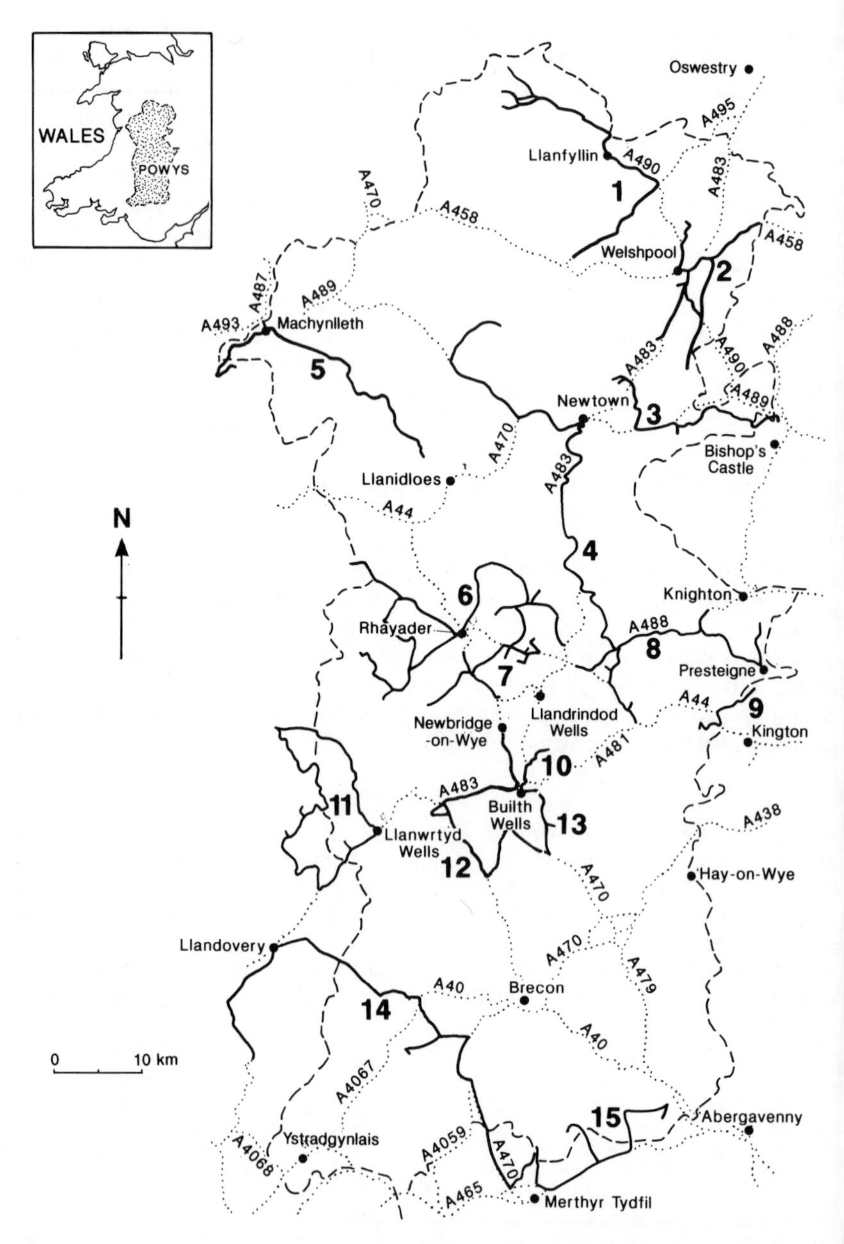

Location and coverage of excursion itineraries in this publication; thick solid lines show generalized routes, numbers 1–15 correspond with the appropropriate chapters.

CONTENTS

A CODE FOR GEOLOGICAL FIELDWORK

Editorial note. This code was first issued by the parent body of the Geologists' Association, with the support of leading geological organizations in the UK.

A GEOLOGICAL 'CODE OF CONDUCT' has become essential if opportunites for fieldwork in the future are to be preserved. The rapid increase in field studies in recent years has tended to concentrate attention upon a limited number of localities, so that sheer collecting pressure is destroying the scientific value of irreplaceable sites. At the same time the volume of fieldwork is causing concern to many site owners. Geologists must be seen to use the countryside with responsibility; **to achieve this, the following general points should be observed.**

1. Obey the Country Code, and observe local byelaws. Remember to shut gates and leave no litter.
2. Always seek prior permission before entering private land.
3. Do not interfere with machinery.
4. Do not litter fields or roads with rock fragments which might cause injury to livestock, or be a hazard to pedestrians or vehicles.
5. Avoid undue disturbance to wildlife. Plants and animals may inadvertently be displaced or destroyed by careless actions.
6. On coastal sections consult the local Coastguard Service whenever possible, to learn of local hazards such as unstable cliffs, or tides which might jeopardise excursions possible at other times.
7. When working in mountainous or remote areas, follow the advice given in the pamphlet 'Mountain Safety', issued by the Central Council for Physical Education, and, in particular, inform someone of your intended route.
8. When exploring underground, be sure you **have the proper equipment**, and the necessary experience. **Never go alone**. Report to someone your departure, location, estimated time underground, and your actual return.
9. Do not take risks on insecure cliffs or rock faces. Take care not to dislodge rocks, since other people may be below.
10. Be considerate. By your actions in collecting, do not render an exposure untidy or dangerous for those who follow you.

Collecting and Field Parties
1. Students should be encouraged to observe and record but not to hammer indiscriminately.
2. **Keep collecting to a minimum**. Avoid removing *in situ* fossils, rocks or minerals unless they are genuinely needed for serious study.
3. For teaching, the use of replicas is recommended. The collecting of actual specimens should be restricted to those localities where there is a plentiful supply, or to scree, fallen blocks, and waste tips.

4. Never collect from walls or buildings. Take care not to undermine fences, walls, bridges or other structures.
5. The leader of a field party is asked to ensure that the spirit of this Code is fulfilled, and to remind his party of the need for care and consideration at all times. He should remember that his supervisory role is of prime importance. He must be supported by adequate assistance in the field. This is particularly important on coastal sections, or over difficult terrain, where there might be a tendency for parties to become dispersed.

Health and Safety at Work Act
Since the introduction of this Act, safety measures are more strictly enforced on sites, including quarries. Protective clothing, particularly safety helmets, must be worn by employees, so visitors are expected to observe the same precaution, often as a condition of entry. Suitable helmets are readily available, cheap to purchase, and should be part of the necessary equipment of all geologists. **THEY MUST BE WORN AT ALL TIMES IN QUARRIES**.

Visiting Quarries
1. An individual, or the leader of a party, should have obtained prior permission to visit.
2. The leader of a party should have made himself familiar with the current state of the quarry. He should have consulted with the manager as to where visitors may go, and what local hazards should be avoided.
3. On each visit, both arrival and departure must be reported.
4. In the quarry, the wearing of safety hats and stout boots is recommended.
5. Keep clear of vehicles and machinery.
6. Be sure that blast warning procedures are understood.
7. Beware of rock falls. Quarry faces may be highly dangerous and liable to collapse without warning.
8. Beware of sludge lagoons.

Research Workers
1. No research worker has the special right to 'dig out' any site.
2. Excavations should be back-filled where necessary to avoid hazard to men and animals and to protect vulnerable outcrops from casual collecting.
3. Do not disfigure rock surfaces with numbers or symbols in brightly coloured paint.
4. Ensure that your research material and notebooks eventually become available for others by depositing them with an appropriate institution.
5. Take care that publication of details does not lead to the destruction of vulnerable exposures. In these cases, do not give the precise location of such sites, unless this is essential to scientific argument. The details of such localities could be deposited in a national data centre for geology.

Societies, Schools and Universities

1. Foster an interest in geological sites and their wise conservation. Remember that much may be done by collective effort to help clean up overgrown sites (with permission of the owner, and in consultation with the Countryside Council for Wales).

2. Create working groups for those amateurs who wish to do fieldwork and collect, providing leadership to direct their studies.

3. Make contact with your local County Naturalists' Trust, Field Studies Centre, or Natural History Society, to ensure that there is co-ordination in attempts to conserve geological sites and retain access to them.

Notes for Landowners

Landowners may wish to ensure that visiting geologists are familiar with this Code. In the event of its abuse, they may choose to take the name and address of the offenders and the Institution or Society to which they belong.

Enquiries may be addressed to The Geologists' Association, Burlington House, Piccadilly, London WIV 9AG.

ADVICE ON SAFETY AND BEHAVIOUR FOR INDIVIDUALS AND PARTIES CARRYING OUT GEOLOGICAL FIELDWORK

Editorial note. This set of guidelines on safety and behaviour is based on a document directed at students and issued by the Committee of Heads of University Geology Departments in the UK. The advice contained here is based therefore on many years experience of directing student parties and as such is applicable to all persons carrying out fieldwork.

SAFETY

A. To: *All persons attending geological field courses*

Geological fieldwork is an activity involving some inherent special risks and hazards, e.g. in coast exposures, quarries, mines, river sections, and mountains. Severe or dangerous weather conditions may also be encountered at any season especially on mountains or the coast.

In accordance with the Health and Safety at Work Act, field leaders are advised that they should follow certain safety precautions and should take every reasonable care concerning the safety of members of their parties. However, the potential dangers make it imperative that everyone should co-operate by behaving responsibly in order to reduce the risk of accidents. Each individual is responsible for his or her own safety.

You are specifically asked to:

1. Observe all safety instructions given by party leaders or supervisors. Anyone not conforming to the standards required may be dismissed from the field course.

Stay with the party, except by clear arrangement with the leaders. Assemble where requested (e.g. outside a quarry) in order to receive specific instructions regarding likely hazards.

Observe instructions for reporting after completion of work.

Report any injury or illness.

2. Wear adequate clothing and footwear for the type of weather and terrain likely to be encountered. Shirt, loose-fitting trousers, warm sweater, brightly-coloured anorak with hood, are normally desirable in the UK. A woollen hat (in addition to the hood of the anorak) is useful in winter or on high ground. Cagoule and waterproof over-trousers are desirable for wet weather. Jeans are generally unsuitable because they do not give sufficient protection when wet and are subjected to a cold wind, but can be all right if waterproof over-trousers are also worn.

Walking boots with rubber mountaineering soles are normally essential. Sports shoes are unsuitable for mountains, quarries and rough country.

Wellingtons are generally best reserved for walking through shallow water, peat bogs and the like.

Leaders may be advised to refuse to allow ill-equipped persons on their field courses, since they have a responsibility to see that individuals observe the provisions regarding personal safety.

3. Wear a safety helmet (preferably with chin strap) when visiting old quarries, cliffs, scree slopes, etc., or wherever there is a risk from falling objects. It is obligatory to do so when visiting working quarries, mines and building sites.

4. Wear safety goggles (or safety glasses with plastic lenses) for protection against flying splinters when hammering rocks or chisels.

Do not use one geological hammer as a chisel and hammer it with another; use only a soft steel chisel.

Avoid hammering near another person, or looking towards another person hammering.

5. Take special care near the edges of cliffs and quarries, or any other steep or sheer faces, particularly in gusting winds.

Ensure that rocks above are safe before venturing below. Quarries with rock faces loosened by explosives are especially dangerous.

Avoid working under an unstable overhang.

Avoid loosening rocks on steep slopes.

Do not work directly above or below another person.

Never roll rocks down slopes or over cliffs for amusement.

Do not run down steep slopes.

Beware of landslides and mudflows occurring on clay cliffs and in clay-pits, or rockfalls from any cliffs.

6. Avoid touching any machinery or equipment in quarries, mines or building sites.

Never pick up explosives, or detonators from rock piles; if found, inform the management immediately.

Comply with safety rules, blast warning procedures, and any instructions given by officials.

Keep a sharp look-out for moving vehicles, etc.

Beware of sludge lagoons.

7. Do not climb cliffs, rock faces or crags, unless this has been approved as an essential part of the work.

Take great care when walking or climbing over slippery rocks below high water mark on rocky shores.

More accidents to geologists, including fatalities, occur along rocky shorelines than anywhere else.

8. Beware of traffic when examining road cuttings.

Avoid hammering, and do not leave rock debris on the roadway or verges.

Railway cuttings and motorways are not open to geologists, unless special permission has been obtained from the appropriate authorities.

9. Do not enter old mine workings or cave systems unless it has been approved as an essential part of the work. Only do so then by arrangement, with proper lighting and headgear, and never alone. Ensure that someone on the surface knows your location and expected time of return. Be sure to report after returning.

Leaders of parties should follow the general guidance contained in: *A Code for Geological Fieldwork*, issued by the Geologists' Association (see p.9); *Mountain Safety – basic precautions*, published by Climber & Rambler; *Guidelines for visits to quarries* – laid down by the British Quarrying and Slag Federation.

B. To: *All persons undertaking geological fieldwork alone, in pairs, or small groups*

All the provisions in Section A also apply to independent fieldwork. However, since the nature of the training involves an important element of self reliance and the ability to cope alone, students in this category are necessarily responsible for their own safety in the field, and the following further advice is offered:

1. Discuss likely safety problems or risks, and check equipment, with experienced geologists before departure or commencement of work.

2. Plan work carefully, bearing in mind experience and training, the nature of the terrain and the weather.

Be careful not to overestimate what can be achieved.

3. Learn the mountain safety and caving codes, and in particular the effects of exposure.

Rock-climbing, caving and underwater swimming may be useful in research activities, but are dangerous for the untrained or ill-equipped. They should only be undertaken with the prior approval of people experienced in the particular techniques.

4. Do not go into the field without leaving a note and preferably a map showing expected location and time of return.

Never carelessly break arrangements to report your return to local people. Camp near habitation if possible.

5. Check weather forecasts. Keep a constant look-out for changes.

Do not hesitate to turn back if the weather deteriorates.

6. Know what to do in an emergency (e.g. accident, illness, bad weather, darkness).

7. Carry at all times a small first-aid kit, some emergency food (chocolate, biscuits, mint cake, glucose tablets), a survival bag (or large plastic bag), a whistle, torch, map, compass and watch.

8. Avoid getting trapped by the tide on inter-tidal banks or below sea cliffs.

Obtain local information about tides and currents. Pay particular attention to tidal range. For sea cliff study, local advice can be obtained from HM Coastguards.

Always wear footwear when wading in rivers, lagoons or on the shore.

9. Know the international distress signal: *a*, six whistle blasts, torch flashes or waves of a light-coloured cloth; *b*, one minute pause; *c*, another three blasts (flashes, waves) at 20-second intervals.

10. Always try to obtain permission to enter private property, and follow the recognized procedure for visits to quarries, etc.

Be careful to report after completion of work.

11. Take special precautions when working off-shore; small boats should normally be used only with an experienced boatman or colleague. Always wear a life-jacket. Aqualung equipment should only be used by experienced divers.

12. Ensure that you are conversant with the particular safety and health requirements if you work in a new environment.

GENERAL BEHAVIOUR

All participants in geological field courses, or undertaking independent field-work, are expected to observe sensible standards of behaviour, to conduct themselves with consideration for others, particularly in hotels or other accommodation, and not to damage property in any way (e.g. by climbing over walls, leaving gates open, trampling crops).

Please do not disturb the environment more than is absolutely necessary.

Do not collect specimens unless required for serious study.

Do not hammer outcrops casually or indiscriminately.

Do not disturb living plants and animals.

Do not leave litter, including rock chippings.

Observe conservation requirements. Remember that public access is an acute problem in the countryside and especially in areas designated as National Parks.

THE GEOLOGY OF POWYS: AN INTRODUCTION

by N. H. WOODCOCK *and* M. G. BASSETT

THE county of Powys forms the heart of Wales, elongated north to south through the districts of Montgomery, Radnor and Brecknock (Fig. 1). Its dominant geological grain is north-east to south-west (Figs. 3, 4), across which Powys provides an informative oblique cross-section. The excursions in this guide each concentrate on an element of this cross-section and its local role in the geological history of Powys. This introductory chapter provides a wider geological perspective, on the county as a whole and on its context within southern Britain. A 'Geological Outline' is provided for readers preferring a short synopsis of sequence, outcrop and structure, followed by a guide to geology and scenery. More detailed sections follow on structure and depositional environments, before a concluding synthesis of the regional setting and geological history of Powys.

Powys has hosted at least one and a half centuries of focused geological research, reported in a large scientific literature. The more enduring contributions on each specific area are mentioned in connection with the relevant excursion and listed in the consolidated reference list at the end of the volume. This introduction to the geology makes no attempt to recite this literature, but only to quote recent review articles and discussions of controversial points. Reference is made to the relevant excursion [as a number in square brackets] for detailed information.

However, it would seem too rooted in the present not to acknowledge the pivotal role in Powys geology of two people, R. I. Murchison and O. T. Jones. Murchison's *Silurian System* (1839) provided the stratigraphical and faunal foundation for later work, particularly in south-east Powys and the Welsh Borderland. Jones (1909, 1912) deciphered the stratigraphy and structure of north-west Powys, using Charles Lapworth's graptolite biostratigraphical ideas. Jones (1938, 1956) also provided the first views of Lower Palaeozoic Wales as an integrated depositional basin.

GEOLOGICAL OUTLINE

The solid geology of Powys comprises Precambrian and Palaeozoic rocks, represented here on a time-stratigraphical column (Fig. 2) and as a

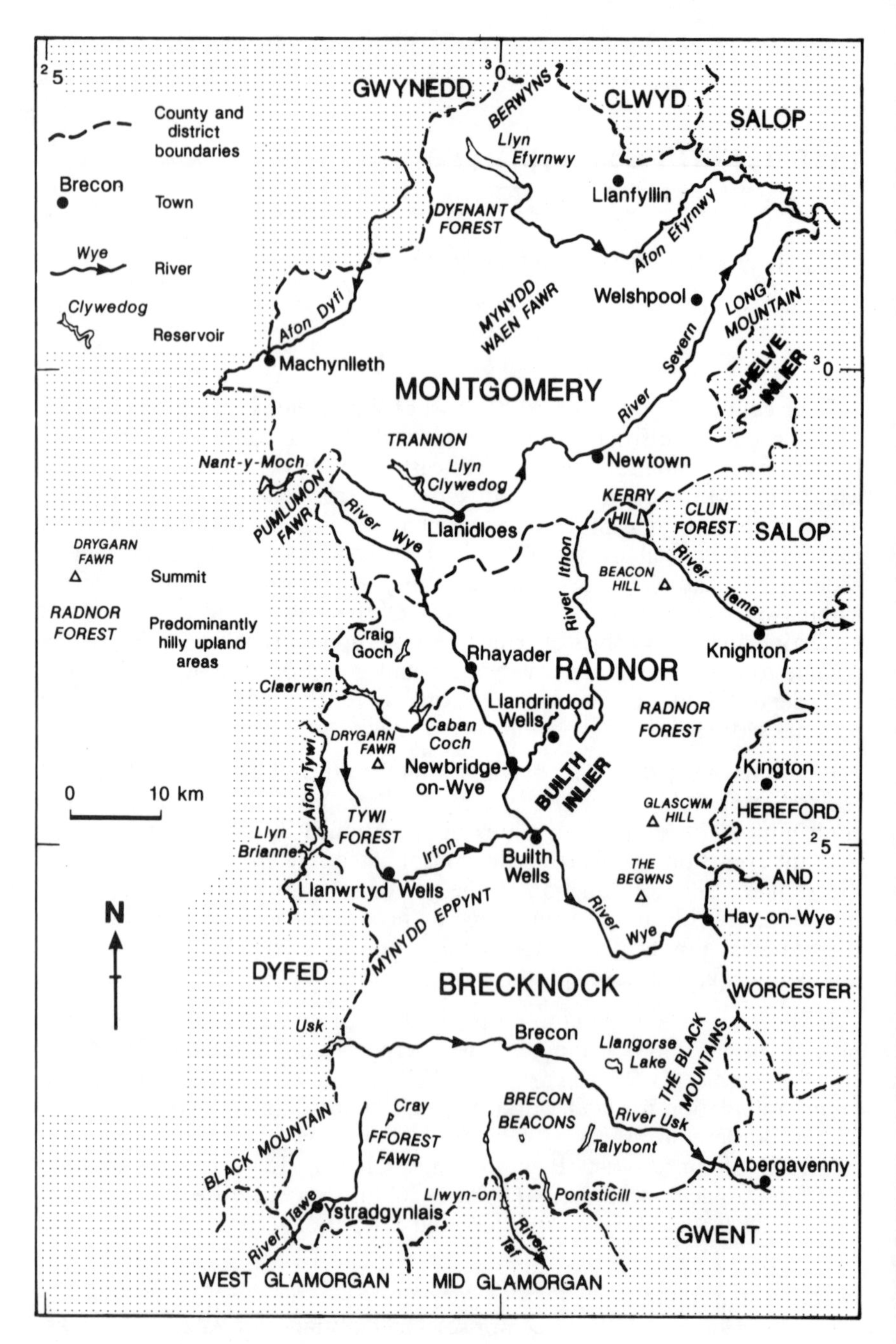

FIG. 1. Selected geographical features of Powys.

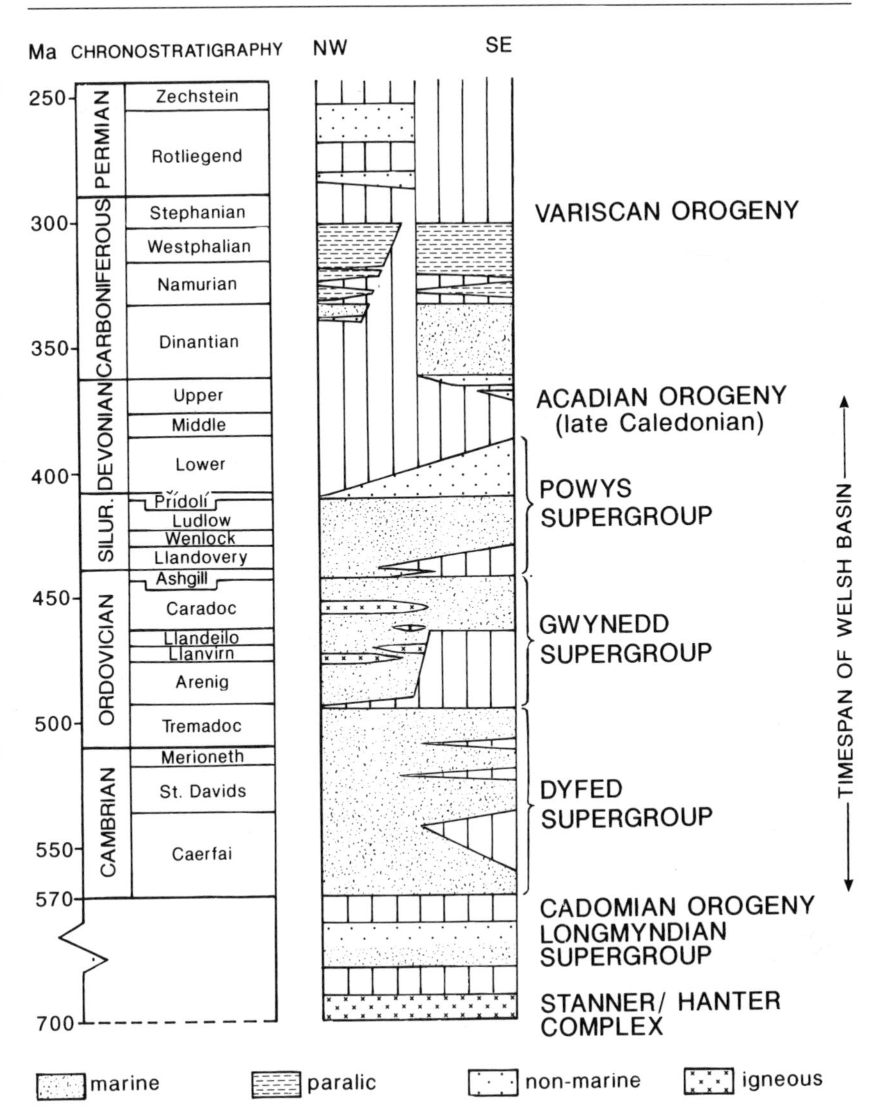

FIG. 2. Stratigraphical column of rocks exposed in Powys and immediately surrounding areas.

lithostratigraphical map (Fig. 3). At least six major sequences of rock can be distinguished (Fig. 2), separated by widespread unconformities. These unconformities record major events in the regional geological history. During three orogenies, the Cadomian, Acadian (late Caledonian) and

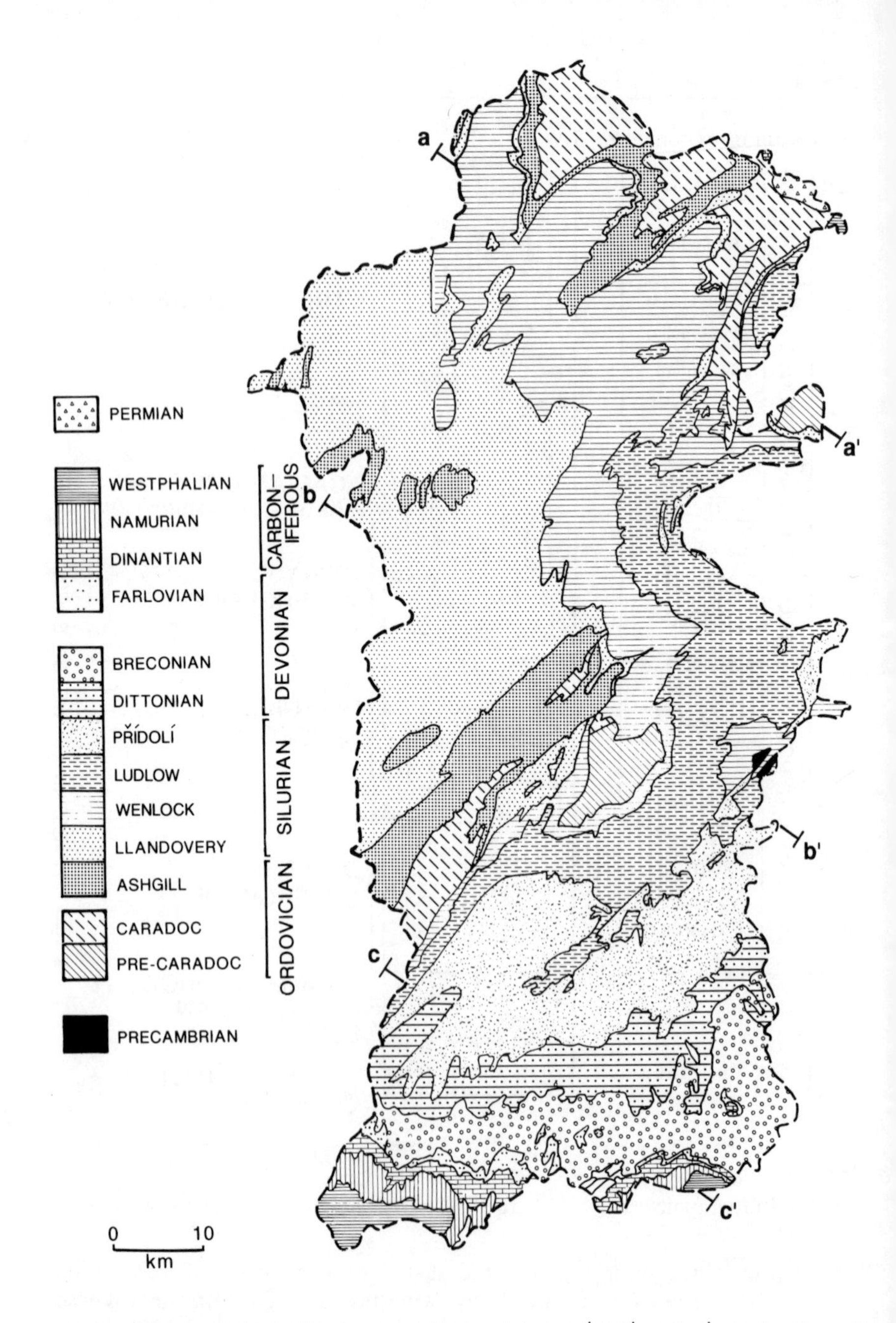

FIG. 3. Chronostratigraphical map of Powys. Lines aa′, bb′ and cc′ are the lines of sections in Figs. 5, 6 and 7.

Variscan, pre-existing rocks were deformed, uplifted and eroded, before deposition recommenced over them after some millions of years. The other two major unconformities record barely less dramatic events. That in early Ordovician time was the rifting away of southern Britain from the major Gondwana continent, as subduction and its related volcanism started up. That in the late Ordovician was the ending of that same prolonged phase of subduction and volcanism.

The distribution of these sequences in map view (Fig. 3) is controlled strongly by the pattern of faults and folds (Fig. 4), themselves mostly a result of the Acadian deformation. The most prominent structures are the three fold/fault belts forming the Welsh Borderland Fault System (Fig. 4; Woodcock & Gibbons 1988). During Cambrian, Ordovician and Silurian times this system marked the boundary between stable continental crust of the Midland Platform to the south-east and the unstable, thinned crust of the Welsh Basin to the west and north-west. Basin-down displacements on these faults allowed more rapid subsidence, leading to the presence of deeper-water facies and greater sediment thicknesses in the basin. The Welsh Borderland Fault System later controlled the extent and character of the Acadian deformation. The thinned crust of the basin was shortened again, developing tight folds, strong cleavage and a weak metamorphism (Fig. 4 inset). To the south-east, the platform was only weakly deformed, with flat-lying or gently folded sequences cut by localized zones of faulting.

Throughout this introduction 'north-west' and 'south-east' Powys refer respectively to the regions north-west and south-east of the Pontesford Lineament, in the centre of the Welsh Borderland Fault System. In general, older rocks crop out in north-west and younger rocks in south-east Powys. The notable exceptions to this trend are where old rocks are brought to the surface locally along faults or in anticlines. For instance, the oldest rocks in Powys are fault-bounded inliers of Precambrian sedimentary and intrusive rocks along the Church Stretton Fault (Fig. 3). These and nearby areas of Precambrian outcrop are small but nevertheless important as clues to the nature of the Precambrian basement that is assumed to underlie the whole of Powys.

The Dyfed Supergroup (Cambrian and Tremadoc) is unrepresented at the surface in Powys, but the next highest sequence, the Gwynedd Supergroup (Ordovician; Arenig to Caradoc) crops out in the north-west. The Shelve and Builth inliers have the fullest sequences of these marine sediments and volcanics. Most other outcrops are of Caradoc age only, as in anticlines along the north-west side of the Welsh Borderland Fault System, and on the south side of the Berwyn Dome.

The Powys Supergroup (Ashgill, Silurian and Lower Devonian) appropriately covers most of the county. In south-east Powys, on the old

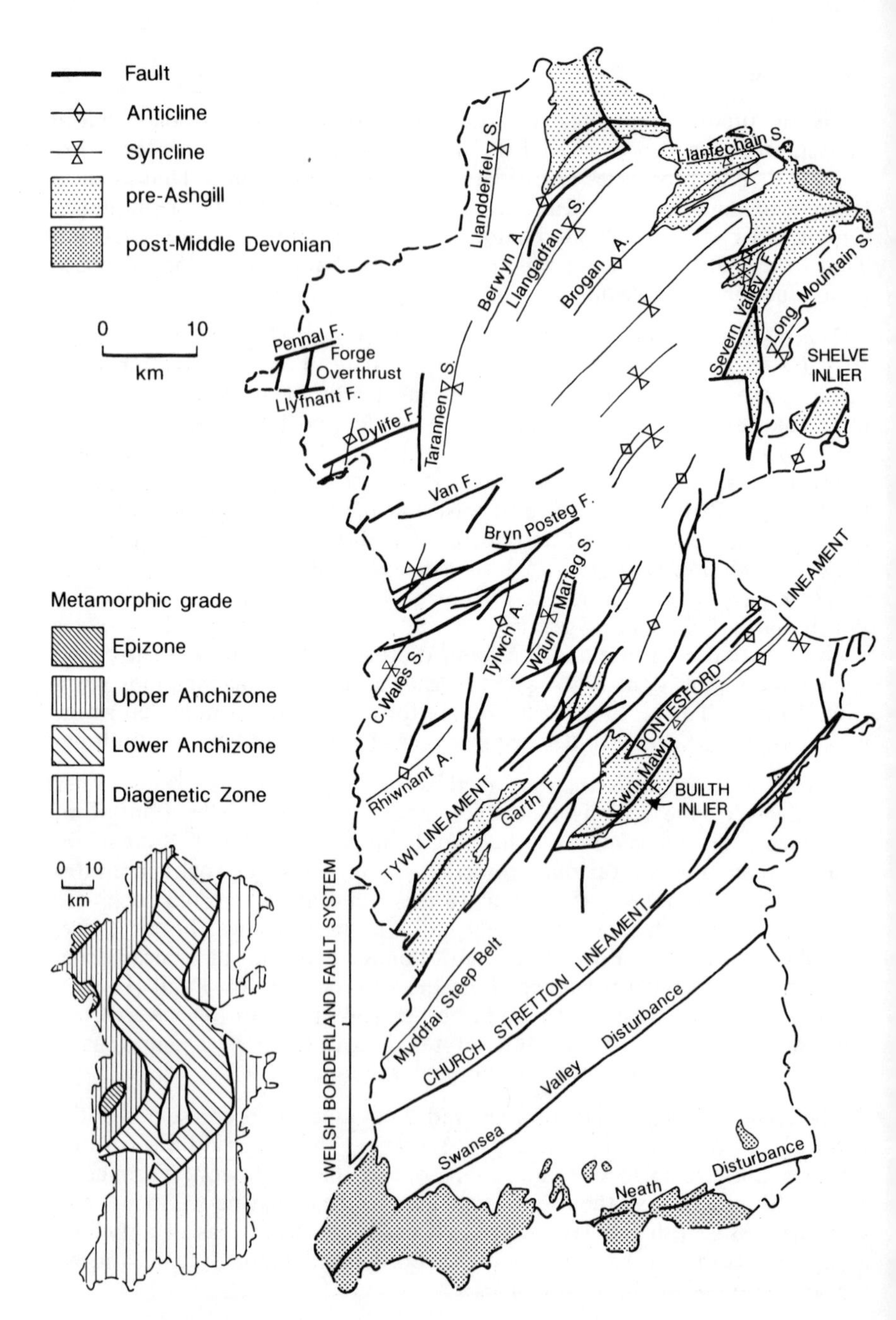

FIG. 4. Structural map of Powys with inset of metamorphic grade variation (from Robinson & Bevins 1986).

Midland Platform, Ashgill and Lower to Middle Llandovery rocks are absent and Upper Llandovery to Ludlow rocks are mostly of marine shelf facies. The Přídolí records a transition from marine to non-marine facies, which then persist through the Lower Devonian. In north-west Powys, in the Welsh Basin, a much thicker and more complete Ashgill to Wenlock sequence is of deeper marine facies, with the exception of shallow marine Ashgill and Llandovery rocks rimming the Berwyn Dome in the north.

Rocks deposited after the Acadian Orogeny are preserved in Powys in two distinct areas. On the southern fringe of the county, Upper Devonian non-marine clastic sediments are overlain by Dinantian marine limestones, and then by the coastal plain deltaic Namurian and Westphalian of the South Wales Coalfield. In north-east Powys the Acadian unconformity is overlain mostly by Permian non-marine sandstones. Outside the county, Westphalian rocks of the Shrewsbury coalfield are preserved below the Permian, to the south-east, and a fuller Dinantian to Westphalian sequence is developed to the north.

GEOLOGY AND SCENERY

The well-consolidated nature of most of its Precambrian and Palaeozoic rocks gives Powys a dominantly upland character, incised by the major river systems of the Severn, the Wye and the Usk (Fig. 1). The lowlands of the Cheshire Plain just encroach into the county along its north-eastern perimeter.

The character of the upland scenery changes across the Pontesford Lineament (Fig. 3), which marks the south-east limit of strong Acadian deformation and uplift. Rocks to the south-east were less deformed or uplifted, with their gentle dips resulting as much from tilting during the later Variscan event. As a result, south-east Powys is characterized by hills with gentle dip slopes and prominent scarps which persist for many kilometres. On the south-eastern rim of the county the main such hills, the Brecon Beacons and the Black Mountains, are formed by resistant units of the non-marine Přídolí to Lower Devonian sequence, which also gives this region its distinctive red soils. Yellow-grey marine Silurian rocks form structurally similar, if lower, flat-topped hills east of Builth Wells and in Radnor Forest, a topography interrupted only by the small, elongate fault-bounded hills along the Church Stretton Fault belt.

The folded marine Silurian rocks in north-west Powys give rise to a more complex topography. Their grossly uniform lithology allowed the development in Mesozoic time of a sub-horizontal peneplain, since uplifted to 500–600 metres. The subsequent dissection of this plateau by river and ice action has, in some parts of north-west Powys, resulted in the production of flat-topped hills superficially resembling those in the south-east. But more often a distinct linear grain to the topography is created by differential erosion. The more resistant rocks are usually thick units of deep marine sandstone. Resistant volcanic and shallow intrusive rocks locally give more rugged topography in the Llanwrtyd, Builth, Shelve and Breidden inliers and in the Berwyn Hills flanking the north of the county.

GEOLOGICAL STRUCTURE

The more recent reviews covering the structural geometry of the area are those by D. A. Bassett (1969), Dewey (1969), Coward & Siddans (1979), and Woodcock (1984*b*). The main component of deformation was Acadian (late Caledonian); that is of early to mid Devonian age. However, several phases of deformation can be demonstrated on some of the fault/fold belts in the south-eastern half of the county, in one or more of the intervals late Precambrian, Ashgill to mid Llandovery, and late Carboniferous.

Folds north-west of the Welsh Borderland Fault System generally trend north-easterly in the east but more northerly in the west. Major folds have wave-lengths in the range of 5–10 km. Those with northerly trends include the Llandderfel and Tarannon Synclines and the Machynlleth and Plynlimon Anticlines [5]. These have steep or upright axial planes and gentle axial plunges. The major NE-trending folds have NW-dipping axial planes and therefore a south-east vergence. In north Powys a train of such folds includes the Berwyn Anticline, the Llangadfan Syncline, the Brogan Anticline and the unnamed syncline–anticline pair to its south-east. Further south, the Central Wales Syncline trends into Powys from the south-west. It probably continues north of the Ystwyth Fault System between the Van inliers (British Geological Survey *in press*). It is flanked to the south-east by the Rhiwnant and Tylwch Anticlines and the Waun Marteg Syncline [6].

Smaller-scale folds usually parallel the major structures, but some have anomalous orientations. East-trending folds in north Powys, for example the Llanfechain Syncline, parallel trends in the north-east Berwyns. They may represent the inner segment of the continuous arcuation of Acadian folds around the Berwyns but their continuity with the NE-trending folds is in doubt. Analogous easterly folds persist down the south-east limb of the Llangadfan Syncline (Cummins 1963) and into the Talerddig area (D. A. Bassett 1955) where they might consistently predate the NE-trending folds (Dewey 1969).

Cleavage north-west of the Welsh Borderland Fault System is approximately axial planar to the major folds. It is generally steep in the upright N–S trending folds of north-west Powys and takes on a more pronounced north-westerly dip in the NE-trending folds in the south-east. Subtle departures from a strictly axial planar geometry are common and

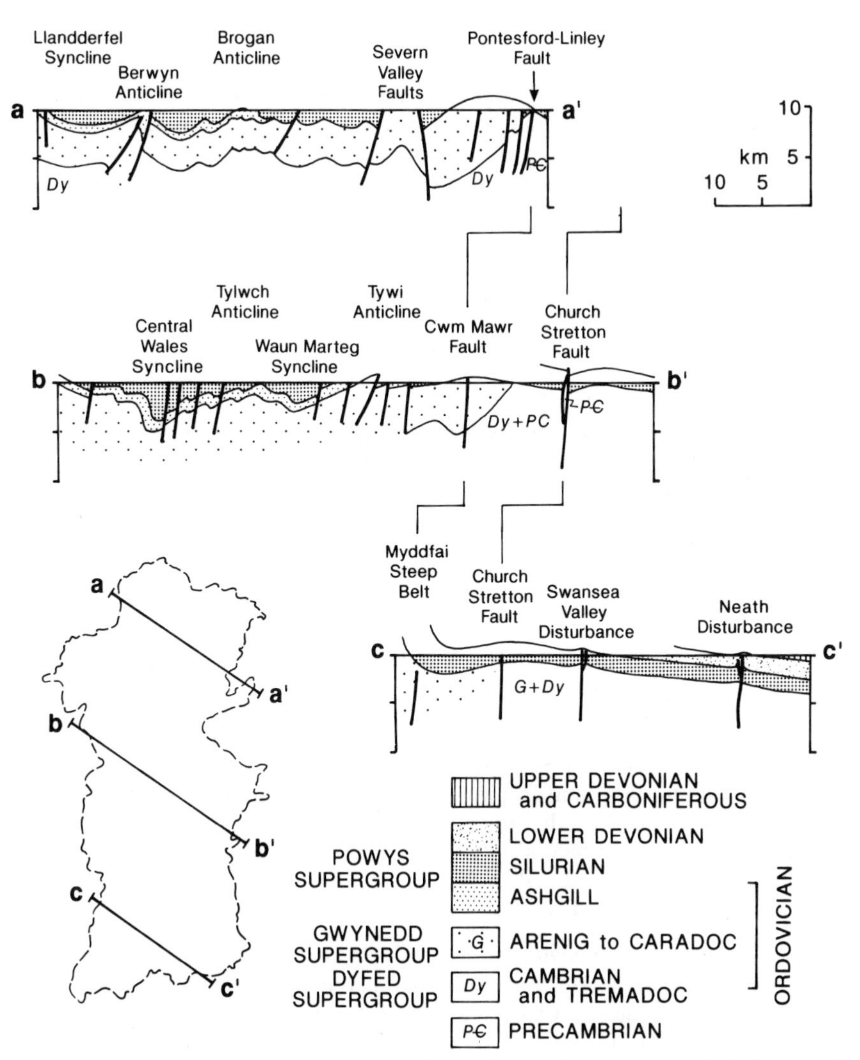

FIG. 5. Generalized structural cross-sections along NW–SE lines aa′, bb′ and cc′ marked on inset and on Fig. 3.

of two kinds. First, the fold profile usually shows refraction of cleavage between beds of differing ductility. Second, the cleavage often transects fold hinges in map view by angles of up to about 20°. Transection is widespread in mid Wales (Woodcock *et al.* 1988). In most examples [e.g. 5], cleavage is clockwise of fold hinges, a phenomenon attributed to a component of sinistral shear as well as shortening during the Acadian deformation.

Faults north-west of the Welsh Borderland Fault System comprise two main sets. One set is related intimately to the Acadian folds, usually taking up locally high strains on the steep limbs of major folds. Examples are the Forge and Brwyno Overthrusts in the Machynlleth Inlier and the fault on the steep limb between the Berwyn and Llangadfan folds. The second class of faults are those with an ENE rather than NE or NNE strike. The main examples in west Powys are the Pennal, Llyfnant, Dylife, Van and Bryn Posteg Faults. These displace the Acadian folds and cleavage, mostly with dip-slip displacements. They host post-Acadian mineralization, but the relative importance of Variscan and Acadian displacements is unclear.

The Tywi Lineament is a NE-trending complex of folds and faults forming the north-western element of the Welsh Borderland Fault System. Where it runs into south-western Powys from its type area in the Tywi Valley, it has a broadly anticlinal structure cored by pre-Ashgill rocks (Fig. 4). This anticline is cut by a number of important strike faults mostly dipping north-westward and with both reverse and normal net displacements [6, 7], (James 1983, 1990). These acted as basin-down faults during deposition of the Powys Supergroup and show a strong control on facies and thickness. Displacements on them were reversed during the Acadian shortening, and the faults influenced the growth of some of the smaller-scale folds that parallel the main Tywi Anticline.

In the Rhayader area the main anticline of the Tywi Lineament plunges to the north-east, causing the Silurian outcrop to close around its hinge [6, 7]. A zone of strong faulting and folding in mid-Silurian rocks seems to continue along the NNE trend of the lineament along the west side of Clun Forest [4] and into the Severn Valley Faults in the Welshpool area [2]. These structures can be considered as a continuation of the Tywi Lineament (Woodcock & Gibbons 1988).

The Pontesford Lineament is the central element of the Welsh Borderland Fault System. It runs south-eastward from the Shelve Inlier to the Builth Inlier, then continues probably along the Myddfai Steep Belt (Fig. 4; Woodcock 1984*a*). In the Ordovician inliers it is represented by steep faults cutting open folds. These faults have a dominant dextral strike-slip component (Lynas 1988; Woodcock 1987*b*) which predates the deposition of the Powys Supergroup. Such faults presumably underlie the intervening Silurian outcrop. Some of these faults were reactivated in the Acadian deformation, localizing upright folds and steep faults in the Silurian cover [8]. Components of strike-slip are suspected (Lynas 1988, Woodcock 1984*a*). Further to the south-west, the Myddfai Steep Belt is the SE-younging limb of a major asymmetric fold, probably formed by basin-up movements on a basement fault continuing the Pontesford line (Woodcock 1987*a*).

The Church Stretton Lineament is the south-eastern element of the Welsh Borderland Fault System. In its type area further to the north-east in Shropshire it is a zone of two or more steep faults, with some displacement as late as post-Triassic. Followed south-westwards it enters Powys near Presteigne (Fig. 4), south-west of which four or more steep strands can be mapped [9], interleaving slivers of Precambrian basement with the Silurian cover. There is a post-Wenlock, probably Acadian, strike-slip component to the fault movements, although with evidence of pre-Wenlock faulting and folding [9]. Further south-west the Church Stretton Lineament is marked by the faulted 'Brecon Anticlinal' of Ludlow through Přídolí rocks [13]. As it leaves Powys it cuts Devonian rocks, and forms the Llandyfaelog or Carreg Cennen Disturbance, a fault belt with a component of Variscan strike-slip displacement (Owen & Weaver 1983).

The Swansea Valley and Neath Disturbances are two narrow fault belts that traverse the south-east corner of Powys. They are sub-parallel to the faults of the Welsh Borderland System, with their important Acadian displacement. However, the faults of the Swansea Valley and Neath belts run up through the Acadian unconformity into Upper Devonian and Carboniferous rocks. The main displacements here are therefore of Variscan (late Carboniferous) age, though the faults probably nucleated on pre-Acadian basement weaknesses. The structure of the fault belts is best known where they cut the the Lower Carboniferous rocks in southern Powys (Owen & Weaver 1983) and comprise steep faults, some with strike-slip components, and upright folds in intervening fault blocks.

The metamorphic grade of Lower Palaeozoic rocks on the Midland Platform does not exceed diagenetic levels, probably indicating temperatures less than 200°C (Fig. 4 inset; Bevins & Robinson 1988). Grade increases rapidly north-westward across the basin margin, through anchizone levels into the epizone – probably produced by temperatures up to at least 300°C.

GEOLOGICAL SUCCESSION AND FACIES

Precambrian rocks of two distinct kinds occupy small fault-bounded inliers near Old Radnor [9]: igneous intrusives dated at about 700 million years old (Ma), and the probably younger sedimentary rocks of the Longmyndian Supergroup (Pauley 1990). More extensive outcrops of the Longmyndian occur to the north-east in Shropshire, where they are faulted against another late Precambrian unit, the Uriconian volcanics. There is also a small outcrop of probable Longmyndian at Pedwardine, bordering Powys to the east of Knighton. Precambrian basement is assumed to underlie the whole of Powys. All that can be deduced given the diversity of the available exposures is that this buried basement is probably of late Precambrian age.

The Dyfed Supergroup (Woodcock 1990) does not crop out in Powys, although it probably occurs at depth below some of the county. Shallow marine representatives occur to the east in Shropshire, and a mixed shallow to deep marine sequence to the north-west in Gwynedd. In Shropshire at least, Cambrian rocks overlie the Precambrian with a strong angular unconformity, representing the deformation and metamorphism of the Cadomian Orogeny (Fig. 2).

The Gwynedd Supergroup is present in Powys, but only its uppermost part is widely represented (Fig. 6). The fullest sequence comprises the marine sediments and volcanics in the Shelve Inlier, into which the county just encroaches. North-west from here, along the north-east fringe of Powys, Caradoc sedimentary and volcanic rocks crop out in the cores of a series of faulted anticlines [2]. In the Berwyn Hills [1] the Caradoc encloses small fault-bounded inliers of Llandeilo carbonates and volcanics. Further south, the Builth Inlier exposes a Llanvirn to Caradoc sequence with important volcanics in the Llanvirn [10]. Caradoc sedimentary rocks also core anticlines in the Tywi Lineament [7] (Fig. 4), and include volcanics near Llanwrtyd Wells [11]. It is probable that a complete sequence of the Gwynedd Supergroup underlies most of Powys north-west of the Pontesford Lineament. However, to the south-east of this line, analogy with outcrops in Shropshire (Fig. 6) suggest that the sequences may be absent or represented only by Caradoc sedimentary rocks. There is a conspicuous mismatch between Gwynedd Supergroup sequences across the major faults (Fig. 6) due to both lateral depositional contrasts and to later strike-slip displacements on the faults.

The igneous rocks of Powys have been reviewed by Allen (1982), Kokelaar *et al.* (1984) and Bevins *et al.* (1984). The predominant volcanic lithologies are acid to intermediate bedded tuffs, commonly reworked by marine processes. More proximal lithologies of agglomerates and volcaniclastic conglomerates suggest volcanic centres in the Breidden [2], Shelve and Builth [10] areas as well as to the east and west of the Berwyn Hills [1]. Abundant basic lavas occur in the Builth area. Basic to intermediate sills occur throughout the Gwynedd Supergroup, and are most probably of Caradoc age. Petrology and geochemistry suggest that the igneous activity of the sequence in Powys represents crustal extension in a marginal basin floored by immature continental crust, behind a subduction-related volcanic arc situated further to the north-west (Kokelaar *et al.* 1984).

Sedimentary rocks of the Gwynedd Supergroup are of two types. Predominant are dark fossiliferous mudstones, their common bioturbation suggesting marine shelf conditions. Subordinate sandstones represent higher-energy shelf conditions. They are often associated stratigraphically and spatially with volcanic centres, which provided both a shallower marine or even subaerial environment and a source of coarse clastic sediment. In contrast to the Powys Supergroup, there was little sediment supplied to the Welsh Basin from beyond its boundaries.

The Powys Supergroup appropriately covers most of its eponymous county. Its base is marked by a major unconformity spanning much of

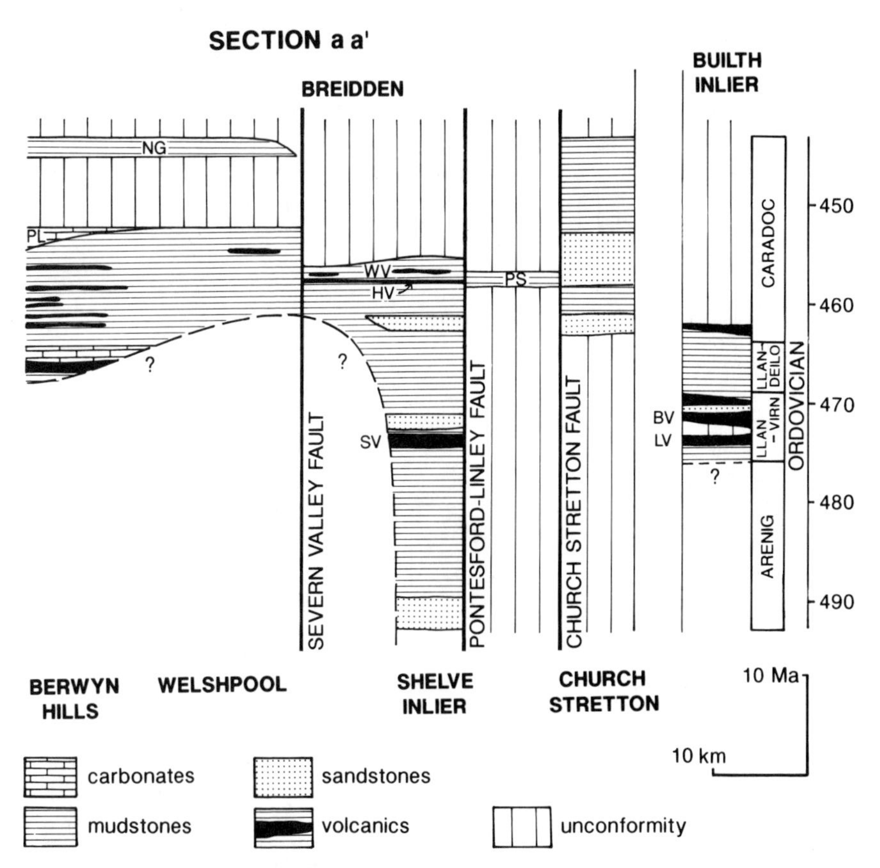

FIG. 6. Chronostratigraphical diagram of the Gwynedd Supergroup along the NW–SE line aa′ marked on Fig. 3, and from the Builth Inlier. Abbreviations of selected units are: BV Builth Volcanic Group; HV Hagley Volcanic Formation; LV Llandrindod Volcanic Group; NG Nod Glas; PL Pen-y-garnedd Limestone; PS Pontesford Shales; SV Stapeley and Hyssington Volcanic Formations; WV Whittery Volcanic Formation.

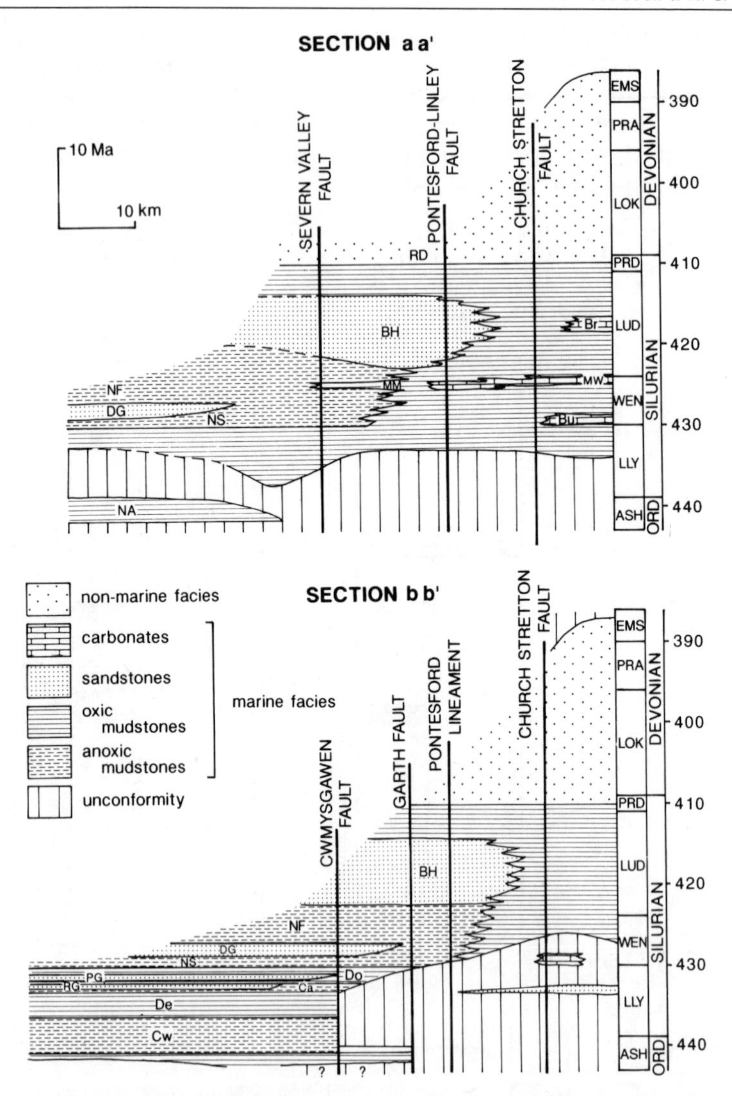

FIG. 7. Chronostratigraphical diagrams of the Powys Supergroup along the NW–SE lines aa′ and bb′ marked on Fig. 3. Abbreviations of selected units are: BH Bailey Hill Formation; Br Bringewood Group; Bu Buildwas Formation; Ca Caerau Mudstone Formation; CW Cwmere Formation; De Derwenlas Formation; DG Denbigh Grits Formation; Do Dolgau Mudstone Formation; MM Mottled Mudstone Member; MW Much Wenlock Limestone Formation; NA Nant Achlas Siltstones; NF Nant-glyn Flags Formation; NS Nant-ysgollen Shales Formation; PG Pysgotwr Grits Formation; RD Red Downtonian; RG Rhuddnant Grits Formation.

Ashgill to mid-Llandovery time (Fig. 7). This persists along the basin margin south-east of the Tywi Lineament (Cwmysgawen Fault on section bb′, Fig. 7) and throughout the area of the old volcanic centres in the north of the county (section aa′). The span of the unconformity reduces rapidly into the basin centre (section bb′). The Powys Supergroup is composed almost entirely of sedimentary rocks, but contains two major facies transitions. One is a lateral transition from shallow marine facies on the margins of the Welsh Basin to deep marine facies in its centre. The other is a vertical transition from marine to non-marine facies in the uppermost Silurian.

The Upper Llandovery to Ludlow rocks around the basin margin are mostly of shelf marine facies [3, 8, 9, 13]. Predominant are bioturbated calcareous mudstones and fine sandstones, in places with relict structures of storm-influenced deposition. These structures are better preserved on the outer shelf where the Bailey Hill Formation (Fig. 7) exemplifies the difficulty of distinguishing storm beds from turbidites [3, 8]. Coarser sandstones occur locally over shoal areas on the shelf [9] and carbonate patch reefs are preserved in the Wenlock and Ludlow of the inner shelf to the east (section aa′, Fig. 7). Holland & Lawson (1963), M. G. Bassett (1974), Cherns (1988) and Tyler & Woodcock (1987) review relevant aspects of sedimentation on the Midland Platform.

Along the basin slope, bioturbated mudstones are replaced at several horizons by laminated hemipelagic mudstones, deposited by vertical fallout of mud and dead micro-organisms [3, 4, 6, 8, 12]. The fine lamination is only preserved if bottom waters were too low in oxygen to support an infauna and epifauna. At times of basin oxygenation, the hemipelagic mud became bioturbated. These anoxic or oxic hemipelagites accumulated continuously in the basin, but were interbedded with the products of intermittent turbidity flows [4, 5, 6, 11]. Small-volume muddy flows formed beds at a scale of millimetres to centimetres, and often make up 50–90 per cent of the basinal mudstone units (Fig. 7). Larger volume muddy sand-flows formed beds typically on the scale of tens of centimetres, commonly stacked in packets hundreds of metres thick. These form the basinal sand units of Fig. 7, each unit corresponding to the deposits of one major turbidite system. The largest systems were of late Llandovery to Wenlock age and were supplied axially from the south-west end of the Welsh Basin. Earlier systems are small localized fans derived from the south-east basin margin. The processes and patterns of basinal sedimentation in the Powys Supergroup have been discussed recently by Cave (1979), Cave & Hains (1986), Dimberline & Woodcock (1987), Dimberline *et al.* (1990) and James (1987).

Přídolí rocks reflect a transition from marine to marginal marine then non-marine facies, the Old Red Sandstone, which then persists through the Lower

Devonian [8, 12, 14]. Non-marine sediments are preserved only in south-east Powys, but their former presence in the north-west is indicated by provenance studies on the Old Red Sandstone of Shropshire (Allen 1962) and by overburden calculations from metamorphic grade (Awan & Woodcock 1991). The facies change was apparently synchronous over both the former Welsh Basin and its margin, implying a rapid shallowing of the basin. This shallowing predated the climax of the Acadian deformation, and may have resulted from oversupply of sediment to the basin rather than to its active shortening and inversion. The ensuing non-marine environment was a muddy low-relief alluvial plain traversed by meandering river channels bearing sand and gravel [14]. Přídolí stratigraphy and sedimentation are reviewed by Bassett *et al.* (1982) and the non-marine sedimentation by Allen (1979, 1985).

Upper Devonian or Carboniferous rocks succeed earlier sequences across the unconformity representing the Acadian Orogeny. In north-east Powys, this unconformity is strongly angular, and spans most of Devonian and Dinantian (early Carboniferous) time. A Dinantian to Westphalian marine and deltaic sequence is preserved above the unconformity north of Powys, but within the county is mostly overlapped by Permian continental sandstones. Westphalian rocks emerge again from beneath the Permian in the Shrewsbury coalfield to the east.

By contrast, southern Powys was beyond the limit of strong Acadian deformation and therefore the unconformity here is weakly angular and spans only mid Devonian time (Fig. 2). Here Upper Devonian non-marine clastic sediments are overlain by Dinantian marine limestones [15], and then by the coastal plain/deltaic Namurian and Westphalian of the South Wales Coalfield. Minor unconformities occur within the sequence.

REGIONAL SETTING

The Lower Palaeozoic geology of Powys is primarily a contrast between the shallow or emergent Midland Platform in the south-east and the more rapidly subsiding trough of the Welsh Basin to the north-west. Both these areas were founded on the continental crust of Eastern Avalonia (Fig. 8). The Welsh Basin was bordered to the north-west on this microcontinent by the small Irish Sea Platform, and may have had marine connections with the Lake District and Leinster Basins to the north and west, and with the Anglian Basin to the east.

Eastern Avalonia was one of a group of microcontinents, including Western Avalonia, Iberia, Armorica and probably Baltica (Fig. 8), that were attached originally to the Gondwana continent. Eastern Avalonia detached itself from Gondwana in high southerly latitudes in early Ordovician time and moved northward towards the Laurentia continent

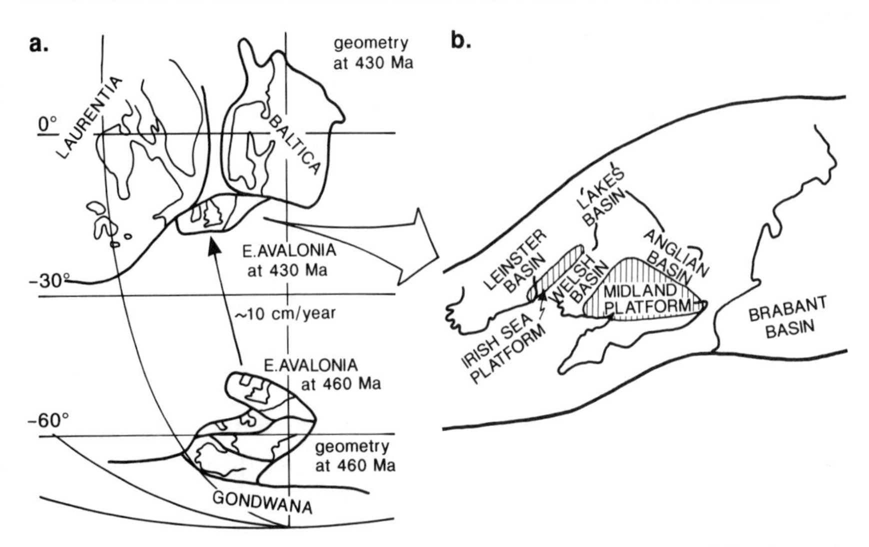

FIG. 8. a) The drift history of the Eastern Avalonian microcontinent (after Soper & Woodcock 1990); b) Position of Welsh Basin and related features on Eastern Avalonia.

lying in low latitudes (review by Trench *et al.* 1991). This northward displacement was related to the closure of the Iapetus Ocean to the north of Eastern Avalonia and the opening of the Rheic Ocean to the south. There is continuing debate about the history of impingement of Eastern Avalonia with other continents. It probably approached Baltica in late Ordovician time, with the obliteration of the intervening Tornquist Ocean, and first impinged on Laurentia in the early Silurian (reviews by Pickering *et al.* 1988; Soper & Woodcock 1990).

The Welsh Basin existed from Cambrian through Silurian time. It and the other basins on Avalonia were inverted and strongly shortened during early to mid Devonian time. This Acadian (late Caledonian) event may have resulted from the impingement of Armorica and the main Gondwana continent on the assembled Laurussia continent (Laurasia + Baltica + Avalonia). After this deformation and uplift the former basins were left as upland areas on Laurussia, bordered to the south by the narrow remnants of the Rheic Ocean. The late Devonian, Carboniferous and Permian sediments of Powys were deposited in newly subsiding basins around the Welsh uplands, formed as a result of active crustal stretching and, in south Wales, by downbending of the crust in advance of Variscan thrust sheets advancing from the south.

GEOLOGICAL HISTORY

The local geological history of Powys can now be outlined against the regional perspective described above. The vertical stratigraphical column (Fig. 2) provides a good template for the main events (Fig. 9). Each major unconformity records a major tectonic event and marks a change in the setting of the Welsh Basin. Palaeogeographical maps (Fig. 10) illustrate the geological history of the Ordovician to early Carboniferous interval best represented in Powys.

The Cadomian Orogeny occurred when Eastern Avalonia was still attached to the Gondwana continent (review by Taylor & Strachan 1990), and is recorded by the unconformity below the Dyfed Supergroup. This supergroup itself records the mainly non-volcanic sedimentation which later covered the eroded Cadomian mountain belt, still probably near a continental margin facing the ancestral Iapetus Ocean to the north.

Near the end of Tremadoc (early Ordovician) time, the Rhobell Volcanic Group to the north-west of Powys accumulated, marking the onset of arc volcanism (Kokelaar 1988), which is presumed to record renewed subduction beneath the Gondwana margin. The subsequent development in the Arenig of small fault-bounded basins (Traynor 1988, 1990), and transition to marginal basin volcanism (Bevins *et al.* 1984), suggests rifting of the Gondwana crust, and developing faunal contrasts (e.g. Vannier *et al.* 1989) point to separation of Eastern Avalonia from Gondwana at this time.

The northward movement of Eastern Avalonia towards Laurentia was promoted by continuing subduction of Iapetus Ocean crust under its northern margin. This is recorded in Powys by the volcanically dominated Gwynedd Supergroup. The Welsh Basin was a marginal basin behind the subduction arc. Volcanic centres developed near the basin margin at Llanwrtyd, Builth, Shelve, the Breidden Hills, and east of the Berwyns (Fig. 10a). More major centres were located north-west of Powys, most notably in Snowdonia, and further centres could be buried beneath the cover of Powys Supergroup in central Wales. Sediment was derived mainly from the volcanic centres themselves during this period; the Eastern Avalonian continent, isolated between Gondwana and Laurentia, was too small to provide much non-volcanic terrigenous sediment at this time.

During Caradoc (late Ordovician) time, volcanism in Wales waned rapidly, suggesting that subduction beneath it was coming to an end. This volcanic shutdown could record a continental collision, for instance with Baltica, but more likely it records the impingement of the Iapetus spreading ridge with Eastern Avalonia to give a California-type margin (Woodcock 1990). The Californian analogy is enhanced by the strike-slip tectonics that ensued in Ashgill time.

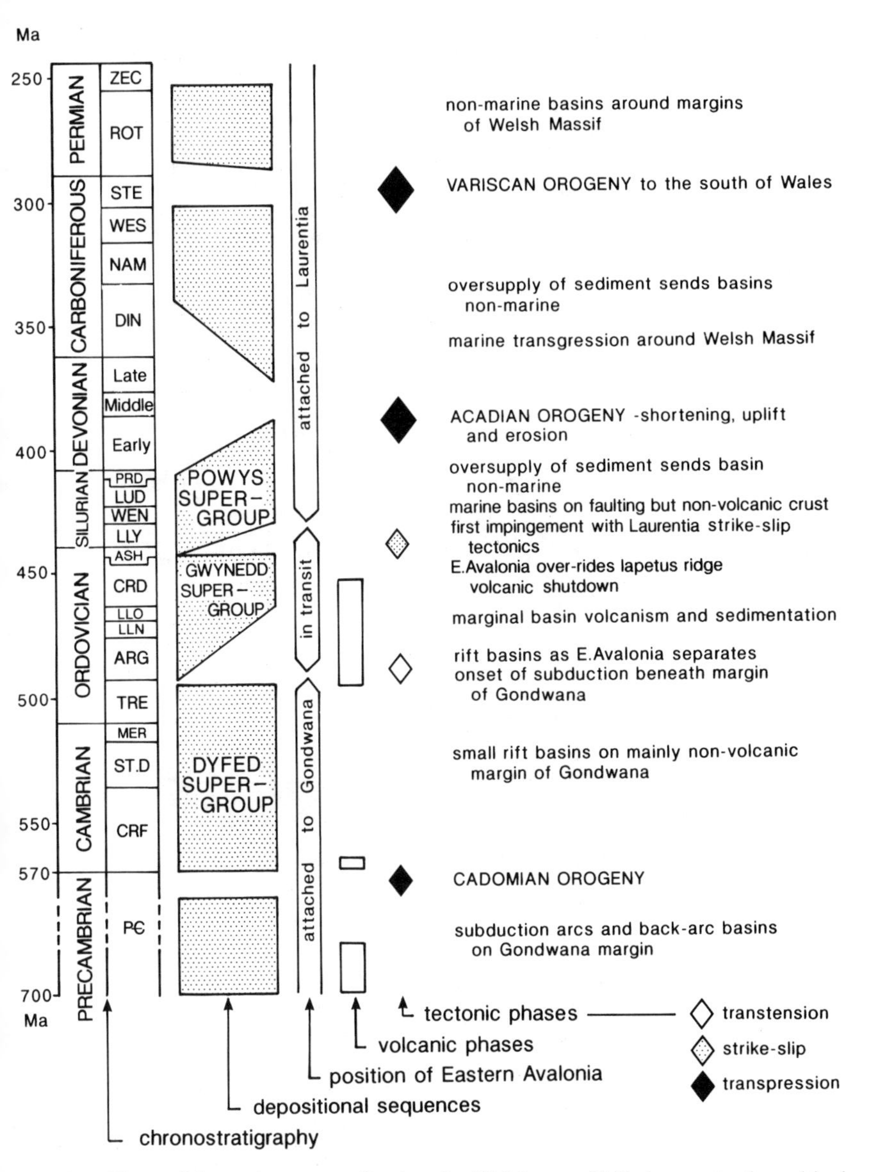

FIG. 9. Chart of the main events affecting the Welsh area. Full chronostratigraphical names and lithostratigraphy are shown on Fig. 2.

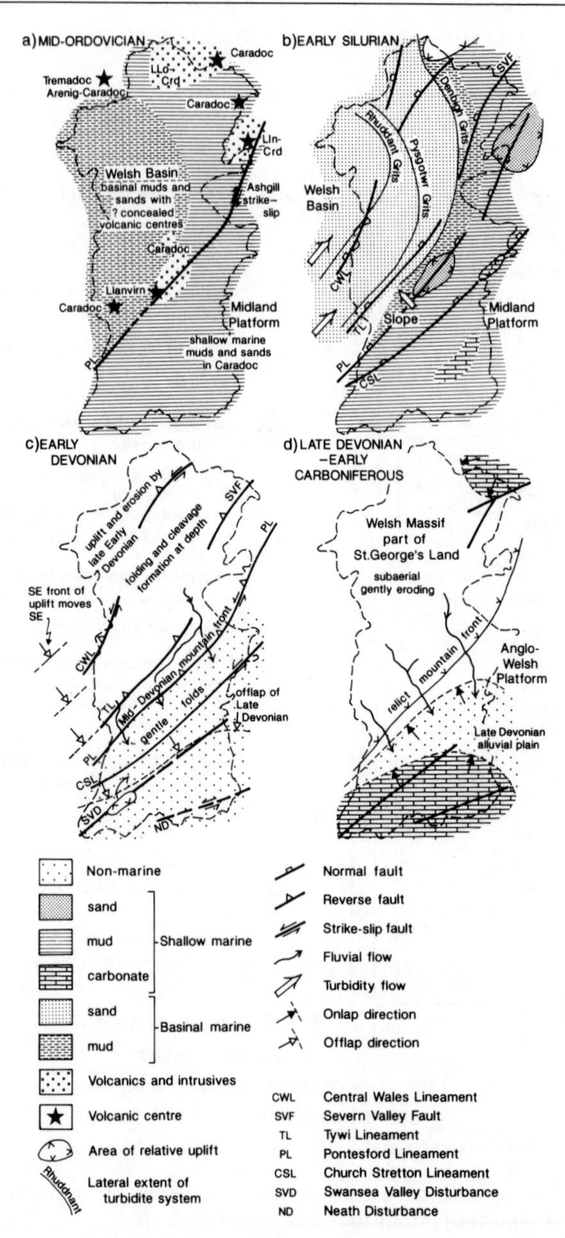

FIG. 10. Palaeogeographical sketches of Powys at intervals through Ordovician to Carboniferous time.

Powys Supergroup deposition occurred away from most proximal volcanism, but in a tectonically active setting. The old faults of the Welsh Borderland Fault System marked a well-defined basin margin, with sediment initially supplied mainly from the Midland Platform (Fig. 10b). The eustatic lowering of sea level due to the late Ashgill glacial event produced a brief but important shallowing of shelf environments and an influx of eroded debris into the basin (Brenchley & Newall 1984). A major influx of turbidite sediment from the south-west in late Llandovery and Wenlock time (Fig. 10b) may have been derived from newly uplifting source areas related to the first impingement of Eastern Avalonia with Laurentia (Soper & Woodcock 1990).

As this continental interaction proceeded, the Welsh Basin became a transpressional collision-zone basin. The basin shallowed as it was oversupplied with sediment, and marginal marine environments gave way to fluvial deposition. The basin was shortened strongly in a NNE–SSW direction during early Devonian time (Woodcock *et al.* 1988). The basin was uplifted, so that fluvial drainage now supplied sediment from the old basin towards the platform (Fig. 10c). Folds and cleavage formed at depth and old faults were reactivated with reverse and strike-slip displacements. By late Lower Devonian time north-west Powys was undergoing strong erosion at the climax of the Acadian Orogeny, restricting deposition of the Lower Old Red Sandstone rocks to the area south-east of the Welsh Borderland Fault System.

The whole of Powys underwent subaerial erosion throughout mid Devonian time, with the former Eastern Avalonian crust now part of the amalgamated continent of Laurussia. The narrow Rheic Ocean existed to the south, with marine conditions as far north as the Bristol Channel. In late Devonian time a pulsed transgression began, allowing first fluvial sediments and then, by early Carboniferous times, shallow marine sediments to accumulate around the margins of the upland remnants of the Welsh Massif (Fig. 10d). The Carboniferous marine basins were sustained by gentle crustal extension, but the resulting subsidence was insufficient to contain the increasing volume of sediment being shed into South Wales from the early uplifts of the Variscan Orogeny to the south, and into North Wales from uplift in northern Britain. The sediment supply overfilled the basins and the alluvial plain and deltaic conditions of the Namurian and Westphalian were established. Deposition was terminated by the weak shortening and uplift ahead of the northward advancing front of the Variscan deformation belt.

The Mesozoic depositional history of Powys can only be guessed from events recorded outside its margins. A review is given by Cope (1984), concluding that Jurassic deposition may have encroached into north-east

Powys from the Cheshire Basin, and that much of the county may have been covered by the late Cretaceous Chalk seas.

It was in South Wales, including southern Powys, that A. C. Ramsay in 1846 recognized that hill summits accord to a gently dipping planar surface and proposed that such planation surfaces were cut by wave action. This enduring hypothesis formed the basis of Brown's (1960) synthesis of the Cenozoic uplift and denudation history of Wales, involving vertical movements of sea-level and of a coherent Welsh structural massif. More recent studies (Battiau-Queney 1984) emphasize the possibilities of subaerial planation by intense subtropical weathering and of differential uplift and subsidence of fault-bounded crustal blocks.

The Quaternary history of Powys must remain outside the scope of this review. A comprehensive guide for the whole of Wales is provided by Campbell & Bowen (1989).

1. THE ORDOVICIAN OF THE SOUTH BERWYN HILLS

by P. J. BRENCHLEY

Maps *Topographical:* 1:50 000 Sheet 125 Bala and Lake Vyrnwy
Geological: 1:50 000 Sheet 136 Bala
1:63 360 Sheet 137 Oswestry

THIS excursion is designed to demonstrate the middle and upper Ordovician rocks of the south Berwyn Hills in northern Powys. The Berwyn Dome has gentle outward dips on its northern and western flanks, but dips are usually steep on the southern flank which is cut by the major Tanat Valley Fault. South of the fault the rocks are folded into periclines with a NE–SW axial trend and a typical wavelength of a few kilometres (Fig. 1).

Ordovician rocks occupy all the central region of the Berwyn Dome and the Silurian rocks occur only on the outer flanks. South of the Tanat Valley Fault, Ordovician occurs in broad NE–SW trending antiforms with Silurian in the intervening synforms (Fig. 1).

The Ordovician rocks within the excursion area are of Caradoc and Ashgill age, although there is a small fault-bounded inlier of Llandeilo rocks nearby in the core of the Berwyn Dome. The thick Caradoc sequence is composed of mudstones and fine sandstones with a varied benthic fauna belonging to different life zones, deposited in a range of shelf environments. Interbedded are two major tuff formations, the Cwm Clwyd Tuff and the Swch Gorge Tuff, the latter of which includes welded ash flows, indicating, very probably, subaerial deposition (Brenchley 1969).

The Ashgill is a rather monotonous sequence of mudstones and siltstones, except near the top where more varied facies were developed in response to the late Ordovician regression.

Within the Ordovician sequence there are three important horizons representing disconformities (Fig. 2). At the top of the Pen-y-garnedd Limestone (Longvillian) there is a concretionary phosphate horizon immediately below the black shales of the 'Nod Glas'. At least two Caradoc stages, the Marshbrookian and Actonian are missing at this level. This is a regional disconformity in North Wales reflecting firstly the end of

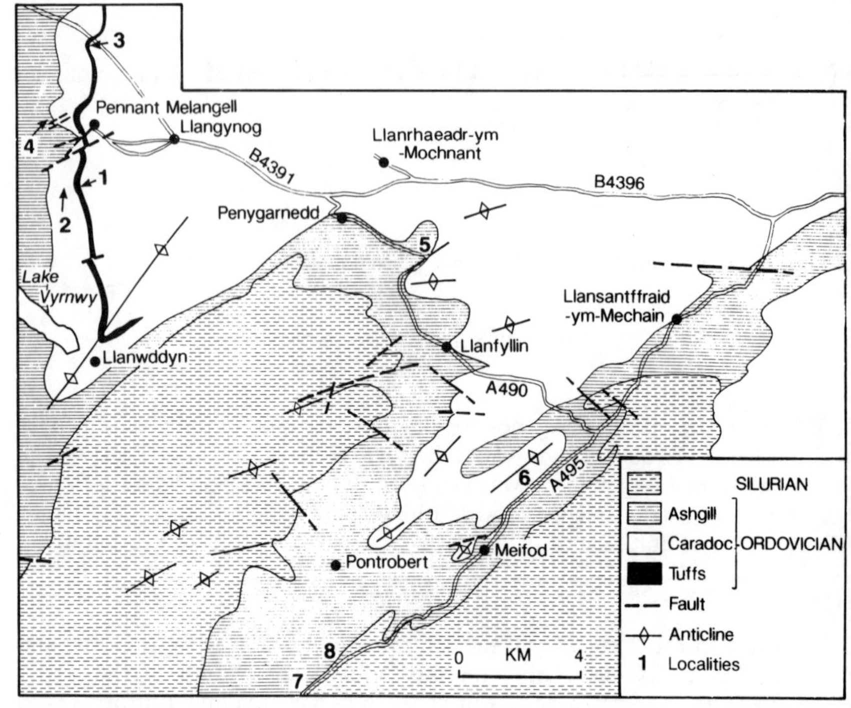

FIG. 1. Simplified geological map of the west and south Berwyn Hills showing Localities 1–8.

volcanicity, which had hitherto supplied much of the sediment to the Welsh Basin, and secondly a transgression which flooded the marginal shelves.

There is a second stratigraphical break at the top of the Nod Glas, immediately below the Ashgill, where there are again phosphate nodules. At several other places in North Wales there is an unconformity at this level, and Woodcock (1990) regards this as a major sequence boundary separating the Gwynedd Supergroup below from the Powys Supergroup above. In terms of basin evolution the sequence boundary is seen as separating a phase when the basin was an active back-arc basin from a later period when the basin was non-volcanic but still in an active margin and was possibly in a collision zone.

The third stratigraphical break occurs near the top of the Ordovician, where late Ashgillian (Hirnantian) sandstones lie with a sharp erosional contact on Rawtheyan mudstones. Traditionally this contact was seen as an unconformity between the Silurian and underlying Ordovician (King 1928), but has subsequently been shown to be a response to a glacio-

			WEST BERWYNS	SOUTH BERWYNS
SILUR-IAN	LLAN-DOVERY	RHUDDANIAN		
ORDOVICIAN	ASHGILL	HIRNANTIAN	Nant Achlas Siltstones ~ 550m	Graig Wen Sandstone 0-10m
		RAWTHEYAN		
	CARADOC	? PUSGILLIAN	Pen-y-garnedd Shales	(Nod Glas) 0-20m
		ONNIAN	phosphate horizon	
		LONGVILLIAN	Pen-y-garnedd Limestone ~ 35m	Pen-y-garnedd Limestone
			Cwm Rhiwarth Siltstones 450m	Cwm Rhiwarth Siltstones: upper >20m, lower
		SOUDLEYAN	Swch Gorge Tuff 0-45m	SGT
			Pen-plaenau Siltstones 320m	Allt Tair Fynnon Beds
		?	Cwm Clwyd Tuff 0-15m	CCT
			Llangynog Formation	

FIG. 2. Stratigraphy of the Ordovician rocks of the west and south Berwyn Hills.

eustatic lowered sea-level during the Hirnantian (Brenchley & Newall 1980). The regression caused widespread erosion on shelf-areas and produced a cover of varied shallow marine facies. The subsequent end-Hirnantian transgression blanketed most places in black graptolitic muds of early Silurian age.

The excursion first shows the characteristic Caradoc succession of the west Berwyns (Brenchley 1978), and the lower two disconformities (localities 1 to 4). It then moves to the south Berwyns to demonstrate one example of facies changes between regions (locality 6) (Pickerill & Brenchley 1979), and finally visits localities showing erosional contacts within the topmost Ordovician (localities 7 to 8).

The Caradoc faunas to be found on this excursion are listed in Brenchley (1978). Many of the brachiopods are figured in Williams (1963) and the trilobites in Whittington (1962–1968). The bivalves are shown in Tunnicliffe (1987) and the trace fossils in Pickerill (1977).

ITINERARY

This excursion is not suitable for visits by large parties in a coach as roads in the west Berwyns are very narrow. **Parking spaces are commonly limited to about three vehicles.**

Start the excursion at the village of Llangynog (SJ 053 262) on the B4391 road to Bala. The village was a busy mining and quarrying centre in the past. Immediately to the south of the village, where there is now a large feldspar-porphyry quarry, was the Llangynog mine where the lead vein was said to be of pure galena, as much as five yards wide. The vein had an E–W strike on a fault which truncated the feldspar-porphyry intrusion (Bick 1978; Williams 1985). Other nearby cross-veins were mined, and there were lead mines on the north and south faces of Craig Rhiwarth, the mountain which dominates the village on its north side. The highest lead output was in the eighteenth century but mining continued throughout the nineteenth century with varying degrees of success and finally ceased in 1912.

Slate mines on the steep south face of Cwm Rhiwarth and other quarries nearby were active throughout the eighteenth century and the first half of the nineteenth century, but declined towards the second half of the century in competition with Caernarfon slates. The opening of the Tanat Valley Railway in 1904 temporarily revived the quarrying, but nevertheless all activity had ceased by the Second World War. The railway line from Llangynog ceased its passenger service in 1951 and goods services in 1952 (Wren 1968).

1. Cwm Llech

This locality shows sediments and fossils of the Pen-plaenau Siltstones and Cwm Rhiwarth Siltstones. Take the small side-road beside the chapel at the south end of Llangynog village and follow it for 2 km to the west to the farm of Rhyd-y-felin (SJ 0330 2541) where there is parking space. From the farm follow a metalled road uphill into **Cwm Llech** (Fig. 3). Follow the track until a major rock step with a waterfall is in sight. About 150m before the waterfall a smooth joint face to the right of the track (SJ 0213 2459) shows one of the deeper facies of the Pen-plaenau Siltstones. Here, thin (< 1 cm), mainly parallel laminated siltstone beds are interbedded with mudstone and there are a few silt-filled burrows. Because of the sequence context in which they occur (see locality 2) the siltstone beds are interpreted as storm layers deposited below storm wave base (Fig. 4). Trackside exposures in the next 150 m show similar sediments, but with more bioturbation.

There is a large difference here between the strike of the bedding (165°–75°) and the strike of the cleavage (030°), typical of the western flank of the Berwyns.

The prominent crag above the waterfall (SJ 0210 2458) shows the facies of the upper Pen-plaenau siltstones. Here siltstone/fine sandstone beds are up to 40 cm thick and have erosional bases and gradational or erosional tops. Internally, they may be parallel laminated, have low angle dips to the laminae or show hummocky cross-stratification. Straight-crested,

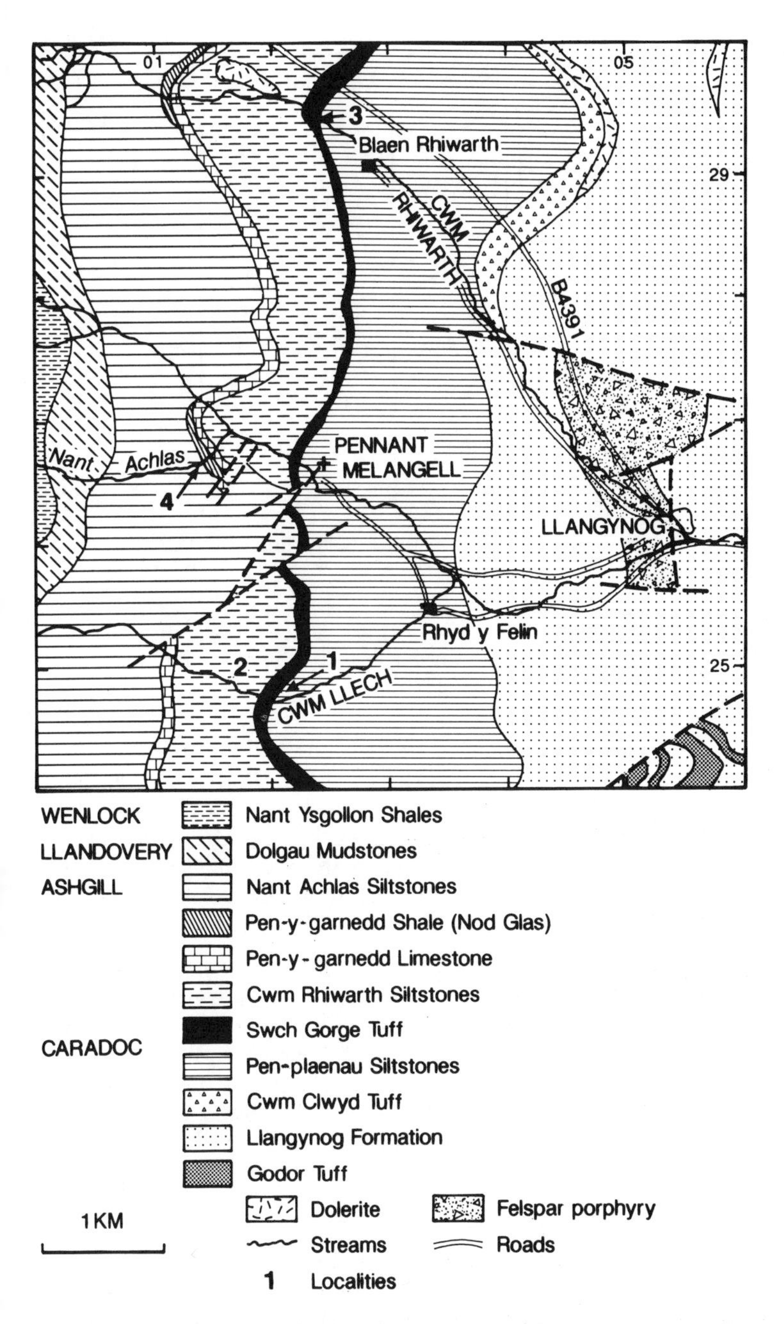

FIG. 3. Geological map of the Llangynog area showing localities 1–4.

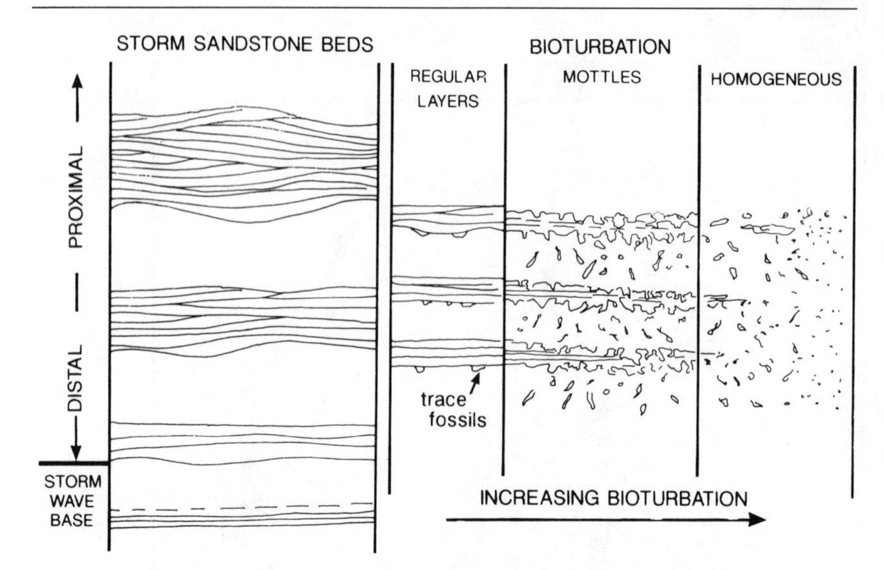

FIG. 4. Typical Caradoc facies of the Berwyn Hills. These represent a spectrum of storm facies from proximal to distal, and varying degrees of bioturbation that appear to vary more with the rate of sediment supply than depth.

bifurcating, symmetric wave ripples occur on some bedding planes. Trace fossils include *Planolites, Skolithos, Teichichnus,* and *Monomorphichnus* (Pickerill 1977). The sandstones are interpreted as storm sandstones, deposited by storm-generated currents and reworked by storm waves (Fig. 4). The facies is best seen on the north face of the prominent crag and on the crag 30 m to the north side of the track. Hummocky cross-stratification is seen in a bed under a prominent overhang on this crag.

Standing on the viewpoint above the waterfall and looking north, the Swch Gorge Tuff is seen in prominent crags at the top of the valley side. The tuff occurs in small exposures by the track 100 to 150 m west of the viewpoint. The light grey flinty rocks are welded tuffs but apart from some loose blocks which show black flattened pumice (fiamme) they look like rhyolites and their ash flow origin is not obvious. The Swch Gorge Tuff is well exposed at locality 3, but even here the welded texture is not clear in the field.

2. Upper Cwm Llech (SJ 0172 2477)

Follow the yellow 'track marks' on to a green track to the scree from a small quarry at the side of the 'forest track' above. The Cwm Rhiwarth Siltstones here are massive and so intensely bioturbated that most original lamination has been destroyed. The scree and quarry yield a fauna with common brachiopods including *Howellites, Sowerbyella, Reuschella horderleyensis, Heterorthis retrorsistria,* and trilobites including *Broeggerolithus broeggeri* and *Brongniartella sp.*, indicating a Soudleyan age. *Dinorthis* is often found with this association which is regarded as a relatively shallow marine *Dinorthis* Community in spite of the fine grained sediment in which it is found (Pickerill & Brenchley 1979).

The Cwm Rhiwarth Siltstones are exposed intermittently until the head of Cwm Llech is reached. Near the top of the formation there are more interbedded siltstones forming beds up to 40 m thick, seen above the waterfall at the head of the valley. Details of the sediments are not well preserved here.

3. Cwm Rhiwarth

The upper part of the Pen-plaenau Siltstones, the Swch Gorge Tuff and the Cwm Rhiwarth Siltstones are all well exposed almost continuously in Cwm Rhiwarth. However, details of the sediments in the Pen-plaenau Siltstones are less clear than in Cwm Llech.

The locality is included because the Swch Gorge Tuff is fully exposed. Follow the road along Cwm Rhiwarth towards the farm of **Blaen Rhiwarth** (park at the start of the private track, or **with permission** at the farm). Follow the marked path beyond the farm until it starts to rise up the north side of the valley. Diverge left from the path towards the entrance to a river gorge. Crags near the entrance to the gorge (SJ 0242 2938) expose the upper Pen-plaenau Siltstones, and the Swch Gorge Tuff is exposed in the gorge. The irregular basal contact of the tuff is exposed by the fence at the entrance to the gorge. The tuff is a massive vitric crystal tuff with some black pumice fiamme. The rock is deformed and although some pumices are flattened it is not clear whether this is original welding or the result of deformation. However, thin sections show some portions of the tuff to have a eutaxitic, welded texture.

Above the Swch Gorge Tuff are the massive bioturbated mudstones of the Cwm Rhiwarth Siltstones. They yield varied Soudleyan faunas at several levels. Approximately 200 m upstream from the Swch Gorge Tuff is the Pen Bwlch Tuff, a thin unit (1–2 m) of lithic lapilli tuffs. These might be of air-fall origin or according to Schiener (1970) were possibly deposited from subaqueous ash flows.

4. Nant Achlas (SJ 0150 2675)

Park by the church in Pennant Melangell, and follow the road west. Just past the farm of Blaen y Cwm the track turns left up to the deserted farm of Tyn y

Cablyd. Soon after this farm cross **the stream of Nant Achlas** and 30 m onwards start to ascend the wooded hillside (SJ 0150 2675). Immediately **above the track are crags** with air-fall tuffs up to 1 m thick and reworked tuffaceous horizons which contain fossils. These beds belong to the upper part of the Cwm Rhiwarth Siltstones. Climb and traverse back towards the stream. Locate a large oak overhanging the north bank of the stream. Below the tree are typical sediments of the upper Cwm Rhiwarth Siltstones with storm beds showing parallel and low-angle cross-lamination and some synsedimentary deformation structures. About 20 m upstream, and 3 m below an oak which grows outwards then upwards from the top of the incised steam valley, is the contact between the Pen-y-garnedd Limestone and the Cwm Rhiwarth Siltstones (**this locality is perilous** at the best of times and **dangerous when the ground is wet**). Here the topmost few centimetres of the Cwm Rhiwarth Siltstones have yielded an abundant fauna of *Howellites* and *Kiaeromena*. The basal beds of the Pen-y-garnedd Formation are tuffaceous and have pebbles of felsite and phosphate together with fragmental fossil material. The pebbly nature of the lowest beds, the presence of phosphate pebbles and an abrupt lithological change suggest there might be a disconformity here. The sequence through the Pen-y-garnedd Formation is as follows.

BLACK GRAPTOLITIC SHALES OF THE PEN-Y-GARNEDD SHALES (NOD GLAS)

~1.5 m	Bioclastic limestone and calcareous siltstone with phosphate nodules and ooids.
2.5 m	Bioclastic limestones, partially dolomitized and grey calcareous siltstones.
17.5 m	Massive siltstones in beds < 1.5 m thick, interbedded with a few thin tuffaceous sandstones and calcareous siltstone.
10.00 m	Interbedded dark grey silty limestones and tuffaceous siltstones.
1.00 m	Tuffaceous calcareous siltstone.
3.00 m	Pebbly tuffaceous silty mudstones. Clasts of felsite and phosphate together with fossil debris.

A rich and varied fauna can be collected from decalcified beds in the side of the stream gorge, particularly in the lower part of the formation. The fauna belongs to the *Nicolella* Community and includes *Nicolella actoniae, Skenidioides costatus, Dolerorthis duftonensis, Flexicalymene planimarginata*. In the well-bedded limestones the fauna is less abundant, but includes the trilobites *Chasmops cambrense, Estoniops alifrons* and *Deacybele arenosa*. The faunas in the Pen-y-garnedd Limestone in the west Berwyns have many species in common with the Bala area, but differ from those further east, found in the north and south Berwyns. However,

Chasmops cambrense occurs throughout the whole area and suggests the *Dalmanella indica* zone of the Longvillian is present everywhere.

The thick-bedded siltstones in the upper part of the formation have yielded only a limited fauna, but a large species of *Kjerulfina* has been collected from the uppermost beds exposed on the south side of the stream.

Cross the stream above the stepped waterfall (about 7 m high) and locate an **old adit** with spoil of black shale about 20 m upstream. The adit was excavated along phosphate beds at the junction between the Pen-y-garnedd Limestone and the black shales above. Phosphate at this horizon was mined at several places in the Berwyn area in the last century (Davies 1875). The phosphate, which is found in the uppermost metre of beds exposed in the stream, occurs in nodules associated with ooids and spastoliths which were probably formed of chamosite but have subsequently been altered to chalcedony and clay minerals. Two stages of the Caradoc are not recognized at this horizon (Cave 1965) and are either missing owing to erosion or are condensed within the phosphate beds.

Poorly preserved graptolites occur in the black shales in the spoil tip and at some points in exposures upstream. The contact between the Nod Glas and the overlying Ashgill is exposed 250 m upstream near the top of a 5 m waterfall. Phosphate concretions occur in the topmost few centimetres of the black shales and lowest few centimetres of the Nant Achlas Formation. *Tretaspis moeldenensis moeldenensis* from near the base of the Ashgill sequence indicates a Pusgillian or Cautleyan age (Price 1984), suggesting that no more than one Stage is missing at this level. This is the 'unconformity' at the base of the Powys Supergroup, though here there is no evidence of discordance nor of subaerial emergence. The 'unconformity' here seems to be a submarine horizon of erosion and/or non-deposition.

5. South-east of Pen-y-garnedd

From Llangynog drive towards Llanfyllin. The road goes through Pen-y-garnedd, type locality of the Pen-y-garnedd Limestone and Shale. The quarry in which they were formerly well exposed is now largely overgrown. At a distance of 3 km south-east of Pen-y-garnedd stop at the roadside (SJ 1230 2240) and look north-east to Allt Tair Ffynnon. An anticline, typical of the folds in this tract of country, is seen in the hillside.

6. Bron y main

Continue through Llanfyllin to the A495 road, turn south-west to a **roadside quarry opposite the farm of Bron y main** (SJ 1660 1448). Parking is available about 600 m north-east of the quarry opposite a stone construction retaining the bank above the road.

Longvillian rocks equivalent to the top part of the Cwm Rhiwarth Siltstones of the west Berwyns are exposed in this quarry. The sandstone facies here represents an environment nearer shore, though not much shallower, than that further west.

In the quarry face opposite the farm are phosphate nodules followed by tuff beds up to 20 cm thick. Stratigraphically above are about 8 m of thick-bedded sandstone exposed between the quarry face and the road. Straight-crested wave ripples are present on two surfaces. Coquinas in the sandstones yield *Dalmanella horderleyensis* and *Sowerbyella sericea* and occasional *Broeggerolithus* cf. *nicholsoni*. Trace fossils include *Skolithos, Planolites* and *Palaeophycus*. About **60 m north-east of the farm entrance** tuffs drape over an irregular megarippled topography on the underlying siltstones. Immediately above are beds with abundant stick bryozoa, and bedding plane assemblages of *Dalmanella* and *Sowerbyella.*

7. Graig Wen Quarry

Continue on the A495 for 6 to 7 km south-west of Meifod, 1.6 km south-west of the Tan House Inn (SJ 0965 0900). The uppermost part of the Ordovician is exposed here in **a roadside quarry**. Rawtheyan mudstones of zone 7 are exposed in the back wall. Small phosphate concretions have yielded a rich fauna of small fossils, often with several species to each concretion. The fauna of the nodules comprises *Tretaspis sortita, Orthambonites* sp. and other trilobites and brachiopods together with gastropods, bivalves and ostracods (Brenchley & Cullen 1984). The fauna in the enclosing mudstones is sparse which raises interesting questions about the taphonomy of the fauna.

At the south-west end of the quarry Hirnantian sandstones lie on the mudstones with an erosion surface. King (1928) figured this face, showing the sandstone lying in steep-sided erosional hollows, some with a concentration of phosphatic nodules at the base. The sandstone itself yielded *Hirnantia sagittifera* and *Hindella crassa* typical of the end Ordovician *Hirnantia* fauna. This locality reflects the major glacio-eustatic lowering of sea-level (estimated at about 70 m) which occurred during the lower Hirnantian as ice caps grew on Gondwanaland.

8. Roadside quarries

About 600 m to the north-east (SJ 1013 0940) there is a second locality at the same horizon (there is limited parking 100 m north-east of the locality). **Old quarries** extend about 120 m alongside the road **south-west from the entrance to a farm**. The quarries are rather overgrown and dark in summer, but the important features are still exposed. Bedding in the quarries is nearly vertical so the rocks are viewed from their upper surface. The relationships

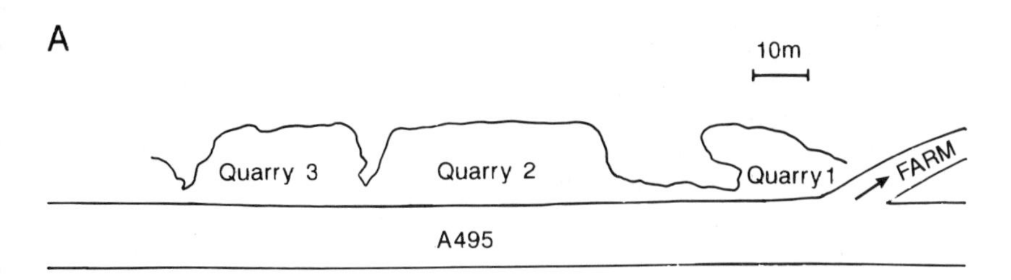

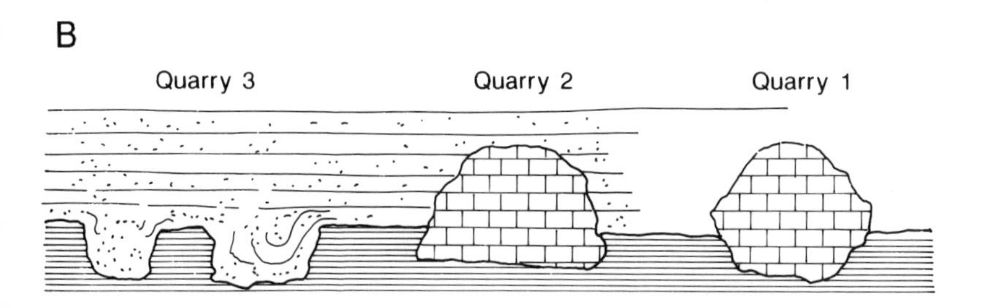

FIG. 5. The contact between the Graig Wen Sandstone (Hirnantian) and Rawtheyan Shales with large limestone 'blocks' at the top. *A*) plan view of the quarries beside the A495; *B*) an interpretation of the stratigraphical relationships between the shales, limestone and sandstone with a reference to the quarry where the relationships are best seen.

here are quite complex. Rawtheyan shales have large rounded limestone blocks or mounds in their uppermost levels. In the small quarry nearest the track into the farm (Fig. 5, quarry 1), shales abut against the side of the limestone blocks. The base of the limestones is not seen. The limestones are mainly crinoidal-bryozoan packstones with some corals, gastropods, crinoids and brachiopod fragments. Tubular fossils filled with fibrous cement are believed to be algae, and there are oncolitic coatings to many grains. The interstitial material is mainly biomicrite, now seen as microspar, but there are patches of coarser granular calcite and sparry calcite fills spaces within the fossils. In places the limestone is transected by narrow fractures filled with fine sand or sparry calcite, and commonly stylolitic at their margin. It is not clear whether the limestones are part of an original sheet which has been broken up into a series of blocks which are scattered on, and embedded within, the shales, or whether the blocks are small bryozoan mounds, analogous to small mud mounds. If the latter is the case, then some at least have been disrupted from their original position.

The overlying sandstones enclose the sides and cover the tops of the limestones (seen in quarry 2, 50 m south-west of the farm entrance) and cut down and load into the shales, with vertical contacts often intersecting at high angles to each other (exposed in quarry 3). About 8 m of sandstone have been quarried out. It is not clear whether these thick-bedded sandstones were deposited immediately after the erosion, during the regression, or were deposited during the subsequent transgression. The overlying beds, not exposed here, are blue-black micaceous shales (King 1928) reflecting the rise in sea-level which accompanied the melting of Gondwanan ice.

2. THE ORDOVICIAN AND SILURIAN OF THE WELSHPOOL AREA

by R. CAVE *and* R. J. DIXON

Maps *Topographical*: 1:50 000 Sheets 126 Shrewsbury
137 Ludlow

1:25 000 Sheets SJ 10, SJ 20, SJ 21
SJ 31, SO 29

Geological: 1:63 360 Sheets Old Series 60 SE, 60 NE

THIS itinerary describes selected exposures of Ordovician (Caradoc) and Silurian (Llandovery) rocks bordering the River Severn near Welshpool, first on the west side, then on the east. No formations older than Caradoc are exposed in the area, whilst the Caradoc, Ashgill and Llandovery series present the stratigraphy with some large non-sequences, including a major unconformity between Ordovician and Silurian rocks.

In the main these rocks are sandstones and mudstones, and were the deposits of fairly shallow marine conditions, in an environment that was mostly oxic. Only in the late Caradoc did it become anoxic, or dysaerobic, for any length of time, though thin layers of dark grey graptolite-rich mudstones reveal that there were short periods in the early Caradoc and during the late Llandovery (Telychian Stage) when poorly oxygenated bottom conditions prevailed.

Igneous activity, both intrusive and extrusive, was important in the late Ordovician, probably all within Caradoc times. In the Silurian it was restricted to distant or mild volcanism, so that the only local expressions are thin layers of pale fawn clay ('bentonite') and a nearby acid vitric tuff of late Wenlock age. Bentonite is a commercial term for a swelling clay that has resulted from very early decomposition of vitric volcanic dust. It is a mixed-layer clay with varying proportions of illite and smectite (montmorillonite), illite becoming dominant when the metamorphic grade rises.

Stratigraphy

Ordovician, Caradoc Series

Harnagian Stage: The oldest rocks in North Powys are those of the Hyssington Volcanic Formation (Llanvirn) in the Shelve Inlier, but the oldest strata near Welshpool probably belong to the Harnagian stage of the Caradoc Series. They are medium to dark grey friable mudstones, of the Trelydan Shale Formation, in the core of the Guilsfield anticline north of Welshpool, and of the Stone House Shale Formation of the Breidden Hills (Fig. 1). The Trelydan Shale gives way upwards to slightly coarser and more blocky mudstone of the Trilobite Dingle Shales and the Middle House Mudstone Formation.

Soudleyan and Longvillian Stages: The Caradoc sequence thereafter coarsens upwards into the top of the Lower Longvillian (Pwll-y-glo Formation and Gaer Fawr Formation, Fig. 1), as silt and sand become increasingly dominant over mud. This sequence is palaeontologically and lithologically comparable with a part of the classic River Onny section in Shropshire and corresponds to the Llewelyn Volcanic Group, the Cwm Eigiau Formation and the Snowdon Volcanic Group in Snowdon (Howells *et al.* 1981).

The top part of the Lower Longvillian is markedly finer and consists of about 12 m of grey-brown mudstone with occasional layers of limestone. The remaining 12 m of Gaer Fawr Formation is within the Upper Longvillian and is a return to siltstones and fine sandstones which are rather calcareous.

During the Soudleyan and Longvillian, submarine and partially subaerial volcanoes were active over West Shropshire and North Wales. Formations composed largely of volcanogenic detritus are well exposed in the Breidden Hills (e.g. the Soudleyan rhyolitic tuffs of the Middletown Formation and andesitic conglomerates of the Bulthy Formation), in Snowdonia (e.g. the Soudleyan Llewelyn Volcanic Group and the Longvillian Snowdon Volcanic Group) and to the north, in the Berwyn Hills, where there are several thin lithic and crystal tuffs (Brenchley 1978). It is strange therefore that no similar, dominantly volcanic, formations are present in the Soudleyan and Longvillian sequence of Welshpool. Most of the arenaceous input which caused the coarsening upwards of the sequence is probably redistributed volcanogenic detritus, but the only direct volcanic contributions are beds of feldspar crystal sandstone, some lithic tuff material and layers of bentonite, mainly in the Lower Longvillian.

Marshbrookian and Actonian Stages: Marshbrookian and Actonian strata have not been recorded here or over the rest of North Wales. It is one of the striking differences between the sequences of Shropshire and North Wales, that in the latter, wherever age has been established, black shales and mudstones conformably overlie Longvillian rocks. These black rocks constitute the Nod Glas Formation, but for reasons of unconformity, either by Ashgill rocks or Llandovery rocks, there are places where the Nod Glas has been removed from the sequence.

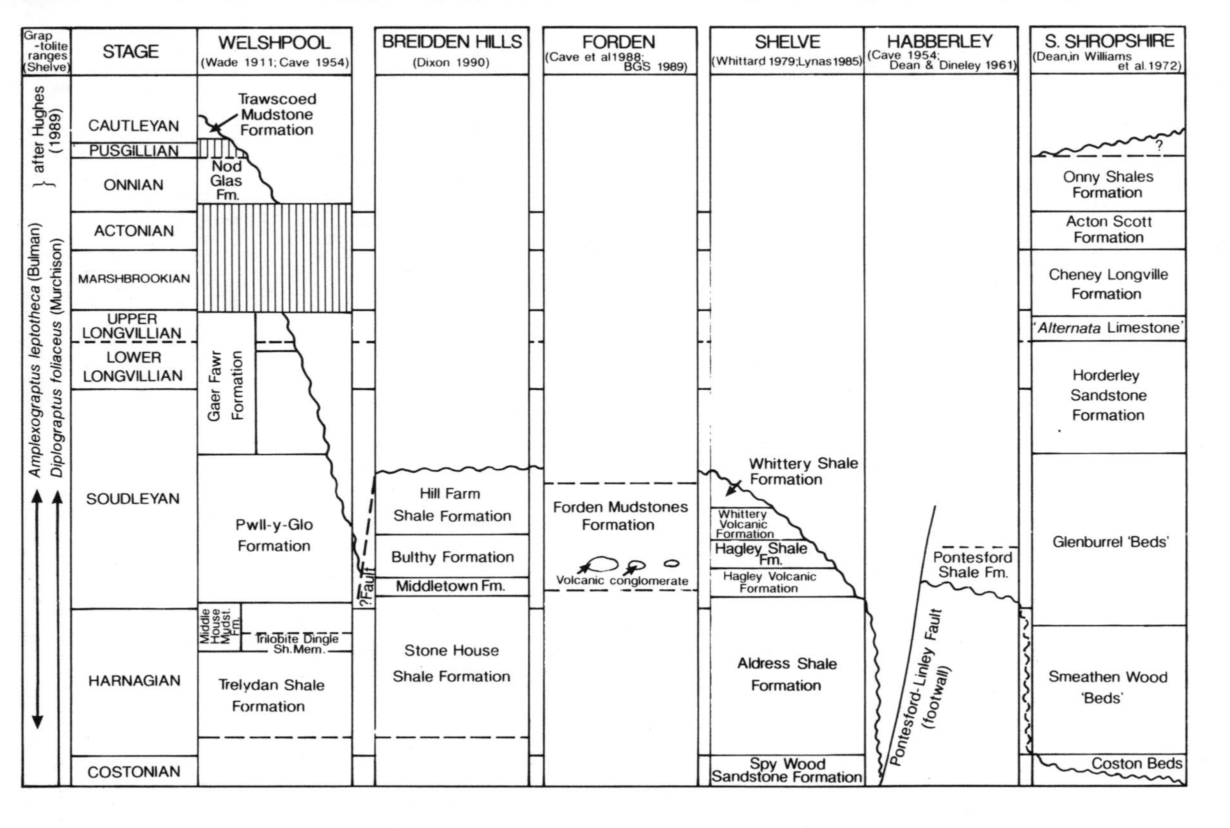

FIG. 1. Correlation of upper Ordovician formations from Welshpool to east Shropshire.

Onnian Stage: Near Welshpool the Nod Glas is between 12 m and 14 m thick consisting of black shales with nodular phosphatic layers in the lowest metre or two and a basal phosphatic limestone. This division holds true for much of the southern Berwyns area, the two members comprising the Pen-y-garnedd Shale and the Pen-y-garnedd Phosphorite (King 1923; Cave 1965).

The Nod Glas of Gwern-y-Brain, just west of Locality 5, has yielded a rich fauna of brachiopods, ostracodes, trilobites and graptolites indicating an Onnian age in the top part of the biozone of *Dicranograptus clingani* and possibly basal *Pleurograptus linearis*.

The Nod Glas indicates a radical change in the depositional environment after the end of Longvillian volcanicity. Marine conditions persisted but they became poorly oxygenated and the input of clastic detritus almost ceased. The crystalline phosphatic limestone was precipitated under such conditions and may represent the condensed Marshbrookian and Actonian stages, though they may be absent by total non-deposition prior to the deposition of the limestone (Cave 1965). The event that caused this change in depositional environment was of fundamental significance in the history of the Welsh Early Palaeozoic Basin; it was probably an interaction of crustal plates that ended southward subduction of oceanic crust beneath continental Cadomia (or Eastern Avalonia) (Campbell 1984; Cave 1992).

Caradoc Palaeontology

As the early Caradoc deposition coarsened, and the marine conditions shallowed, so the initially dominant graptolite faunas including *Diplograptus foliaceous* and *Amplexograptus leptotheca* of the Trelydan Shale and the Stone House Shales gave way to shelly faunas. First there was a trinucleid trilobite fauna seen in the Trilobite Dingle Shales, Middle House Mudstones, Pwll-y-glo Formation and Bulthy Formation, then a dominantly brachiopod fauna seen in the Longvillian Gaer Fawr Formation. Longvillian rocks are amongst the most fossiliferous of the Ordovician and they indicate that a shallow-marine depositional environment was established at this time over Shropshire and all of North Wales (Brenchley 1978), apart from brief emergences in the very active volcanic areas (Howells *et al.* 1981, p. 30).

Comparison of the Caradoc shelly faunas of Welshpool with those of East Shropshire provides evidence that there was a long pause in deposition at Welshpool during Marshbrookian and Actonian time. Here, characteristic Lower Longvillian fossils such as *Bancroftina typa, Dalmanella horderleyensis, D. indica, Kjaerina jonesi* and *Broeggerolithus nicholsoni*, are separated by some 12 m of sandstones and siltstones from rocks with: *Onniella broeggeri, Onnia gracilis* (Bancroft) and *Onnicalymene onniensis* characteristic of the middle of the Onnian stage. Those 12 m of rocks do not possess obvious Marshbrookian and Actonian fossils such as the pterygometopid trilobite *Chasmops extensa* with its extended pygidium, nor several species of *Onniella*, and it is assumed that the rocks are contained within the Longvillian Stage and overlain within a metre or two by Onnian rocks.

Ordovician, Ashgill Series

Little of the Ashgill succession is preserved in the Welshpool area. What remains has been called the Trawscoed Mudstone Formation (Cave & Price 1978), from its exposure just above the Nod Glas in a gully near Trawscoed Hall, Guilsfield. It is a mudstone, probably calcareous when fresh, yielding a varied fauna of small brachiopods and trilobites. From these its age was concluded to be Cautleyan Zone 2, leaving the probability that the Pusgillian Stage and the basal part of the Cautleyan are not represented. Such a non-sequence is believed to be widespread over North Wales and in places on the western and northern sides of the Berwyn Hills, Ashgill rocks rest with angular disconformity on Caradoc strata (Bancroft 1928; Bassett *et al.* 1966). The nature of the mudstone – fine-grained, calcareous, bioturbated and with a shelly fauna but no bed forms – suggests fairly deep marine shelf conditions some distance from the shoreline. Early Ashgill rocks, at least, were probably present over all of the area, and extended well to the east of the River Severn before late Ashgill and early Llandovery erosion removed them.

Silurian, Llandovery Series

The Silurian Period commenced with the Welshpool area as land. The late Ashgill sea had regressed to the west to re-advance eastwards during the Llandovery. It took the whole of Rhuddanian times (early Llandovery) to return as far as the line of the River Severn, so that west of the Severn Valley a sandstone and conglomerate, the Powis Castle Conglomerate Formation, form the base of the Rhuddanian succession (Locs. 2 and 3). It thus overlies Ordovician rocks with lithostratigraphical overstep and chronostratigraphical overlap (Fig. 2).

It was not until late Aeronian (mid Llandovery) times that the sea advanced east of the Severn Valley to Buttington (Loc. 6) and beyond, but once it did cross the Severn 'barrier' it very rapidly inundated an area extending many miles east to Church Stretton. Deposits of the marine transgression were shelly, sandy and even conglomeratic. These deposits belong to the biozone of *Monograptus turriculatas* and a little lower in some places (Cocks & Rickards 1969).

The succeeding deposits were very widespread and uniform muds eventually forming the Tarannon Shales (or Pale Shales) of North and Central Wales.

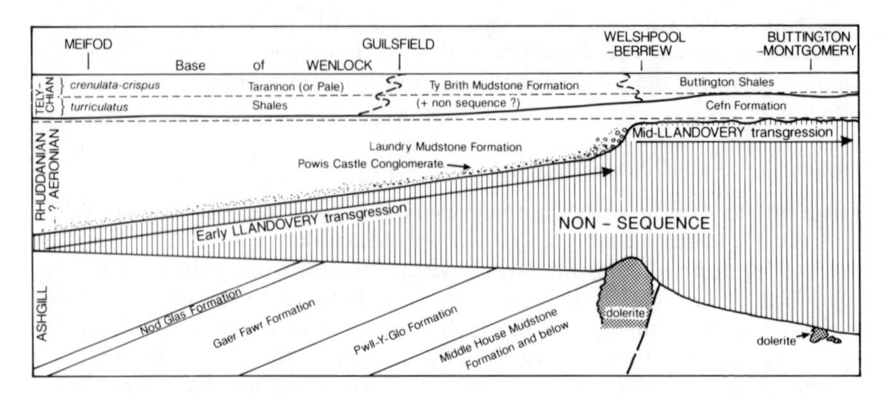

FIG. 2. Diagrammatic NW–SE time section showing Llandovery formations unconformable upon Ordovician formations.

The Tarannon Shales is one of the most distinctive formations of the Silurian by reason of its pale-grey, green and maroon colours and its lithological uniformity. Such beds extend well beyond the confines of Britain into Continental Europe. The formation has received many subsequent and subordinate local names, e.g. Purple Shales, Buttington Shales, Pale Slates and Dolgau Mudstones. The conditions of its deposition must have been likewise uniform over very wide areas. Ziegler & McKerrow (1975) viewed the maroon colour as being due to primary haematite imparted to the detritus while it was undergoing weathering on an adjacent arid landscape. The lateral impersistence of the maroon colour, with passage into green and grey suggests that the colours are more likely to be due to early diagenesis.

The Tarannon Shales are not wholly without variety. Near Welshpool there was a shelly fauna resulting in the local Ty-brith Mudstones while in basinal areas (cf. the River Trannon type-section (Wood 1906, p. 657)) the rocks are shales with innumerable thin (<1 cm) fine sandstones (probably of turbiditic origin) and a sparse graptolite fauna. The main variability however was the date of commencement of the facies. In the basinal type-section the base of the formation is approximately the base of the *crenulata* Biozone although similar pale mudstones are present in some places in the *turriculatus* Biozone. Only 18 miles north-east the base lies within the *turriculatus* Biozone as indeed it might at Buttington. The base of the formation is thus diachronous.

Silurian, Wenlock Series

The Tarannon Shales are generally succeeded by dark grey, shaly mudstones with graptolites belonging to the lowest biozones of the Wenlock Series. Thus approximately at the top of the *crenulata* Biozone *s.l.* (see postscript) there was a sharp depletion of oxygen and the marine sediments became dysaerobic or anoxic. In many sections, e.g. Buttington (Loc. 6) there are a few metres of beds below the oxicity decline which have yielded no date indicator and thus absolute coincidence between this change and the base of the Wenlock has not been established. In Wales this oxygen-depletion of marine deposits continued well into Ludlow times. The change was thus of major importance and explains why most Welsh Wenlock rocks are graptolite bearing, for in poorly oxygenated sedimentary conditions pelagic fallout from the water column does not get consumed by a burrowing infauna or a scavenging epifauna.

Igneous Intrusions

There are two types of intrusive igneous rock: the widely quarried basic 'dolerite' of Breidden Hill, Cefn, Welshpool and other localities in the Shelve Inlier, and the calcalkaline andesite of Moel-y-Golfa. The andesite intrudes Soudleyan rocks and has an extrusive component which confirms a Soudleyan magmatic age. It is doubtful whether any of the dolerites are hosted by strata younger than Harnagian. They could be as late as Rhuddanian but are probably Soudleyan or Longvillian.

ITINERARY

The itinerary is arranged geographically as a roughly circular clockwise tour starting in the south-west at Berriew just west of the main A483 trunk road, five miles SSW of Welshpool (Fig. 3).

1. Rose Hill, Berriew

There are two exposures at Rose Hill, one a quarry within a small wood, and the other within the garden of a private house. They expose similar sequences and both are owned by Mrs M.A.R. Corbett-Winder of Vaynor Park, Berriew, Welshpool, Powys, **who should be contacted for permission** and directions to visit the quarry.

The Rose Hill locality provides a hitherto unrecorded exposure of sandstones and conglomerate of early Llandovery (Rhuddanian) age which is significant in being so far south of other exposures of Rhuddanian rocks in this area. They reinforce the view that Rhuddanian marine deposits are present west of the line of the Severn Valley, but absent to the east. Unfortunately the base of the deposit is not exposed, but the lowest beds are mudstones, which, at the top at least, are mass-flow deposits 1.5 m thick containing sandstone clasts < 14 cm long.

Overlying this is a 5 m packet of thick-bedded sandstones (< 32 cm) with mudstone partings or interbeds. The sandstones are coarse and some are

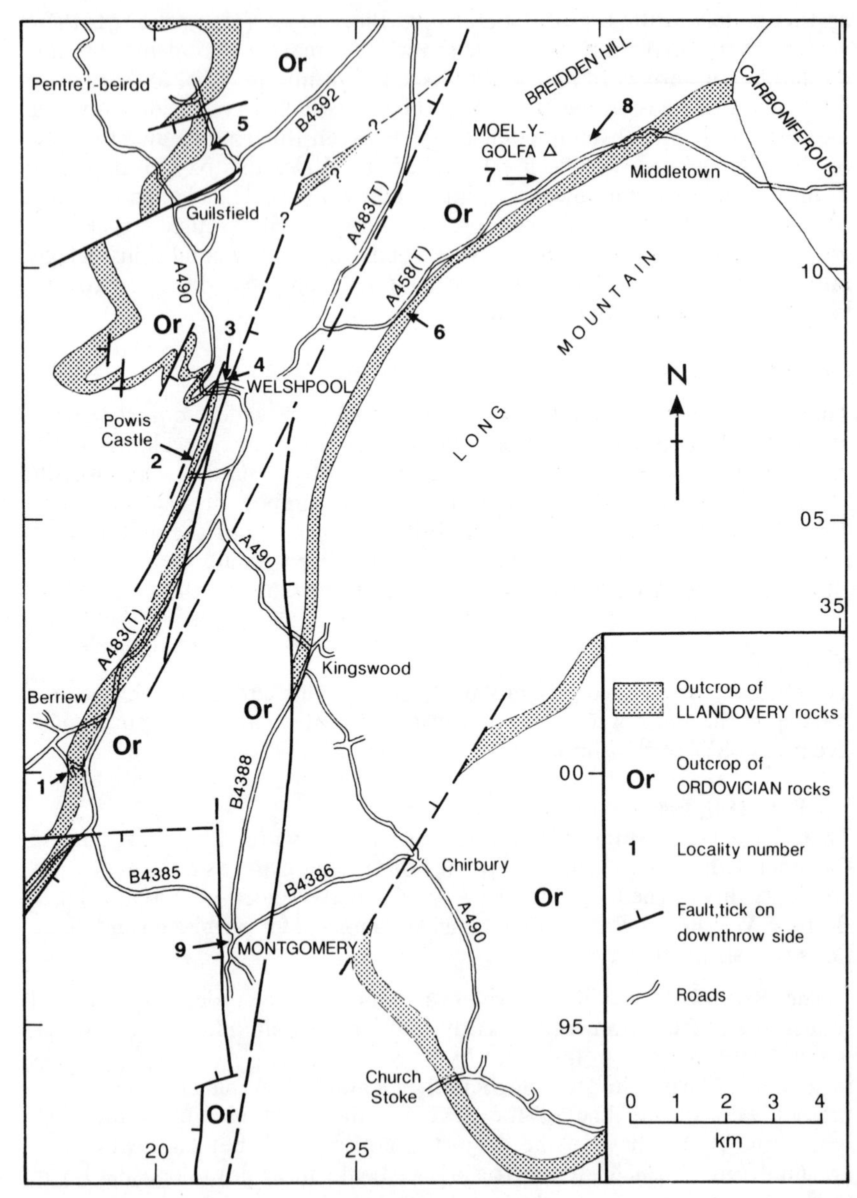

FIG. 3. Itinerary and outcrop of Llandovery rocks.

conglomeratic with clasts (< 2 cm). Most of the clasts are rounded and of quartz, but there are others of shale and igneous rocks. Shell and bryozoan debris is also present.

The most striking feature of these sandstones is their attitude and condition. Some, especially in the lower part, are inverted. Bottom structures, particularly large burrow or grazing-trail casts and some flute-casts, indicate inverted beds. Other beds appear to have a normal attitude and yet others are discontinuous with dismembered parts being enclosed by a mudstone wrap. Although the general attitude of the bedding is the same as that of the overlying strata, the inversions and disruptions indicate that this 'packet' is an allochthonous slide-sheet.

Above the thick sandstones are < 20 m of evenly bedded thinner sandstones (1–20 cm thick) with partings of silty mudstone. Some beds are structureless, bioturbated and thus rather rubbly, others are less argillaceous, more siliceous and harder. They are not unlike the beds of the Laundry Mudstones (early Llandovery) above the Powis Castle Conglomerate near Welshpool and were probably sheet-sands which spread across the deepening shelf sea by storm-generated, current surges.

Shells are fairly common in all parts of the sequence and from above the thick sandstones include: ?*Dolerorthis gwalia, D. sowerbyiana* (Davidson), ?Fardeniid, ?*Hyattidina* ?*Leangella scissa, Leptostrophia* cf. *antecedens, Meifodia* cf. ?*prima, Mendacella*, cf. *Stegerhynchyus pusillus*, *Stricklandia* sp. and a calymenid cheek.

Those from below the thick sandstone include *Clorinda undata, D. sowerbyiana, Plectatrypa tripartita.* These fossils indicate a late Rhuddanian age although *M.* cf. *prima* might suggest a little earlier.

The clasts of the conglomeratic sandstones have an uncertain provenance, but it is most unlikely that igneous clasts came from an easterly sector. The regional evidence indicates that a marine shoreline lay close to the east providing coarse clastic detritus including shell remains and the highly energetic re-sedimentation of this detritus by mass-flow suggests the shoreline might have been fault controlled.

From Rose Hill turn right and rejoin the A483 (T), turning left towards Welshpool. Powis Castle is indicated to the left about one mile south of Welshpool.

2. Powis Castle, park and gardens (216 065)
Access to the castle garden is by **permission of the National Trust** and may have to be arranged beforehand. For access to the Park **permission will be required** from the Powis Castle Estate. There is ample space in the vehicle park adjacent to the Castle (Fig. 4).

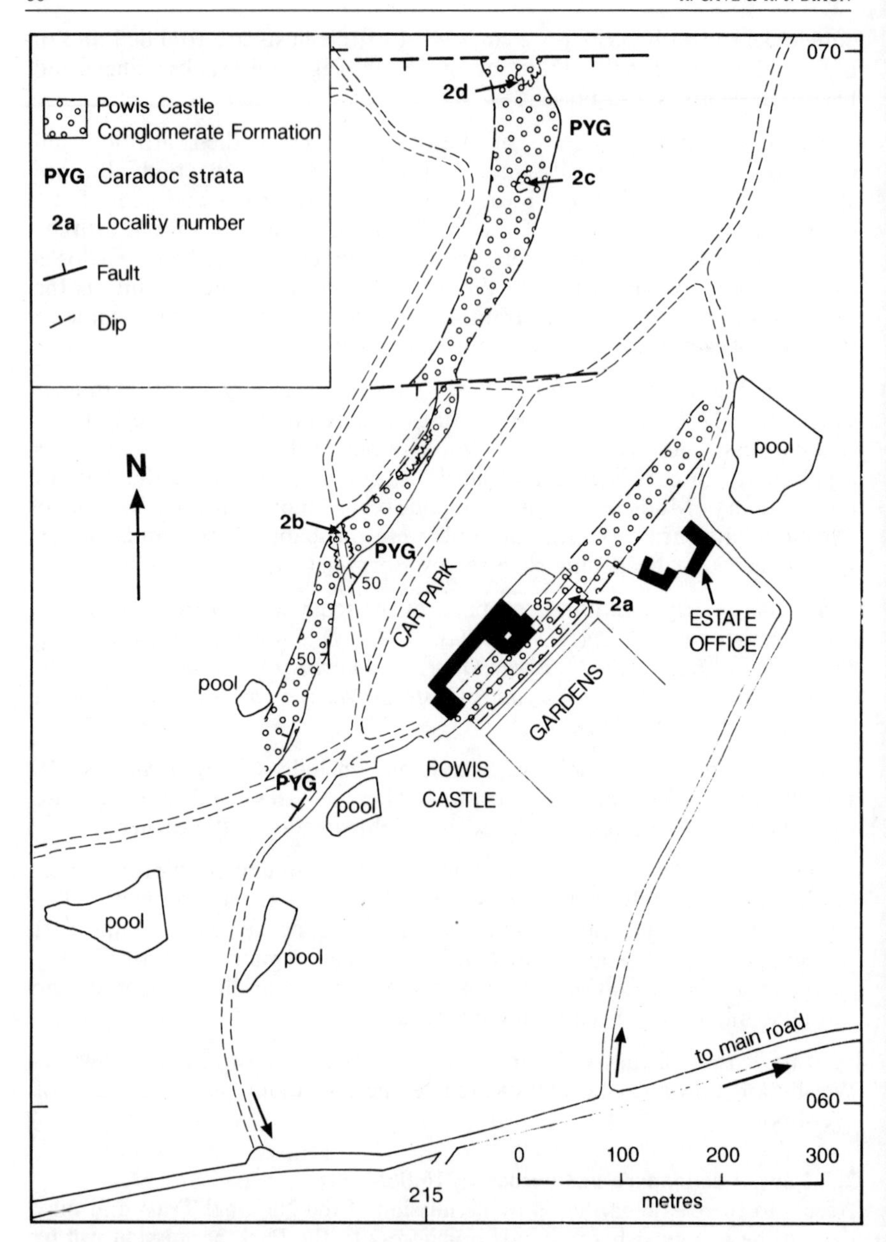

FIG. 4. Sketch map of Powis Castle and Park showing localities 2a–d.

Rocks are exposed near Powis Castle in small quarries and cuttings recorded first by Murchison (1839, p. 304). He viewed the structure there as a sharp anticline, trending NNE–SSW and faulted along its axis (Fig. 5). Thus were produced two outcrops of sandstones and conglomerates – the Powis Castle Conglomerate (Rhuddanian) – one in each limb. Underlying these, and thus in the core of the anticline, are softer, silty mudstones, probably a little higher in the Caradoc than the Trilobite Dingle Shales of Wade (1911).

Murchison's anticline has never been questioned yet the evidence is debatable. The two outcrops (Figs. 4, 5) can be viewed alternatively as a strike-faulted repetition of WNW-dipping beds. The debate centres upon the nearly vertical beds of conglomerate exposed in a vertical face (**Loc. 2a,** 2163 0647) under the north-east corner of the castle on the top terrace of the garden behind Hercules' statue. Should this rock face prove to be the under side of a bed of conglomerate then the main structure is not an anticline. The available evidence comprises linear and parallel ridges on the rock face. If they are erosional groove-casts then the rocks young WNW into the face. An alternative exposure (2172 0654), beyond the confines of the gardens, but of the same beds, occurs c.100 m to the north-east facing the Estate Office (Fig. 4).

The base of the Powis Castle Conglomerate is exposed in the **banks of a park drive** (**Loc. 2b**, 2144 0644). The underlying silty mudstones here, and a little **further north in the park** (close to **Loc. 2d**) have yielded

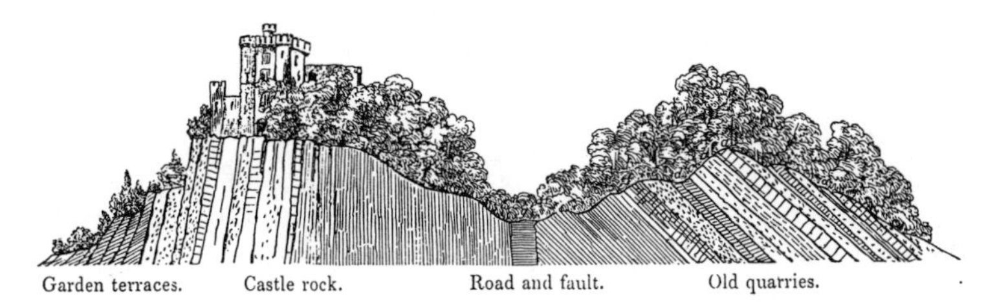

The character of the strata varies much, even in a very short distance; but in a transverse section made near the south end, the quarries present in descending order:

		Feet.
a.	Purplish brown, calcareous grit, with many fragments of encrinites, in beds from eight inches to one foot and a half, with way-boards of red shale, in parts a sandy and gritty, in others an encrinital, subcrystalline limestone (see wood-cut.)	20
b.	Blotchy grit with much red shale	5
c.	Hard, fine-grained, calcareous grit, passing downwards into a mottled, dingy green, and purple impure limestone with irregular traces of way-boards	6
d.	Purple and white limestone with blotches of red shale	4
e.	Hard, fine, thick-bedded, calcareous conglomerate, red where weathered, but greyish where freshly broken. It consists of fragments of encrinites, green earth, chocolate-coloured schist, and a few quartz pebbles, of sizes varying from almonds to small peas	6
f.	The lowest bed visible is a purple and whitish, semi-crystalline limestone with white veins	3

FIG. 5. Section at Powis Castle by Murchison (1839, p. 304)

Broeggerolithus broeggeri, Salterolithus caractaci, Harperopsis cf. *scriptus* and *Orthograptus amplexicaulis*. Although these rocks were considered by Wade (1911) to be Trilobite Dingle Shales their fauna is of a low Soudleyan age and the mudstone is too silty. They may belong to the base of the Pwll-y-glo Formation (Fig. 1).

The conglomerate is not as coarse here as it is further north, but a sandstone block 25 cm long yielded *Howellites* sp., *Skenidioides* sp., *Broeggerolithus* cf. *soudleyensis, Brongniartella* sp., *Piretopsis (Protallinnella)* cf. *salopiensis* and *Lepidocoleus* sp. This fauna signifies a mid to upper part of the Soudleyan Stage and the sandstone suggests that a formation higher than the Pwll-y-glo Formation was undergoing erosion close by. This source could not have been to the west where the Llandovery transgression had already concealed Ordovician strata. In the immediate surroundings mid to high Soudleyan rocks had already been eroded away so the clast must have had a source to the east.

Northward from the road cutting, the Powis Castle Conglomerate gives rise to a sharp ridge in which there are several old quarries. The ridge is broken and shifted westwards by small ENE–WSW faults. The clasts become larger northwards and at **Locality 2c** they are dominantly rounded cobbles of a dark rock that possesses a foliation which simulates bedding. The clasts are in fact of a fine-grained igneous rock with a trachytic texture. The mudstone matrix has yielded *Meifodia* cf. *prima, Stricklandia* sp. bryozoa and colonial corals compatible with the late Rhuddanian age indicated elsewhere at Welshpool. The unconformity with the Ordovician is thus large.

In the **most northerly of the quarries** (**Loc. 2c,** 2161 0695) the conglomerate is made almost entirely of the same igneous clasts but of sizes ranging to boulders 1 m across. It would seem from this that the source of the boulders must be close at hand, yet the underlying formation is still a mid Caradoc silty mudstone.

Return to the A483T and turn left to Welshpool. Continue across the traffic lights and take the first turn left into Brook Street.

3. The Standard Quarry Welshpool (2183 0767)

This quarry is situated in the west end of Welshpool (Fig. 6) and provides ample space for vehicles. The current owner is Miss O'Hare whose agents Harry Ray & Co., Lloyds Chambers, may be contacted for **permission to visit the quarry. Hard hats will be required**.

The quarry has two high walls in splendidly columnar-jointed oligoclase-trachyte (lime-bostonite, Wade 1911) and its texture is most distinctive. In thin section it is immediately evident that this rock was the source of the cobbles and boulders of the Powis Castle Conglomerate in Powis Castle Park.

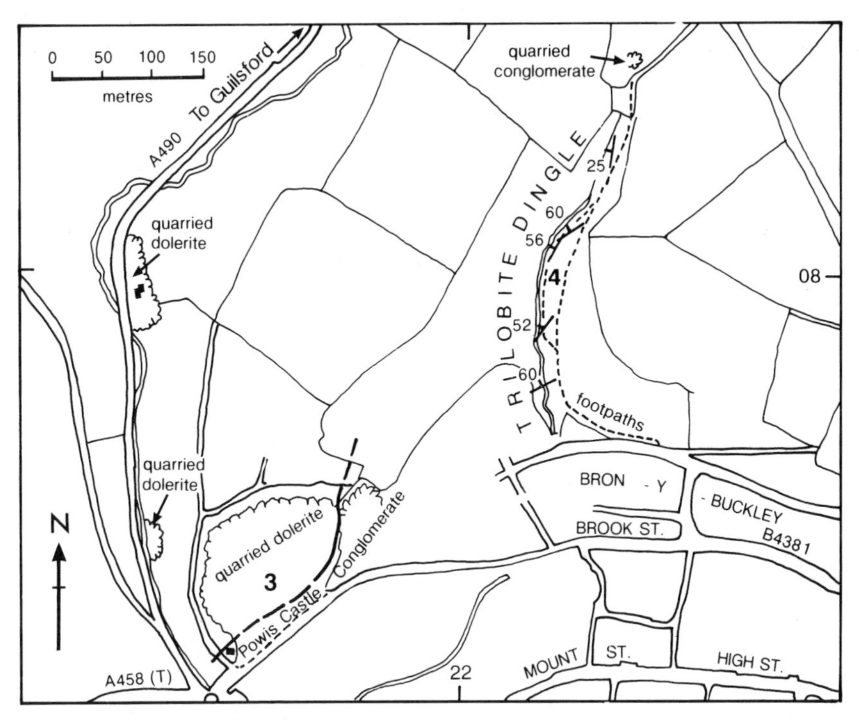

FIG. 6. Sketch map of north-west Welshpool showing positions of localities 3 and 4. Arrows indicate direction and amount of dip of the strata. Broken, heavy line indicates base of Powis Castle Conglomerates.

Many buildings in and around Welshpool were built of this stone, which has a distinctive grey-green colour when weathered. The most striking feature of the rock faces, however, is the columnar jointing. Many are classically hexagonal and pentagonal in section and are several metres long. They plunge at about 25° to the WSW. Like the columnar joints in the Tertiary basalts of Antrim these are the result of shrinkage during cooling and they will have developed with their long axes normal to the cooling surfaces. For columns of such size to have formed, the magma must have cooled slowly and for it to have been a lava the flow would have been thicker than any known in the area. However, rocks of similar composition form intrusions into the nearby early Caradoc strata of the Breidden Inlier so it is reasonable to assume, as did Wade, that the Welshpool rock is intrusive. It is not intrusive into Silurian rocks however, for the Powis Castle Conglomerate, containing much detritus from the intrusion, rests directly upon it in this quarry.

The whole of the **eastern face of the quarry** is composed of Powis Castle Conglomerate (Rhuddanian) and the permission of tenants in the quarry will be required to inspect it (e.g. 2184 0777, Messrs Jones & Greatorex).

The clasts in the conglomerate range up to large boulders, metres across, and are singularly composed of the trachytic intrusive rock. Matrix is sparse, calcitic and bituminous. In places it contains shell fragments and in others is sandy, showing oversteepened cross-bedding. The contact between the intrusive rock and the conglomerate (broken, thick line, Fig. 6) is nearly vertical on this side of the quarry and passes up the north-east corner.

The conglomerate is again visible in the south-west corner of the quarry under Miss O'Hare's bungalow and visible from within the District Council's compound (2176 0766). Here the contact is almost horizontal and since the rocks are not obviously folded it seems likely that the different attitudes of the contact are largely those of the partial burial of an upstanding igneous mass – an island, sea-stack or fault-scarp.

About 300 m north-east of Locality 4 a wooded dingle can be entered (2217 0785) via a footpath from Bron-y-buckley housing estate.

4. Trilobite Dingle (2207 0800)

This locality which exposes the Trilobite Dingle Shales (Harnagian) is under the control of a conservation trust and vehicles can be parked here (Fig. 6). It is well known for the abundance of the trinucleid trilobite *Salterolithus caractaci*, for which it is the type locality (Cave 1957).

The main **exposure follows the stream northwards for some 300 m**, but further exposure has been created recently **beside a new track** on the west of the dingle not marked on Fig. 6. The Trilobite Dingle Shales are, in fact, rather blocky mudstones of grey, buff and dark-red colours. The redness is due to secondary oxidizing agencies, possibly connected with nearby major faults and the sub-Triassic surface, though the effects of possible early Llandovery subaerial weathering cannot be ruled out. About 45 m of beds are exposed, with dips of 25° to 60° towards the west and north-west. They are presumed not to have been inverted, but a count of the attitudes of trilobite cephala was equivocal (Cave 1955).

The fauna comprises *Leptaena* sp., *Onniella avelinei*, *Parabasilicus powisii*, *Salterolithus caractaci*, *Plumulites peachi*, *Diplograptus foliaceus*, *Pseudoclimacograptus scharenbergi scharenbergi*, gastropods and bivalves.

No specimens of *Broeggerolithus* have been found here and this is the only exposure in the area where *S. caractaci* occurs abundantly, but alone. Thus the Harnagian age and *Diplograptus multidens* Biozone seem well diagnosed; it is strange however, that no similar fauna occurs in the Shelve sequence at about the level of the Aldress Shale or in the Breidden sequence at the level of the Stone House Shale.

Continue to the western end of Brook Street, turn right at the roundabout and travel to **Guilsfield**. Sarn Bridge, 650 m north-east of the village, on the Oswestry road, is **as far as a coach will go**. Smaller vehicles can turn left here and travel to the **top of the hill** where a small quarry will accommodate a minibus or two cars.

5. Gwern-y-Brain (218 125), **Guilsfield**

The Gaer Fawr Formation, in the upper part of the Caradoc Series, is well exposed in **Gwern-y-Brain stream directly below the small quarry** (Fig. 7), 1 km north of Guilsfield. It is a fossiliferous sequence comparable with that in south Shropshire, and many of the zones of Bancroft (1933) can be applied to the Welshpool Caradoc sequence (Cave 1955). Attempts to apply them in areas away from Shropshire have met with mixed success. Brachiopod faunas are perhaps too greatly influenced by environmental conditions to remain diagnostic over long distances, e.g. Bassett *et al.* (1966) for the Bala area, yet the attempt at Welshpool appeared to be effective as did that of Whittington (1938) nearby. The lithostratigraphy on the other hand has a strong north-Welsh flavour as indeed have some aspects of the faunas, so that if reliance on Bancroft's zonation is justified, then the Gwern-y-Brain section is a useful link between Shropshire and North Wales.

The section (Fig. 7) illustrates the Lower Longvillian part of the Gaer Fawr Formation. A series of beds of different hardnesses and composition are present which reflect varying activity in the North Welsh volcanism of the time. In the lower two zones fine-grained sandstone with a high feldspathic content is the main rock type. Beds of soft clayey shale, probably bentonites, divide this sandstone and are themselves feldspathic.

Feldspathic sand is most concentrated, however, immediately above bentonites. There are two thick beds of bentonitic shale, each at the base of a brachiopod zone at *a* and *c* and the latter has caused instability on the south side of quarry *g*. There are also three hard beds of fine sandstone which are calcareous and fossiliferous. Above the topmost of these, at *o*, there follows a rather dark grey-brown mudstone with thin blue-grey limestones which coincides with the zone of *Bancroftina typa*. *Scopelochasmops cambrensis*, which is common in the underlying strata, is not found higher, but the trilobite *Flexicalymene* cf. *caractaci* appears in place of *F. planimarginata*.

A thin limestone (< 1 m) at locality *t* caps these mudstones. It is very fossilferous and yields a variety of brachiopods including *Heterorthis alternata*. It is approximately equivalent to the Cymerig Limestone (Bassett *et al.* 1966), once called the Bala Limestone or Gelli Grin Limestone, near Bala (Elles 1909; Whittington 1962; Williams 1963). This and the beds above are considered to belong to the Upper Longvillian

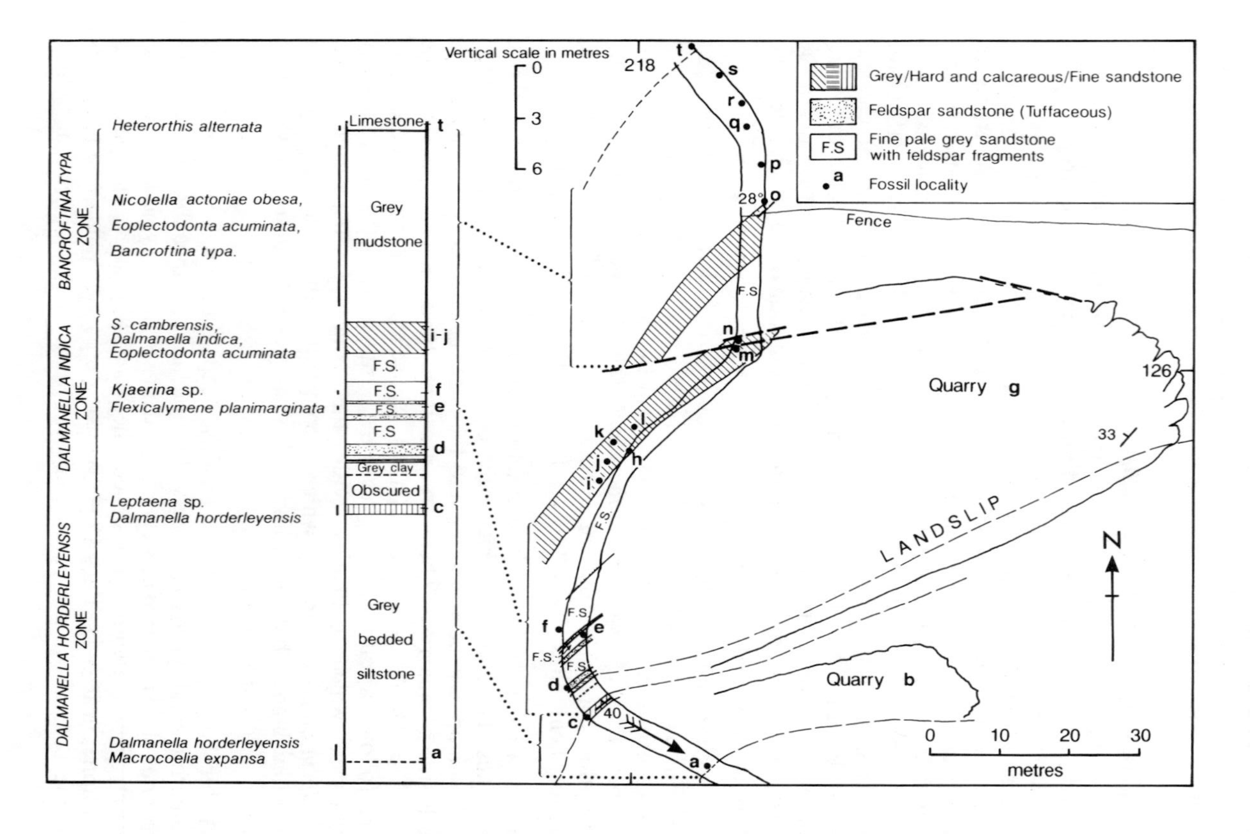

FIG. 7. Sketch map of the Lower Longvillian (Bancroft 1933) part of the Gaer Fawr Formation in Gwern-y-Brain, Loc. 5.

Substage, with fossils including *Dolerorthis duftonensis, Eoplectodonta rhombica, Leptestiina oepiki, Nicolella actoniae obesa, Platystrophia sublimis, Strophomena* sp. and *Estoniops alifrons* (McCoy).

Common fossils from the illustrated section (Fig. 7) include:

a	*Dalmanella horderleyensis, Macrocoelia expansa.*
b	*D. horderleyensis, Howellites antiquior, Kjaerina jonesi, M. expansa, Sowerbyella soudleyensis, Trematis siluriana, Brongniartella* sp., and *Kloucekia apiculata.*
c	*D. horderleyensis, H. antiquior, Leptaena* sp., *Bicuspina spiriferoides, Broeggerolithus* sp., and *F. planimarginata.*
d,e	*B. spiriferoides, D. indica, D. duftonensis, H. antiquior, Kjaerina* sp., *Skenidioides* sp., *Broeggerolithus nicholsoni, F. planimarginata, K. apiculata, Parabasilicus powisii*, and *Tentaculites* sp.
h–l *m,n*	*Dalmanella indica, Eoplectodonta acuminata, H. antiquior, Kjaerina* sp., *Leptaena* sp., *Brongniartella* sp., *S. cambrensis, K. apiculata* and lichid fragments.
p–s	*Bancroftina typa, E. acuminata, E.* cf. *rhombica, L. oepiki, P. sublimis, Skenidioides* sp., *S. soudleyensis, Lingulella* cf. *ovata, Brongniartella* sp., *F.* cf. *caractaci* and lichid fragments.

From Sarn Bridge **continue north-east to Arddleen, turn right onto the A483(T)** and travel south for 4.5 miles; **turn left onto the A458(T)** road towards Shrewsbury. The entrance to **Buttington Brick Pit (Loc. 6c)** is on the right after about 1.3 miles and can be identified by the brick chimney stack. This quarry is owned by Butterley Brick Ltd. **Hard hats are essential**.

6. Buttington Brick Pit (6c) (265 100) **and adjacent quarry (6b)** (2638 1004)
The bricks are fired from crushed mudstones of the Tarannon Shales Formation, or Buttington Shales of Wade (1911). The formation consists of about 71 m of red, green and grey mudstones which are very soft and crush easily into clay. Unlike the Cefn Formation below there are few sandstone layers and, unlike the Trewern Brook Mudstone Formation above, it is not calcareous; both of these factors inhibit brickmaking. The beds dip south-east at *c*.88° and have a strike of 054°. Quarrying thus has left a trench to the depth of the adjacent stream with a north-west 'footwall' formed by the top of the Cefn Formation and a south-east wall of basal Trewern Brook Mudstones. The north-east face depicted in Fig. 8 in its 1986–9 position, is illustrated in detail in Fig. 9.

The oldest beds in the vicinity are exposed in **two defunct roadside quarries, Loc. 6a** (2631 1000) and **Loc. 6b** (2638 1004). Here, medium grey, very friable shales of mid Caradoc age are exposed. They have the appearance of the

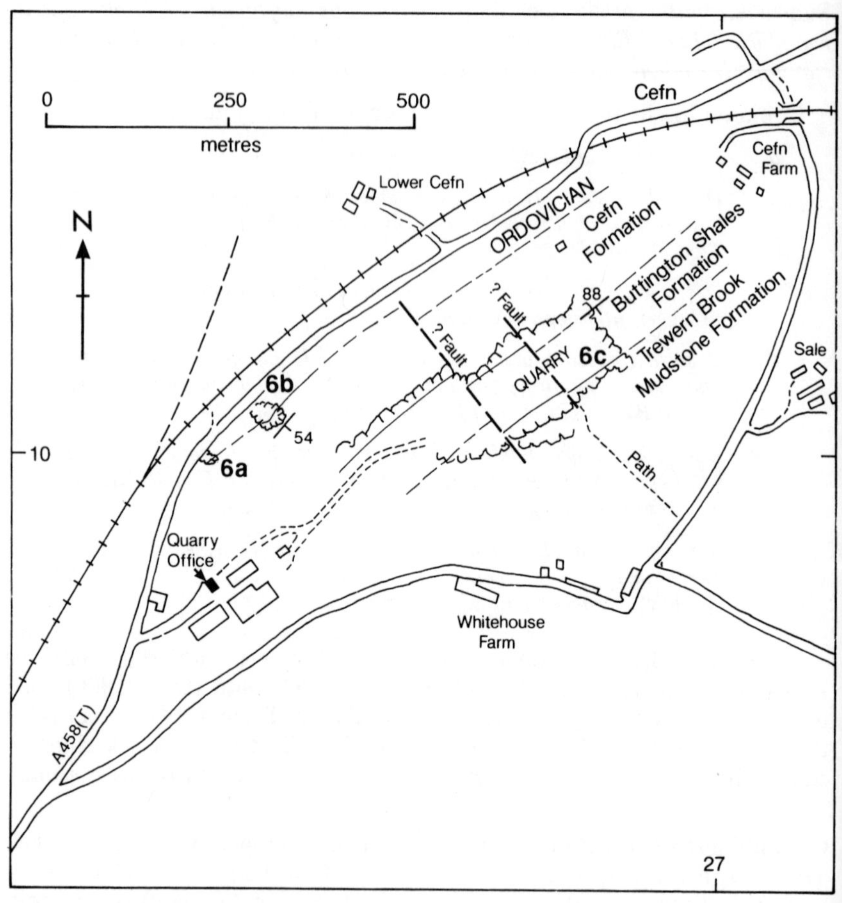

FIG. 8. Sketch map of Buttington Brick Pit and environs, Loc. 6.

Trelydan Shales north of Welshpool, of similar agé, and are possibly equivalent to the Hill Farm Shales Formation of the Breidden Hills (Dixon 1990) and the Forden Mudstones Formation to the south (Fig. 1). These shales have been intruded by a dolerite possessing vesicles filled with calcite and chlorite.

The Ordovician rocks are overlain unconformably by the Cefn Formation (Wade 1911) of Upper Llandovery age. Its type section, with base, is in the roadside quarry (**Loc. 6b**) and it is some 100 m to 110 m thick, mostly of the Telychian Stage, but with the basal part possibly Aeronian. The basal metre or so contains abundant pentamerid brachiopods and pebbles and

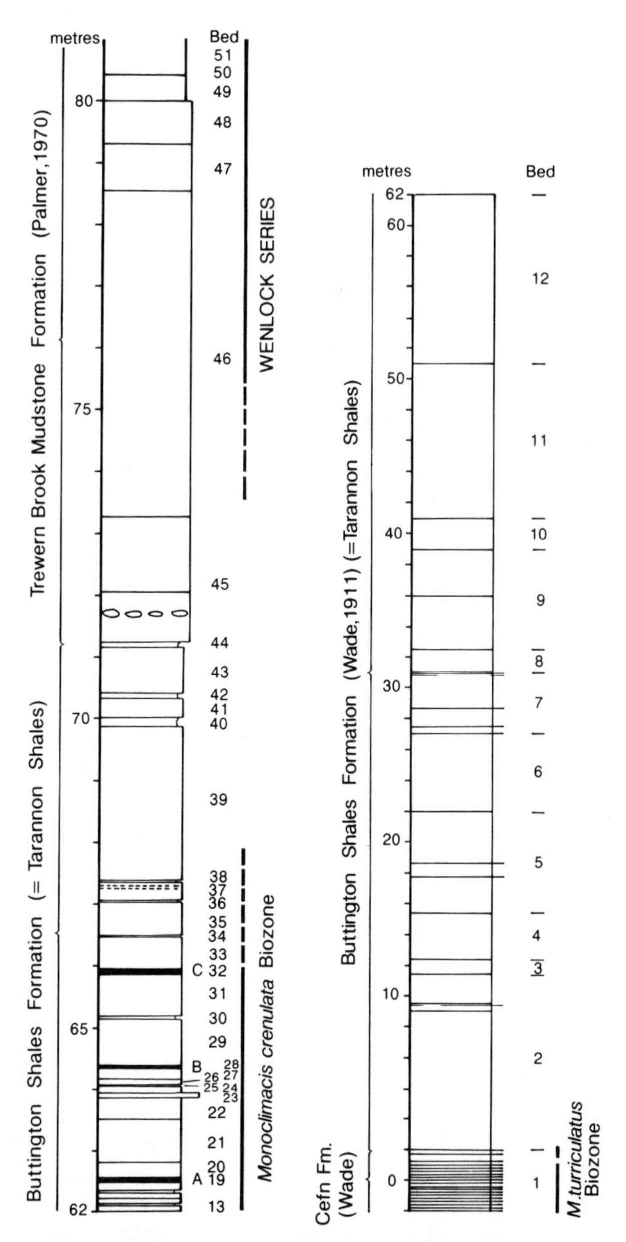

FIG. 9. The sequence in Buttington Brick Pit (bed descriptions in text), Loc. 6c.

cobbles of vesicular dolerite eroded from the Ordovician rocks below. This is a relationship very like that in Welshpool between the Powis Castle Conglomerate (Rhuddanian) and the underlying Ordovician rocks. There is no evidence that the dolerite intruded the Cefn Formation as stated by Wade (1911, pp. 417, 435).

Apart from these basal beds the Cefn Formation is a sequence of thinly interbedded greenish-grey, fine-grained sandstones < 8 cm thick and rather pale grey-green silty mudstones which are very bioturbated. When weathered the mudstones crumble readily into small dice. The beds are poorly fossiliferous but *Monograptus turriculatus* has been found in sandstones near the top (J. Zalasiewicz personal communication 1989). It is possible, therefore, that the whole of the Cefn Formation belongs to *turriculatus* Biozone although Ziegler *et al.* (1968) placed the basal beds in the Aeronian Stage based on the presence of *Stricklandia lens* cf. *intermedia.* The top part of the formation is exposed in the **north-west wall of Buttington Brick Pit (Loc. 6c)** where the surfaces of thin sandstones cover large areas displaying excellent ripples. Near the top of Bed 1 (Fig. 9) some of the sandstones are dark grey and very hard. One of these, 12 cm thick, is intensively ripple bedded, the small-scale ripples being outlined by pale buff-coloured laminae. The pale parts appear to have the same grain size as the dark suggesting that the ripples were wave generated.

The initial deposits of gravel, sand and mud were the shoreline and shallow marine products of the marine transgression of the area from the west. This sea deepened rapidly and the many layers of sandstone were probably introduced by periodic storm-surges. The lack of shell detritus in these suggests, however, that the currents were not strong and that the shoreline had migrated rapidly far to the east. The overlying Buttington Shales are a formation of mudstones rather than fissile shale. They contain several thin films and layers of plastic clay and soft mudstone, usually cream-coloured or pink. These are bentonites and are thickest at the top where beds 40, 42 and 44 are 15 cm, 10 cm and 7 cm thick respectively. Where thick enough such beds are a valuable resource, i.e. Fuller's Earth or Walker's Earth. Similar material of Ordovician age was seen at **Location 5.**

Palaeocurrent indicators are sparse, but sole marks on a 30 cm sandstone in Bed 5 are groove casts aligned 050° indicating currents from the south-west or north-east – probably the latter.

The most varied and interesting part of the formation is the top 10 m. Apart from the bentonites there are three layers of dark-grey mudstone – Beds 19, 28 and 32 (or A, B, C) 7 cm, 5 cm and 8 cm thick respectively. They are graptolitic (see descriptive log) and represent very brief periods when the sea bed was anoxic or dysaerobic, probably the latter for the layers are mottled with pale grey spots (the trace-fossil *Chrondrites*), the product of burrowing organisms which probably required some oxygen. All three layers yield *Monograptus spiralis* and beds B and C yield *Retiolites geinitzianus* and *R. angustidens* respectively, indicating the Llandovery *Monoclimacis crenulata* Biozone s.l (see also Postscript, p.84).

Only 4.03 m of red and green mudstones lie above these Llandovery graptolites and it is unlikely that such mudstones extend into the Wenlock Series here (cf. Cocks & Rickards 1969).

Bed 23 is an 8 cm buff-coloured, non-graded, very fine clastic rock. In appearance it is similar to a nearby Wenlock acid vitric tuff. It is, however, a calcareous very fine-grained sandstone, perhaps the first product of an influence which introduced the overlying Trewern Brook Mudstone Formation.

The base of the Trewern Brook Mudstone Formation is put at the base of bed 45 on top of the 8 cm bentonite. The formation marks a change to more calcareous deposits. Initially the sea-bed conditions were oxic, for the lowest 9 m or so are silty mudstones, highly bioturbated, including darker grey, long (< 5 cm) sub-bedding-parallel burrows < 3 mm wide. A shelly fauna of brachiopods and trilobites is present in the upper half, i.e. top of bed 46 – bed 48. Graptolites are sparse and fragmented but include *R. geinitzianus*.

Palmer (1970, 1972) made a large collection of graptolites which revealed that the lowest two Wenlock biozones of *Cyrtograptus centrifugus* and *C. murchisoni* (in order of ascent) are present in this quarry in beds probably equivalent to beds 46 to 49. Bed 49 is a sharp change to grey, silty, calcareous, largely anoxic, hemipelagic shales with abundant graptolites including cyrtograptids which have a plane-spiral stipe with externally radiating cladia. Bed 49 is only 44 cm thick and is capped by a 1 cm thick bentonite.

Bed		*Thickness in metres*
	TREWERN BROOK MUDSTONE FORMATION	
51	Shales, grey, calcareous with much hemipelagite	Seen
50	Clay, yellow Bentonite	0.01
49	Shales, grey, calcareous, weathered buff, graptolitic	0.44
48	Mudstone, calcareous, pale yellow-green when weathered – brachiopods	0.70
47	Mudstone, buff coloured, silty and shaly, bioturbated	0.75
46	Mudstone, buff, massive, tough and silty. Calcareous and grey when fresh. Trilobites, brachiopods and graptolites in top part where Palmer (1972) recorded: *Glassia* sp., *Skenidioides lewisii,*	

	Sowerbyella sp., *Cyphoproetus binodosus* and *Dalmanites* sp.	5.30
45	Mudstone, buff, very crumbly. Ovoid calcareous concretions < 15 cm across, 40 cm to 50 cm from base. Uneven layer of rubbly clay 2–3 cm thick *c*.80 cm above base	2.00
	BUTTINGTON SHALES FORMATION (=TARANNON SHALES)	
44	Clay, yellow and white. Bentonite	0.08
43	Mudstone, buff	0.75
42	Clay, buff, shaly. Bentonite	0.10
41	Mudstone, grey-green very fractured	0.30
40	Clay, cream-coloured. Bentonite	0.15
39	Mudstone, buff, green, and red, very crumbly. Three layers of cream-coloured clay < 1 cm each	2.50
38	Mudstone, pink, soft. Bentonite	0.03
37	Mudstone, green-buff, crumbly. Two 1 cm layers of red mudstone	0.30
36	Clay. Bentonite	0.02
35	Mudstone, mainly buff coloured or olive-green, but patchily maroon	0.55
34	Clay, grey-purple. Bentonite	0.02
33	Mudstone, buff and olive-green	0.55
32	Mudstone, dark grey, with graptolites: *M. priodon, M. spiralis, Mcl. vomerinus, Retiolites angustidens* and *M. vesiculosus*	0.08
31	Mudstone, mainly grey-buff. Clay layer < 1 cm *c*. 55 cm from base	0.70
30	Clay, pink to purple. Bentonite	0.04
29	Mudstone as 31. Two purple layers < 1 cm each. 53 cm and 64 cm from base	0.73
28	Mudstone, dark grey, small pale-grey spots common. (*Chondrites*). Graptolites: *Diversograptus rectus, Mcl. continens*?, *Mcl. griestoniensis*?, *Mcl. vomerinus. M. anguinus, M. nodifer, M. parapriodon, M. priodon, M. spiralis, P. prantli, R. geinitzianus*	0.05
27	Mudstone, buff-grey	0.16
	Clayey silt parting, pink. ?Bentonite	–
26	As 27	0.12
25	Clay, pink. Bentonite	0.02
24	Mudstone, green-buff	0.10
23	Siltstone, or fine sandstone, calcareous, buff	0.08
22	Mudstone, mainly green-buff, conchoidal fracture	0.35
21	Mudstone, mainly red/purple with green bands	0.60
20	Mudstone, buff-coloured, conchoidal fracture	0.25
19	Mudstone, dark grey with abundant pale olive-green spots (*Chondrites*) in lower half. Graptolites: *M. parapriodon, M. priodon* and *M. spiralis*	0.07

Bed	Description	Thickness (m)
18	Mudstone, greenish, soft, bioturbated	0.10
17	Clay parting, white (bentonite lamina)	< 0.01
16	Mudstone, maroon	0.09
15	Mudstone, grey-green, bioturbated	0.09
14	Mudstone, very soft, pink, ?Bentonite	0.04
13	Mudstone, green-buff, spotted – bioturbated	0.10
12	Mudstone, red, few very thin sandstones	11.00
11	Mudstone, mainly red, with green bands 0.5–10 cm thick and thin layers of green siltstone/fine sandstone 1–2 cm thick, 10–25 cm apart	10.00
10	Mudstone, red apart from thin (0.5 cm) green silty stripes	2.00
09	Mudstones, red and green (50%–50%) in layers < 20 cm, soft, crumbly. Thin green sandstones above 36 m	5.50
08	Mudstones, red (80%) and green, soft, crumbly	*c.*1.50
07	Mudstones mainly grey-green with frequent thin sandstones and four thicker sandstones (9 cm, 5 cm, 7 cm, 2 cm), with large basal burrows or trails. Low-angle minor faults	4.00
06	Mudstones, homogeneous; layers red < 10 cm, green < 8 cm, 50%–50%. Sandstones 0.2–2.0 cm common in green mudstones only. (*c.*10 per 1 m), fine-grained, non-laminated, grey-green	5.00
05	Mudstones green (80%) and red, soft crumbly. Two sandstones (2–3 cm), the lower with basal burrow/trail casts and groove casts 050°. Uneven, low-angle minor faults	6.35
04	Mudstones, red 40%, green 60%, very crumbly/diced with thin (*c.*1 cm) grey-green sandstones in the green mudstone, bases load-cast	3.00
03	Abrupt change to red mudstone, soft, very crumbly/diced	1.00
02	Mudstone, medium-pale grey homogeneous, crumbly. Sandstones, buff and dark grey, 1–2 m apart. One 4 cm thick, cross-bedded, with a thin film of white clay 34 cm below and a 7 cm white clay 9 cm above (bentonites)	*c.*11–15
	CEFN FORMATION	
01	Mudstone, grey, homogeneous, crumbly. Sandstones, generally < 2 cm, *c.*6 per 1 m, fine-grained, hard, siliceous, rectangularly jointed. A dark grey, strongly ripple-laminated sandstone occurs near the top with burrow casts on the base and two other sandstones (5 cm and 3 cm respectively) in the overlying 70 cm. *M. turriculatus* occurs in these beds	Seen 6.00

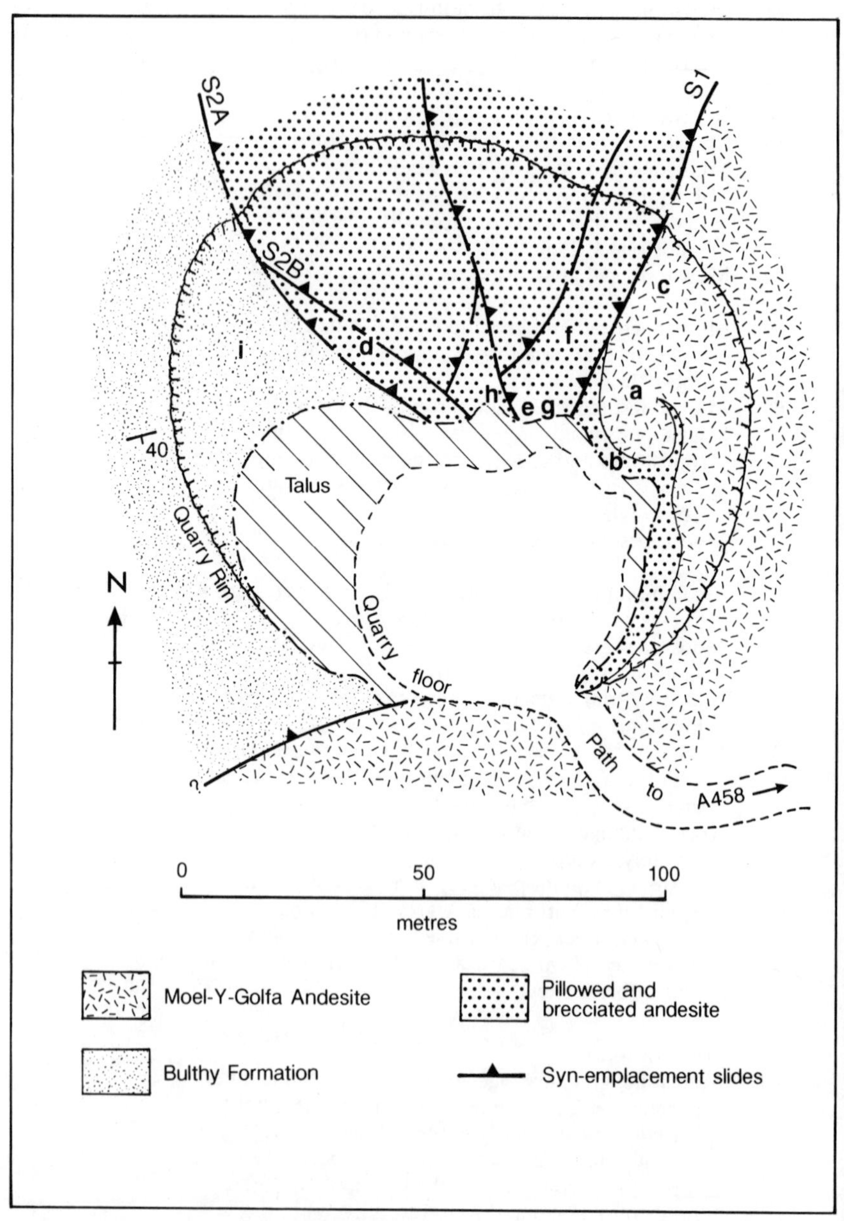

FIG. 10. Map of Moel-y-Golfa Quarry

Take the A458 (signposted to Shrewsbury) north-eastward through Cefn and Trewern. Access to **Moel-y-Golfa Quarry** is by the first track on the left as the road starts to climb. Off-the-road **parking is available for one or two small vehicles** at the end of the track. **Do not block the private driveways**. A footpath leads south-westward from the parking place to the quarry (Fig. 10).

7. Moel-y-Golfa Quarry (2875 1192) provides a section through the Moel-y-Golfa Andesite and its host sediments, the volcanic conglomerates of the Bulthy Formation (Soudleyan). The quarry is no longer worked and presents a single, precipitous, curving face some 20 m high in its central part. A scree of fresh-looking debris testifies to the occasional instability of the central part of the face, and the need for care below it.

The intrusion and host sediments are similar in their mineralogy and weathering characteristics. Before approaching the quarry face it is worth establishing the nature of the large-scale geological features viewed from the quarry entrance. They are described below from east to west (Fig. 10).

Moel-y-Golfa Andesite

The north-eastern aspect of the quarry face is dominated by the intrusive margin of the Moel-y-Golfa Andesite (a & c, Fig. 10), displaying a large-scale, lobate geometry. The andesite that forms the large lobe (**Loc. 7a**) is not easily accessible. However, when viewed from the **foot of the quarry face**, it displays columnar joints and crude surface corrugations, the latter probably reflecting slight withdrawal of magma during the final stages of intrusion (cf. Kokelaar *et al.* 1985).

Contact Zone

The margin of the andesite has a veneer of pillowed and brecciated andesite (e.g. **Loc. 7b**) and is separated from the thick contact zone of pillowed and brecciated andesite by a steeply dipping slide (S1, Fig. 10). From the quarry entrance the S1 slide and its complementary (antithetic) slides (S2A & B) are seen as prominent features of the **main quarry** face. These surfaces have been interpreted as syn-emplacement slides (Dixon 1990), their association with pillowed and columnar jointed andesite within the contact zone (**Locs. 7c and 7d**) suggesting synchroneity with andesite intrusion. Other planar features **in the central face** (Fig. 10) may also represent early slides, though slickensiding also indicates some later tectonic movement.

The contact zone is also characterized by the spectacular andesite pillows and/or tubes. Some pillows are isolated in three dimensions (**e.g. Loc. 7e**), whereas others are seen as circular or ovoid sections higher in the quarry face (**e.g. Loc. 7f**). The pillowed andesites also enclose an enclave of the host conglomerate (**Loc. 7g**) and are overlain by a zone of altered host sediment that is cut by numerous vertical to subvertical veins. The veins are infilled by fragments of the host sediment mixed with blocks of andesite (**best examined at Loc. 7h** and in the large fallen blocks at the **foot of the main face**).

The sequence of events envisaged in this complex contact zone is as follows:

1. Intrusion of andesitic magma into partly lithified, but water-saturated host sediments (Bulthy Formation conglomerates).

2. Intrusion, assisted by the presence of abundant steam generated by magma/pore fluid interaction, dramatically reduced the yield strength of the host sediments, facilitating the loss of cohesion between clasts and the formation of andesite lobes, pillows and tubes. The andesitic magma probably intruded as a series of pulses (Dixon 1990), the intrusive pillows being quenched and brecciated upon contact with the vapour-saturated host sediments.

Fluidizing streams of vapour (steam) generated during rapid quenching and carrying large volumes of entrained debris were explosively injected through the adjacent host sediments and are now preserved in the network of veins seen at **Location 7h**.

3. Gravitative collapse of the pillowed and brecciated contact zone along syn-emplacement slides (S1 & S2, Fig. 10) also took place at this time probably as a result of a loss of cohesion at the steep margin of the intrusion (no doubt enhanced by the presence of fluidized sediments, cf. Kokelaar 1982).

The original upper contact of the Moel-y-Golfa Andesite is not exposed and it is not possible to be sure whether the andesite was totally intrusive or partially extrusive (Dixon 1990). The abundance of andesitic debris contained within the host volcanic conglomerates however, strongly suggests that similar magmas were locally extrusive or that high-level intrusive magmas were subject to penecontemporaneous erosion.

Host Sediments (Bulthy Formation)

The **south-western quarry face** is dominated by thickly bedded massive volcanic conglomerates of the Bulthy Formation (e.g. **Loc. 7i**). Their dip and strike is uncertain here but mapping has shown that they dip at approximately 40° to the south-east and that the Moel-y-Golfa Andesite is markedly discordant to this trend.

Although the conglomerates have not yielded fossils at this locality, trilobite fragments have been recorded from a similar facies elsewhere in the Breidden Inlier and suggest a Soudleyan age. The Bulthy Formation has been interpreted as a turbiditic slope apron sequence (Dixon 1988, 1990). The conglomerates are oligomict (containing only andesite clasts), the well-rounded nature of the clasts and the presence of a 'shelly' fauna implying that the provenance of the turbidites was an andesitic volcanic island.

Travel north-eastward along the A458 to Middletown where ample parking is available at the Breidden Hotel car park. Cross the A458 and **take the road to the right of the telephone box**. You are now walking up the dip slope formed by the Llandovery (Aeronian) Cefn Formation; continue to the junction and **take the right-hand fork onto the quarry access road**. The latter essentially follows the break in slope formed by the Ordovician/Silurian unconformity, with small outcrops to the left of the road comprising thinly bedded, argillaceous and bioturbated sandstones of the Cefn Formation. The unconformity is crossed at the tight hairpin bend, the outcrops on the right hand side of the road up to the quarry being formed by andesitic volcanic conglomerates of the Bulthy Formation (Soudleyan).

8. Middletown Quarry (2990 1289) is a working quarry and **permission is essential for any visit** (for authorization contact Border Hardcore Ltd., Middletown, Powys).

The quarry provides an excellent section through the rhyolitic tuffs and interbedded basinal mudstones of the Middletown Formation. The geological structure of the quarry area is complex, the Ordovician and Silurian sequences being strongly folded and cut by several faults, some of which are mineralized (Fig. 11).

Middletown Quarry itself is located in the faulted hinge zone of a ENE–WSW trending anticline/syncline fold pair. Most of the quarry faces are in the faulted southern limb of the syncline, characterized by steep dips (60°) to the north. They are cut by two approximately E–W trending faults both of which downthrow to south, but which have a component of strike-slip indicated on the most northerly fault by sub-horizontal slickensides. The latter fault is well seen in the uppermost quarry face when standing near the quarry entrance and looking northward.

The syncline is separated from the anticlinal structure by a prominent NE–SW trending fault zone (**Loc. 8a**) that dips to the south-east at 40° and displays large sub-horizontal slickensides.

The folding in the Breidden Inlier is attributed to the Acadian phase of deformation (early to mid Devonian), the Middletown folds being part of an arcuate belt of deformation that also affects the Silurian sequences of the Long Mountain (Palmer 1972).

The most complete stratigraphical section through the Middletown Formation is located in the faulted southern limb of the syncline, and is best described with reference to the detailed map (Fig. 11) and graphic sedimentological log (Fig. 12). The Middletown Formation can be subdivided into three members (Fig. 12). The main working quarry-faces are all within the Lower Rhyolitic Tuff Member and are easily accessible.

The Graptolitic Shale and Upper Rhyolitic Tuff Members, however, are

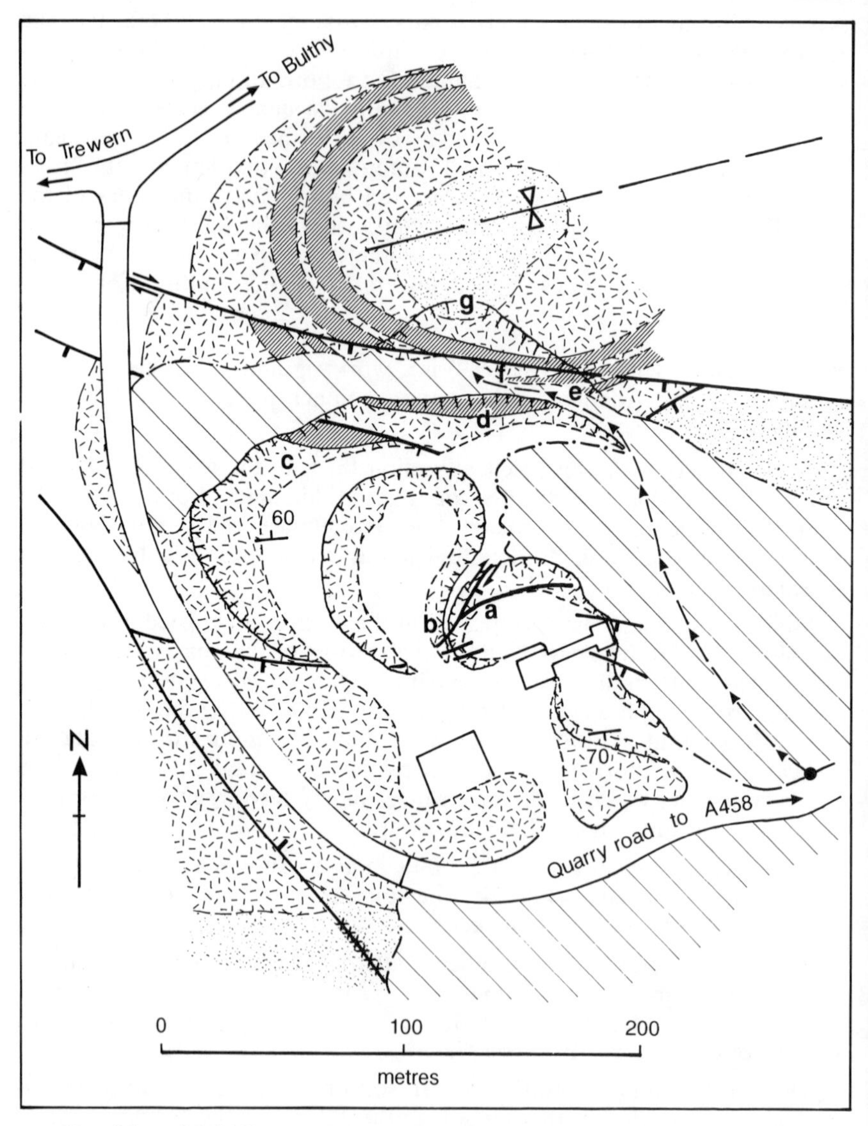

FIG. 11. Map of Middletown Quarry. (See Fig. 12 for key.)

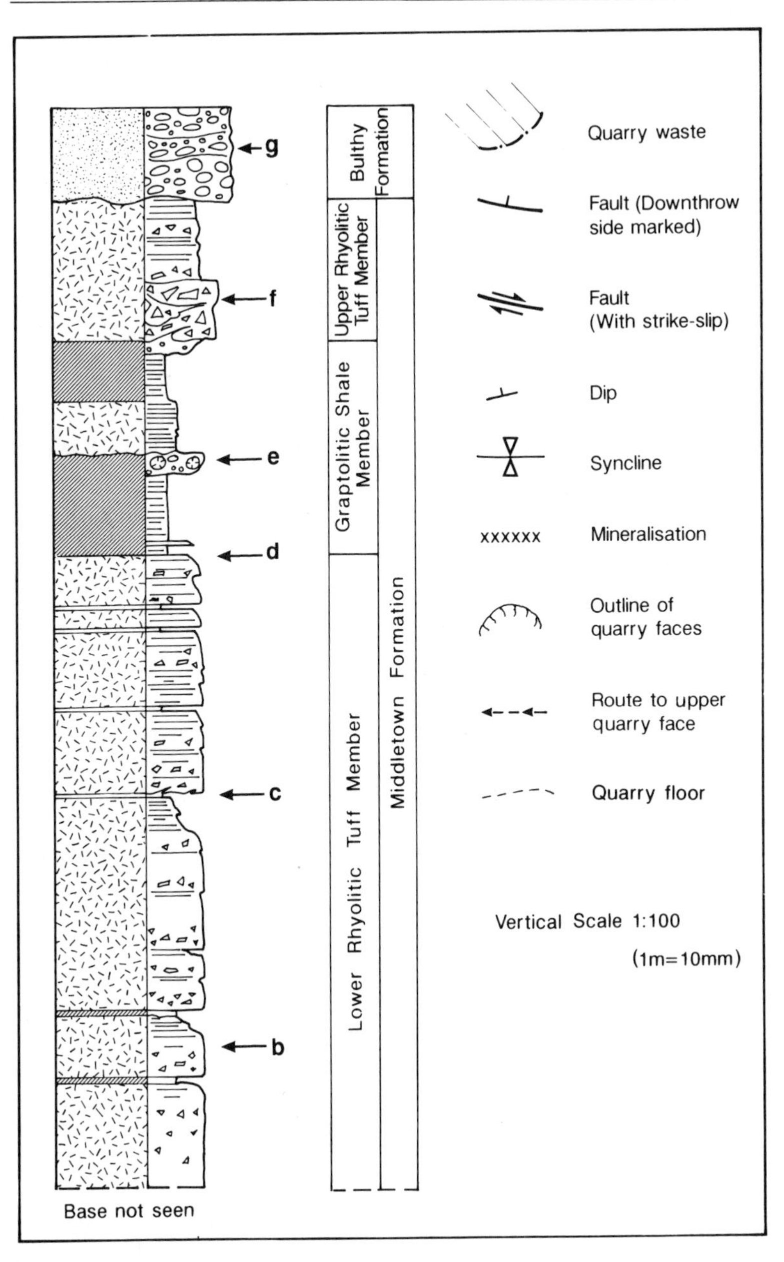

FIG. 12. The section at Middletown Quarry.

exposed only in the **disused upper quarry level**, access to which is gained by walking up the levelled slope that forms the eastern flank of the quarry (route marked on Fig. 11).

Lower Rhyolitic Tuff Member

This unit comprises rhyolitic tuff-breccias, lapilli-tuffs and bentonites with subordinate interbedded black mudstones. The quarry section begins with thickly bedded, rhyolitic tuff-breccias with occasional interbedded mudstones. A typical bed seen at **Location 8b** shows the characteristic sedimentary structures of a coarse grained pyroclastic facies. The bed is graded (fining upward via planar to wavy bedded lapilli-tuff into a coarse-grained lithic tuff) and capped by a fine to medium grained lithic tuff with a rippled top. This sequence of sedimentary structures indicates turbiditic sedimentation, pyroclastic debris being supplied to the turbidity currents either directly from a submarine eruption column or by the resedimentation of unstable coarse debris that accumulated on the flanks of the volcano.

In addition to the size grading noted above, many of the tuff-breccia beds also show evidence of density grading, the upper part of many beds being rich in pumice debris. Because the pumice in the Middletown Formation has been replaced by chloritic clay minerals (notably chlorite/smectite and illite/chlorite, mixed-layer clays), the pumice-rich horizons are often distinctly green in colour.

The presence of pumice within the pyroclastic sequence provides direct evidence of explosive magmatic activity (significant volumes of pumice are unlikely to have been reworked from older sequences) and in the marine environment may give some indication of the emplacement temperature of pyroclastic deposits ('cold' pumice tends to float and thus may be widely dispersed, whereas 'hot' pumice rapidly sinks and thus accumulates close to the source vents) (Whitham & Sparks 1986).

Interbedded with the coarse-grained pyroclastic facies in the upper, more thinly bedded part of the Lower Rhyolitic Tuff Member are several horizons of yellow, bentonitic clay (e.g. **Loc. 8c**, Figs. 11 and 12). This clay is the alteration product of very fine ash, the final depositional phase of the eruptive sequence that presumably blanketed the flanks of the volcano and the nearby seafloor. Large clasts of bentonitic clay are also present in the basal parts of the coarse-grained pyroclastic facies, incorporated within the turbidity currents during their downflank transport.

Graptolitic Shale Member

These beds are exposed in the **upper part of the main working quarry-face**, but can only be examined in detail in the uppermost (disused) face (Fig. 11). The lower boundary of the member is sharp and planar, with black, pyritic shales resting conformably on the underlying lower Rhyolitic Tuff Member. The contact can be examined by scrambling up the shale scree at **Location 8d**. The basal part of the Graptolitic Shale Member contains several thin (less than 20 cm) beds of rhyolitic lapilli-tuff, but soon passes upward into a thick shale interval (Fig. 12). The thin basal tuffs reflect either the gradual waning of volcanicity or a shift in the locus of coarse-grained pyroclastic deposition.

The black, pyritic shales are well laminated and have yielded the graptolites *Climacograptus brevis* and *Diplograptus foliaceous*. Deposition below storm wave-base under anoxic conditions seems likely. This facies is also typical of the Caradoc mudstone successions in the Shelve and Montgomery inliers and has been interpreted as a hemipelagic, basin-fill sequence (Dixon 1988). Interbedded with the basinal mudstones is a distinctive sequence of white, very fine to fine grained, lithic tuffs (Figs. 11 and 12) interpreted as a series of water-lain ash-falls from a subaerial eruption column.

At the base of the ash-fall sequence a lenticular pebbly mudstone bed is developed locally and exposed sporadically at **Location 8e**. The pebbly mudstone contains large (up to 1 m) blocks of rhyolite. It is notable for containing the tabulate coral *Foerstephyllum simplissimum*, indicating that the volcanic centre periodically built up into shallow water and was mantled by coral reefs. The pebbly mudstone bed is interpreted as a mud-flow deposit. Its contained rhyolite blocks and overlying ash-fall sequence suggest eruption-related instability or slope failure, the resultant mud-flows sweeping reef debris down the flanks of the volcano into deeper water.

The Upper Rhyolitic Tuff Member

The Upper Rhyolitic Tuff Member commences with a spectacular rhyolitic breccia, best seen at **Location 8f**. The breccia mainly comprises angular blocks and lapilli of banded rhyolitic lava, though pumice blocks and lapilli are also present, as are numerous clasts of black mudstone.

Both clasts and matrix are extensively silicified giving the rock a flinty appearance. The breccia is interpreted as a debris-flow deposit, the abundance of lava debris suggesting that it was derived by the collapse and/or explosive disintegration of a submarine rhyolitic lava flow or dome. Crudely developed columnar joints, and *in situ* brecciation of rhyolite clasts suggest that the debris-flow (or at least some of its components) was deposited at high temperature.

The breccia is overlain by a sequence of bedded rhyolitic tuff-breccias and lapilli-tuffs similar to those described from the Lower Rhyolitic Tuff Member.

The Upper Rhyolitic Tuff Member of the Middletown Formation is overlain by massive, andesitic volcanic conglomerates belonging to the Bulthy Formation. These may be seen at the top of the quarry face (**Loc. 8g**, Fig. 11), but are best examined on the quarry access road. They are similar to those seen at Moel-y-Golfa Quarry (**Loc. 6**).

From Middletown **return south-west along the A458(T)** towards Buttington. About one mile beyond the Brick Pit **turn left along the B4388 at Kingswood** and continue southwards to Montgomery.

Admission to the castle is free and **there is a small car/minibus park at the entrance**.

9. Castle Rock (SO 2215 9680), **Montgomery** is an eminence worthy of a visit for the view alone. The Rock rises abruptly from the south side of the Drift covered ground of the Severn and Camlad Valleys and faces the Forden Ordovician 'inlier' in the middle distance to the north, with Welshpool in the far distance. Much of the earlier part of the itinerary is thus visible from here. To the NNE is Long Mountain, a syncline in an almost complete sequence of Silurian rocks from mid-Llandovery to the Přidoli Series, whilst five miles to the east is the Ordovician inlier of Shelve dominated by Corndon Hill, a concordant intrusion of dolerite. Montgomery Castle was built under the direction of King Henry III in the year 1223 once it was realized that the timber-and-earth motte and bailey, about one mile to the north, had become an inadequate defence. The castle was demolished by its owner, Lord Herbert, *c.* 1650.

Castle Rock has been eroded from a large lenticular mass of volcanic conglomerate (Fig. 13), an oligomict deposit similar to the Bulthy Formation and composed of cobbles and boulders of andesite comparable with that of Moel-y-Golfa (Loc. 7). They are packed in a feldspathic sandstone matrix so similar in composition that it is difficult to recognize the outlines of clasts within it. This mass of conglomerate is enclosed within a thick formation of mudstone with thin sandstones, the Forden Mudstone Formation (BGS 1:10 000 sheet SO29NW). It was considered by Sanderson (1975) to have been emplaced here as a lahar, but the authors see no evidence of the high temperature of a lahar and favour normal submarine mass-flow from a volcanic source. The volcanic centre might well have been that examined at Moel-y-Golfa (Loc. 7), but another possibility lies to the north-east in the Shelve inlier. Further details have been described in recent accounts (Dixon 1988, Cave *et al.* 1988).

At the **northern end of the Rock is a quarry** (**Loc. 9a**, Figs. 13 and 14). Here the upper surface of the volcanic conglomerate is overlain by thin sedimentary conglomerates and sandstones. These have yielded fragments of brachiopods, trilobites and even a graptolite, the main species being: *Dalmanella* cf. *salopiensis gregaria, D.* cf. *salopiensis transversa,*

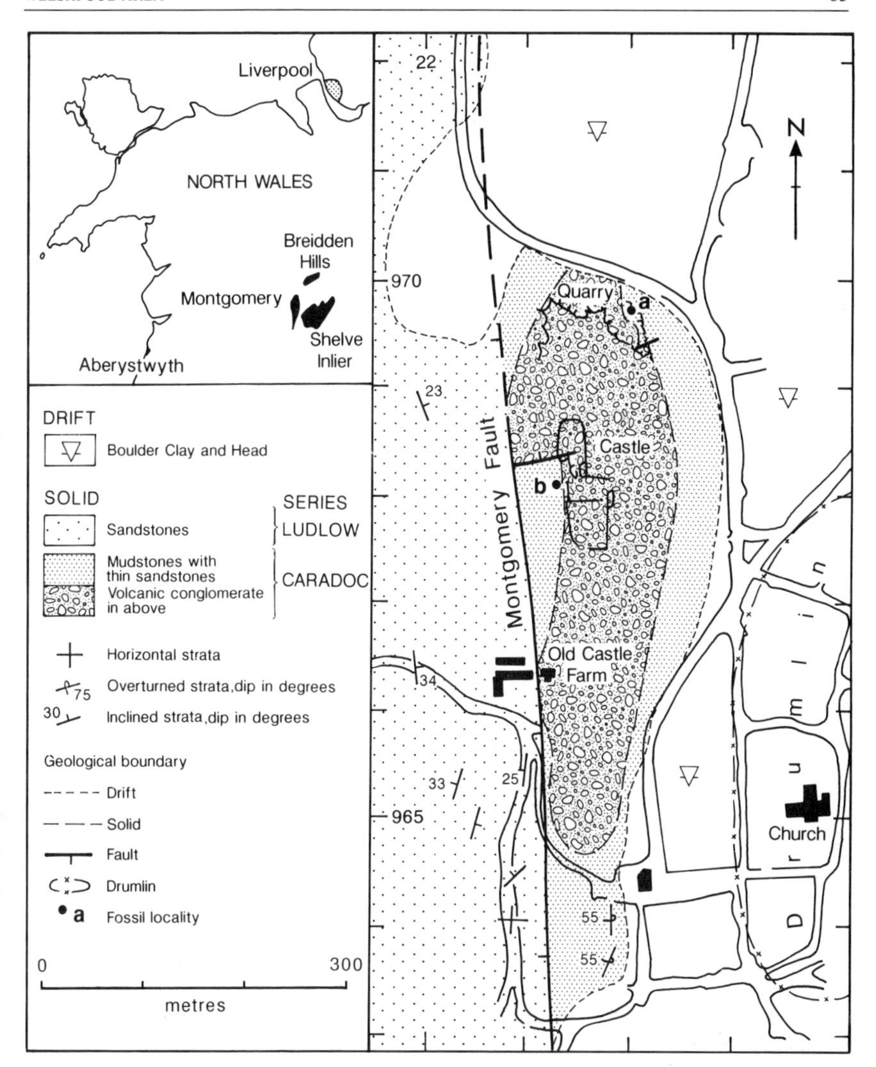

FIG. 13. Geological sketch map of Montgomery including Castle Hill (Cave et al. 1988, fig. 1.).

Skenidioides cf. *costatus, Sowerbyella* cf. *multipartita, Broeggerolithus broeggeri, Salterolithus caractaci* and a variety of acritarchs, chitinozoans and ostracods.

The basal contact of the volcanic conglomerate has a nearly vertical dip and can be seen on the west side of the castle wall. On this underside,

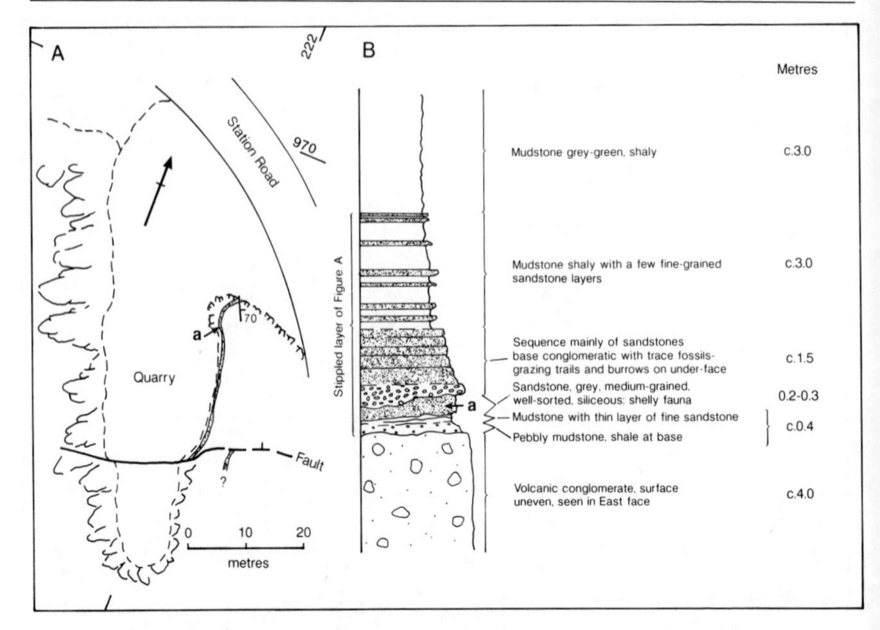

FIG. 14. a) Station Road Quarry and b) the sedimentary sequence overlying the volcanic conglomerate at Loc. 9a in Station Road Quarry (Cave et al. 1988, fig. 2.).

below the Castle Keep (**Loc. 9b**, Fig. 13) a cephalon of *S. caractaci* can be seen impressed into the conglomerate. The conglomerate clearly belongs to the lower part of the Soudleyan Stage.

Postscript.
Since this chapter was written, further collecting of graptolites with Dr D.K. Loydell from the Tarannon Shales of Buttington (Loc. 6) has enabled us to adopt the more restricted Bohemian use of the *crenulata* Biozone. Graptolite bands A, B and C can be assigned to the three Bohemian subzones, *parapriodon, anguinis* and *geinitzi*, respectively, of the *spiralis* Biozone, overlying the *crenulata* Biozone *s.s.* A fourth graptolite band, D, has been discovered near the top of Bed 6, which yielded a fauna indicative of the *crispus* Biozone or basal *griestoniensis* Biozone, below the *crenulata* Biozone *s.s.*

Acknowledgements. The authors are grateful to BP Exploration and the University College of Wales, Aberystwyth, for the use of their facilities, to Dr David Loydell for identifying graptolites from the *crenulata* Biozone, to Myra Bloomfield who typed the account, to Dr Adrian Rushton for his ever helpful discussions, and to Mary Hignett who has given much encouragement to geological studies in the area.

3. THE WENLOCK AND LUDLOW OF THE NEWTOWN AREA

by R. CAVE, B. A. HAINS *and* D. E. WHITE

Maps	*Topographical*:	1:50 000 Sheets 136 (Newtown & Llanidloes) and 137 (Ludlow, Wenlock Edge)
	Geological:	1:63 360 Sheets 60 NE & 60 SE (Old Series) 1:50 000 Sheet 165 (Montgomery) (*in press*)

THE 'Welsh Lower Palaeozoic Basin' is a simplified view of a SSW–NNE tract of sialic crust, on the southern side of the Iapetus Ocean, which accumulated large thicknesses of clastic detritus during the Early Palaeozoic Era. This basin did not always behave as one tectonic unit and probably formed as several sub-basins. However, the concept provides a framework which facilitates discussion of Welsh Early Palaeozoic history, especially during its gerontic, Silurian, phase.

The Arenig to early Caradoc Shelve–Builth sedimentary basin was inverted and deformed during late Ashgill times, whereupon the area became emergent as part of the shelf to the Welsh Silurian Basin. The gradual submergence of the shelf during the Llandovery is a subject of study in Excursion 2, whilst this itinerary examines deposits that followed the Llandovery transgression, including those that prograded eventually into the basin, effectively filling it. Large parts of the margins of this basin are very ill-defined but the eastern margin lies in eastern Wales and its position is fairly well constrained in an abundance of rock exposure.

The early Silurian basin margin of eastern Wales was probably fault controlled along N–S lines. In the north there were two such lines, one through the Berwyn Hills, the other along the Severn Valley and thus the basin margin was broad and diffuse (Cave 1992). Southwards these combined into the 'Towy Lineament' to produce a very sharp basin margin with a steep slope to the west of it.

In mid Silurian times in northern Powys the main basin slope lay further east, probably in a position between Newtown in the west and Church Stoke in the east, thus transecting the line of the fundamental structure known as the Severn–Towy lineament. Outcrops of Wenlock and Ludlow rocks on the south side of the Camlad–Caebitra–Mule valley lie on this transect and allow investigation of formations that were deposited across

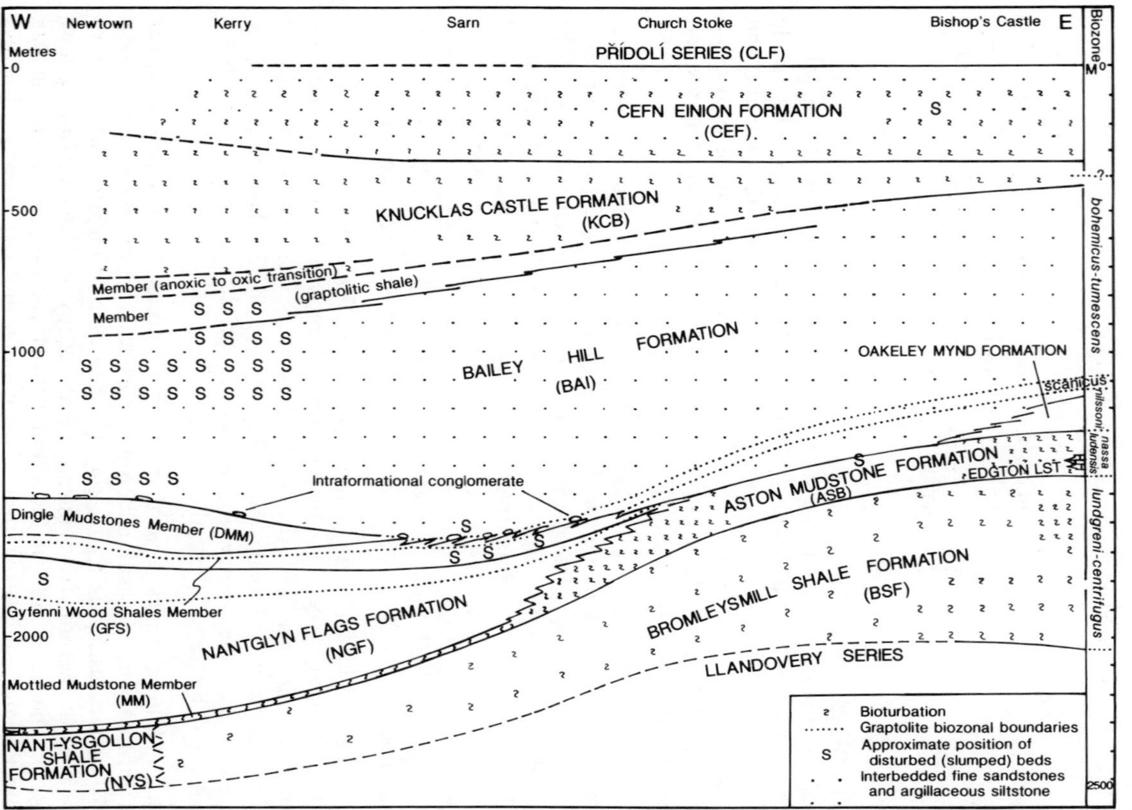

FIG. 1. The Wenlock and Ludlow sequence on a broad, generalized traverse from Bishop's Castle to west of Newtown.

the slope. In both Wenlock and Ludlow times, Bishop's Castle, near the eastern end of the transect, lay within a marine shelf environment, whereas the western end, near Newtown, lay probably within the basin itself at least until early Ludlow times. The 'core' of the itinerary lies in this E–W transect.

ROCK-TYPE ASSOCIATIONS, OR FACIES

1. Dark grey carbonaceous silty laminite (hemipelagite)
This is a facies described very well by Cummins (1959b) as 'laminated muddy siltstone' and is a rock-type for which the term hemipelagite was employed by Natland (1976): 'Hemipelagite sediments are formed by the slow accumulation on the sea-floor of biogenic and fine terrigenous silt particles ... Fossil species in the hemipelagite are indigenous because they lived and died during the deposition of this layer.' Pelagite was a term that probably originated with the Challenger expedition of 1872–6 to describe organic and inorganic deep-sea oozes and fine clay.

Carbonaceous, thinly laminated, hemipelagite accumulated specifically on an anoxic marine bottom for otherwise a benthos would have scavenged the organic debris, burrowing and destroying the depositional laminae. In hand specimen these laminae can be seen as millimetre-thin alternations of dark grey to brown organic debris and pale grey, usually calcareous, silt in small ovoid segregates.

This facies in fact forms the background deposit during much of the period under investigation (Wenlock and early Ludlow) and, in event-dominated formations, appears in the 'quiet' intervals between the event-generated beds. Where event-deposits are minimal it becomes the dominant facies of course, as in the Nant-ysgollon Shale Formation, the Oakeley Mynd Formation and the Gyfenni Wood Member of the Nantglyn Flags Formation (Fig. 1).

Because sea-bottom conditions were unfavourable for the establishment of a benthic fauna, fossils found in this facies are pelagic forms, principally graptolites, although orthocones are also well represented.

2. Calcareous, fine sandstone and argillaceous siltstone event-couplets
This facies has been described fully in recent literature on the area (Lynas 1987, Tyler and Woodcock 1987) and is specific to the Bailey Hill Formation. Fine sandstones form the lower part of a litho-couplet and range from 1 cm to 40 cm in thickness. They are composed of fine, sub-angular, silica sand, with grain-contact and a calcitic argillaceous matrix. Thus when unweathered these sandstones are pale to medium grey, but they become porous and fawn to brown when weathered. The tops and particularly the bases are sharp, the latter commonly being erosive revealing flute-casts, prod-marks and groove-casts. Many beds pinch and swell normally in consequence of the tops being waved, usually symmetrically, with ampli-

tudes up to 5 cm and wave-lengths of 25 cm to 50 cm. Others are asymmetrical and are large current ripples.

The internal bedding fabric is most commonly a sub-parallel wavy lamination; low-angle cross bedding is also common whilst steeper, current-ripple foresets occur, particularly in the ripple-form tops. Convolute bedding is also fairly common, often about the middle of the bed, sometimes grading laterally from low-angle lamination. There are examples where convolutions are truncated by wavy, sub-parallel laminae concordant with, and forming, the waved surface. Within the sandstones there may be many flat or wavy truncation surfaces revealing erosion and reactivation during the deposition.

The higher part of the couplet consists of grey argillaceous siltstone, virtually homogeneous, though faint lamination is sometimes visible. It is an incohesive rock that weathers quickly to rubble. It is also often a rather indistinct entity with thin wisps or layers of fine, laminated sandstone or siltstone dividing it. Thicknesses vary from tens of centimetres to thin partings separating the fine sandstone layers.

Fossils occur as shell-detrital lags; graptolites, commonly current orientated, are also present, especially in the lower part of the formation.

Although the bases of the sandstones commonly display groove-casts and flute-casts (Fig. 12), and vertical grading of grain size is present occasionally, for instance where a coarse shell-detrital lag occurs at the base, the couplets are not so orderly and well defined as the Bouma sequences displayed by the turbidites of basinal Wales; curved, sub-parallel lamination and internal surfaces of discontinuity are much more abundant. Likewise, the argillaceous siltstones are very uneven without the distinctiveness of the Te interval as seen in Silurian turbidites farther west. Nevertheless, the couplets indicate that each was the deposit of a waning current.

The exact nature of the flow regime that delivered these beds is still debated, but storm-generated marine currents with a tractional carpet were considered to be the likely agent of delivery (Tyler 1987). It is also possible to view some bed forms as having been marine-wave influenced during deposition (Brenchley 1985, Duke 1985).

A variant of this facies forms the Dingle Mudstones Member of the Bailey Hill Formation (Fig. 1). It is dominated by argillaceous siltstone with fine sandstones, which are mostly very thin, up to 2 cm, impersistent and wavy. The paucity of sand does not signify a markedly lower input of terrigenous detritus, otherwise graptolitic hemipelagite would have become an important constituent of the facies, especially as inter-event partings, and it is, in fact, inconspicuous. There are, however, horizons of current-orientated graptolites.

3. Fine sandstone/siltstone–mudstone rhythmite

This facies is a fairly uniform repetition of well-ordered couplets of fine-grained sandstone or siltstone, and mudstone. Again this is an event-promoted sedimentary couplet usually associated with laminated hemipelagite as the background deposit which became dominant in the intervals between events. The rhythm is thus that described by Cummins (1959) and as 'ribbon-banded mudstones' by Warren *et al.* (1984), in which the fine sandstone, or siltstone, is subordinate beneath homogeneous mudstone, usually capped by a thin interval of laminar hemipelagite. The fine sandstones are pale grey, usually no more than a centimetre or two thick and have a bedding fabric of parallel lamination and current-ripple cross-lamination. Occasionally there is a coarse shell-detrital lag at the base, a form of grading, so that they conform closely to the Bouma turbidite sequence Tacd or Tcd. The homogeneous mudstone in the upper part of the couplets satisfies Te of the Bouma sequence and indeed they are considered to be low-density turbidites of a basinal environment. Graptolites are common in both the hemipelagite interval and in the turbiditic fine sandstone, where they are orientated (by currents) and usually sorted and monospecific. Graptolites also occur in the Te interval, where they are usually well preserved. Bottom structures are not common, though flute casts do occur.

This facies characterizes the Nantglyn Flags Formation.

4. Bioturbated calcareous mudstone

In the sequence under examination, bioturbated calcareous muds were dominant only in one short interval in the Wenlock. The resultant mudstones are rather silty, very calcareous, massive and unbedded. They are pale to medium grey when fresh but decalcify on weathering to buff and olive-green colours. Burrows are pervasive and in places strikingly obvious. They are orientated sub-parallel with and sub-normal to the plane of bedding. Representatives of benthic and pelagic faunas are preserved – brachiopods of the comparatively deep-water *Visbyella trewerna* Community, trilobites, and sparse, but definitive, graptolites.

Mudstones of this nature form the Aston Mudstone Formation (a name adopted by BGS for the current mapping of 1:50 000 Sheet 165 from Allender 1958) and the Mottled Mudstone Member of the Nantglyn Flags Formation (Boswell & Double 1940; Warren *et al.* 1984).

5. Siltstone laminate

A thinly fissile siltstone facies forms the Knucklas Castle Formation (Holland 1959). Although this itinerary does not include exposures of the Formation, it is significant to the facies study in that it overlies the Bailey Hill Formation and its establishment records the end of the anoxic marine bottom conditions that had mostly dominated the area since Llandovery times. It also records the end of the event-dominated deposition of early Ludlow times here. The rock consists of very closely spaced, calcareous, green-grey siltstone laminae, commonly 2–3 mm thick, but impersistent. It is virtually unfossiliferous; indigenous brachiopods, principally *Dayia navicula*, are rare and allochthonous detritus is too fine to have incorporated shell debris. However, a pelagic fauna consisting of a few orthocones, entomozoan ostracods and small bivalves are also present, but graptolites, so conspicuous in older facies, have not been found. Bioturbation is evident as burrows of various sizes which have disturbed, though not always destroyed, the lamination. Small surface (?grazing) trails are also visible. This facies might be an oxygenated equivalent of laminar hemipelagite-rich older formations and as such might reasonably be expected to have preserved some graptolites, in spite of bioturbation. Their total absence from this facies in the Knucklas Castle Formation and indeed from other (albeit less favourable) facies of similar, mid Ludfordian, age elsewhere in the Welsh Basin and its shelf areas remains an enigma. It is to be noted that such a facies had been developing at the top of the Bailey Hill Formation (the Cwm Mawr Beds of Allender (1958) and the Leintwardinensis Shales of Earp (1938)) just before the shelf here became oxygenated.

6. Minor facies

a) Bentonite. Layers of soft, pale cream-coloured clay, usually < 10 cm thick, are common in the Wenlock–Ludlow sequence. They are the product of submarine decomposition of vitric volcanic dust and consist of illite-montmorillonite mixed-layer clays. Illite is the more dominant in higher metamorphic grades. Bentonites occupy the terrigenous extreme of a range of water column detrital precipitates with biogenic pelagite at the other extremity.

The volcanic centres providing this 'dust' have not been located. They may be very distant, but microgranites of possible Silurian age occur at Owlbury and The Heblands near Bishop's Castle (Sanderson & Cave 1980), an acid vitric tuff of late Wenlock age occurs near Meifod, and a thick volcanogenic turbidite of early Wenlock age occurs near Llanidloes.

b) Slumps. Disturbed beds occur sporadically throughout the Wenlock and lower Ludlow in several rock types. They constitute thin single layers, thick mappable units and, in the Clun Forest area, stratiform alternations of slumped and non-disturbed beds which make up a large thickness of the Bailey Hill Formation (Earp 1938, Woodcock 1973). The reason for the disturbance is soft-sediment instability, though the cause of the instability is less clear. It may have resulted from repeated seismic shock or perhaps build-up of the sedimentary slope to super-critical angles. In the main, slumping is considered to have been down-slope, translatory into mass flow, and at times reconstituted sediment, but some might have resulted from readjustment of reversed density gradients during shock-induced thixotropy more or less *in situ*.

c) Conglomerate. Intraformational conglomerate is a minor but significant lithology in the Wenlock–Ludlow sequence. It occurs in association with a number of rock types; the event-couplet sandstones and mudstones, the turbiditic rhythmites and the bioturbated mudstone – though before disturbance this too might have been turbiditic rhythmite. The clasts vary from round to angular and are ill-sorted. They consist of a variety of contemporaneous rock types – laminated sandstone, laminar hemipelagite, homogeneous mudstone, concretions and also small, rounded, black clasts of ?phosphatic mudstone which does not occur as a primary rock. The clasts are closely packed, often in contact – perhaps due in part to allutriation of matrix during redeposition. The matrix is mudstone, including some bentonite, and in places is very rich in brachiopod shell detritus.

These conglomerates form very lenticular bodies and are of scattered occurrence within the Nantglyn Flags, Mottled Mudstone and Bailey Hill Formation. In the last formation such bodies appear more commonly at, and near, the base of the formation.

7. Lithofacies variation
Figure 1 is a generalized scheme of formational relationships between the Bishop's Castle area in the east and the Newtown area in the west based upon the rock-type associations (or lithofacies) described.

The graptolite biostratigraphy reveals that, overall, some of these lithofacies are diachronous from east to west, others are replacive synchronously. An example of the latter category is the early to mid Wenlock Nant-ysgollen Shales west of the River Severn, of typical laminar hemipelagite. Eastward they become less fissile, silty mudstones in which the carbonaceous laminae are more widely dispersed and discontinuous, or fragmented. This facies has been termed the Trewern Brook Mudstone Formation (Palmer 1970; e.g. SJ 2470 0102 near Rhyd-y-groes) and the Bromleysmill Shale Formation of the current BGS survey of Sheet 165 (e.g. SO 2630 9940 near Walcot). Burrowing is evident and there are sparse occurrences of small brachiopod shells (e.g. SO 2365 9869 near Crankwell Farm) but the scatter of carbonaceous filaments indicates that the deposit was still largely anoxic or, at most, dysaerobic.

In the extreme east, however, at Bromleysmill, the mudstone still possesses bedding-orientated short tabulae of carbonaceous material, including graptolites, but there are no continuous laminae; burrowing is more intensive, and in parts calcareous, bioturbated mudstones with trilobites are interbedded.

STRUCTURE, SILURIAN ROCKS

Folds and cleavage. In the east of the area the rocks are little folded. Gentle dips prevail, but there are widely spaced, N–S aligned zones of intense disturbance which include eastward facing, medium to large, monoclines. Such a fold occurs in the north-facing scarp 300–400 m south-west of Pentre Hall (240 911) and it is interesting that it is not reflected in the topography. Folds of this nature occur quite widely on the eastern flank of the Towy–Severn lineament north of Crossgates.

Approaching the Severn Valley from the east, folds become more regular, and west of the Severn large, fairly open folds are pervasive with a NE–SW trend.

Cleavage in the rocks is sparse. East of the Severn Valley there are local 'patches' of cleavage usually associated with a sharp fold, as at Sarn. West of the Severn Valley a weak spaced cleavage occurs more widely, thus matching the greater intensity of folding.

Faults. The major part of the area has been affected by post-Přídolí faulting. East of the Severn Valley many faults trend approximately N–S though some in the east have a NE–SW trend. They are comparatively straight, persist for up to 5 km and throw down under 100 m, mainly to the west. Their distribution is fairly even, but one in the centre of the area, the Montgomery Fault, is much longer and throws mid Ludlow rocks against Soudleyan, (Caradoc) strata. The north end of this fault appears to be connected with the Severn Valley major structural system.

West of the Severn the fault pattern is very different. It consists of NE–SW strike faults aligned with, and probably associated with, the folds. They are difficult to map, but their most striking characteristic is an almost asymptotic convergence with the Severn Valley lineament, and its northward projection, as far as Llanymynech.

There is no evidence to indicate that Silurian sedimentation was influenced specifically by any of these faults, but they may well reflect late movements within the basement on the same lines that had influenced the Wenlock basin margin.

ITINERARY

This excursion begins in the east and progresses westward (Fig. 2). At least $1\frac{1}{2}$ days should be allowed to cover the whole itinerary.

1. Bromleysmill Pool (SO 3313 9147).

The first locality is **near the junction of the A488 and A489** roads, (SO 332 917) where there is **parking space for cars or minibuses**. About 300 m to the south are a **quarry and stream** exposing many metres of the Bromleysmill Shale Formation of mixed dysaerobic shale and oxic mudstones with graptolites and trilobites respectively.

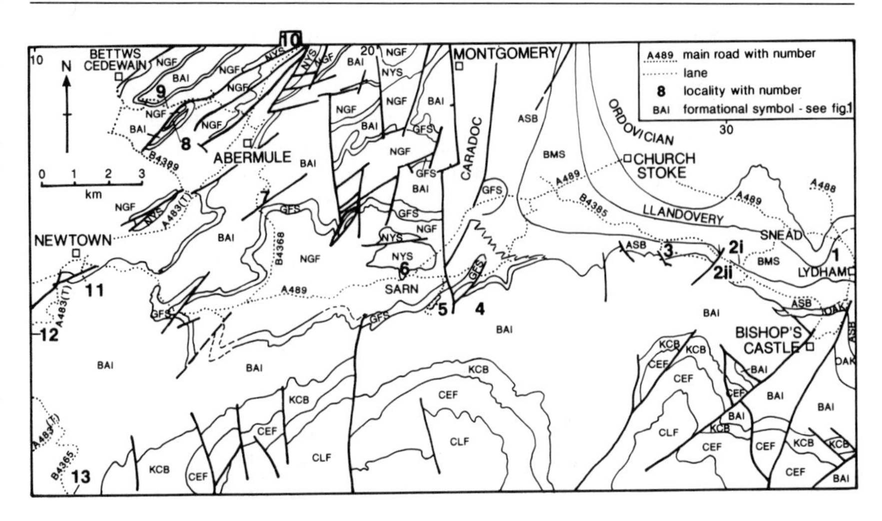

FIG. 2. Geological sketch map and outline of itinerary.

Mudstone, medium, grey, very shaly and friable. Fauna: from top, in stream, *Cardiola* sp., *Monograptus flemingii flemingii*, *M. flemingii* cf. *elegans*; from bottom, in quarry, *M. flemingii flemingii*, *?Pristiograptus pseudodubius*.	13 m
Mudstone, medium grey, very shaly; graptolites and bivalves.	3.25 m
Soft clay, cream coloured. Bentonite.	4 cm
Mudstone, medium grey, shaly at top, passing down to blocky and olive grey, where it is burrowed (mottled) containing ochreous blotches and 'pipes'. Three, possibly four, layers of cream-coloured clay (bentonites) up to 6 cm divide the mudstone. Fauna includes dalmanitinid trilobites.	4.15 cm
Pale cream-coloured clay. Bentonite.	6 cm
Mudstone, olive grey to buff, blocky, calcareous, burrowed and contains ochreous blotches and 'pipes'. A layer of hard calcareous concretions is present 13 cm from top. Fauna mainly shelly: *Glassia?*, orthocone and bivalve fragments, *Dalmanites* sp., *M. flemingii* s.l. and *P.* cf. *pseudodubius*.	0.9 m
Clay, pale cream coloured. Bentonite.	6 cm
Mudstone, grey to buff, rather shaly. Fauna: '*Orthoceras*' aff. *recticinctum*, *Dalmanites* sp., *Monograptus* cf. *flemingii* s.l., *P. pseudodubius*.	2 m
Clay, pale cream coloured in track on east side of pond. Bentonite.	a few cm

The horizon is not dated precisely, but the graptolites indicate a biozone well above the base of the Wenlock, but not above that of *Cyrtograptus lundgreni*. This part of the formation is equivalent to part of the Coalbrookdale Formation of south Shropshire, but it appears to be a more anoxic deposit with few brachiopods. Nevertheless, as the section indicates, the graptolitic shales are much more bioturbated here than they are further west and there are several intervals of highly bioturbated trilobite-rich mudstones. These factors suggest that deposition here must have been in a position on the outer marine shelf where the sea bed was close to the oxygen minimum surface and thus sensitive to its fluctuations.

From Bromleysmill there are two routes. There is the **main road via Bishop's Castle**: B4384 and B4385 past Upper Heblands (SO 3252 9029), where small irregular masses of micrograrite intrude Aston Mudstones (Sanderson & Cave 1980), or there is a more direct route **westward on the A489 to Snead** and thence via a lane past Owlbury Hall (**not suitable for coaches**). This route joins the B4385 where the base of the Aston Mudstones is exposed in an unnamed dingle.

2(i). Unnamed dingle (SO 3038 9094)
The name Aston Mudstones Formation follows Allender (1958), who called the formation the Aston Beds. He did not identify the base of the formation nor, thus, a type section. The base is exposed *c*.5 m **below the foot of a small waterfall** (SO 3038 9094) in the unnamed dingle on the north side of the B4385 as mentioned above. The dingle has been used as a tip for old fencing and farm rubbish, but for those interested in the base of the formation the exposure reveals:

ASTON MUDSTONES FORMATION

Small quarry and dingle S of main road (SO 3037 9091). Mudstone, blocky, calcareous weathered brown. Fauna: decalcified rugose coral and *?Clorinda dormitzeri*.	seen 2 m
To main road and northward, mainly unexposed except at top of waterfall and just above. Mudstone, tough, blocky, calcareous, weathered brown. Fauna: *Glassia* sp., *Jonesea? Strophochonetes* cf. *cingulatus, Visbyella trewerna*, orthocone fragments, *Platyceras* sp., *Dalmanites* sp., beyrichiacean, crinoid columnals and *P*. cf. *pseudodubius*.	*c*.20 m
Face of waterfall except base. Mudstone as above. Fauna: middle of waterfall, *Leangella segmentum, ?Ananaspis* aff. *communis, Gothograptus nassa, P*. cf. *dubius ludlowensis*.	2.5 m

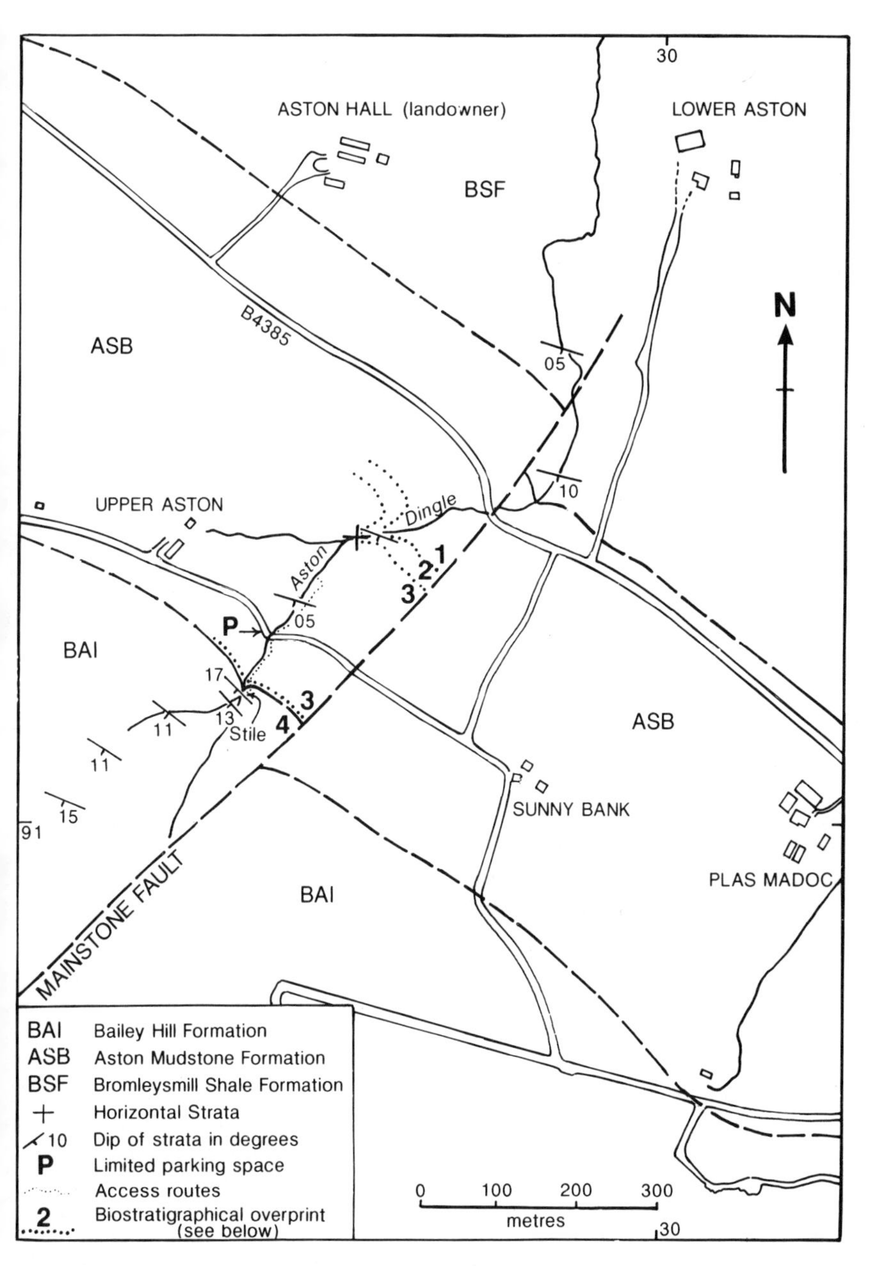

FIG. 3. Geological map of Aston Dingle with biostratigraphical overprint. 1, *G. nassa* Biozone – *G. nassa* and *Pristiograptus dubius* only. 2, *M. ludensis* Biozone — large monograptids and sparse *G. nassa*. 3, *M. ludensis* Biozone — large monograptids. 4. *N. nilssoni* Biozone — slim monograptids including spinose forms. See Fig. 6.

At base of waterfall. Mudstones, grey, calcareous weathering buff, more fissile than above. Fauna: *Jonesea?, S.* cf. *cingulatus, G. nassa, P.* cf. *dubius ludlowensis*, P. *jaegeri.*	seen *c*.1 m
Downstream of waterfall. Mudstones, poorly exposed. Fauna: *Jonesea grayi, G. nassa, P. dubius?* s.l., *P.* cf. *dubius ludlowensis.*	*c*.5 m

BROMLEYSMILL SHALE FORMATION

Mudstones, slightly silty, grey, very fissile in parts. Fauna: *J. grayi, Ludfordina pixis, Nuculites?* and graptolite fragments.	*c*.2 m
Unexposed.	*c*.8 m
Mudstones, as above. Fauna: *G. nassa, P. dubius* s.l., *P. pseudodubius?*	*c*.2 m
Unexposed.	*c*.2 m
Mudstones, grey, shaly. Fauna: *'Orthoceras' argus?, 'O'.* cf. *recticinctum, M. flemingii elegans, P.* cf. *pseudodubius.*	seen

The smaller brachiopods, including *Visbyella trewerna*, which are commonly associated with the graptolites in the lower beds of the Aston Mudstones Formation, indicate the presence of a *Visbyella trewerna* Community. This is a relatively deep-water community, with a comparatively low diversity, which may be related to low oxygen values (Cocks & McKerrow 1978, p. 121).

The approaches to the top end of Aston Dingle from the B4385 are **lanes, not suitable for coaches**, but two cars or a car and a minibus can be parked (Fig. 3).

2(ii). Aston Dingle (SO 2950 9122)
The rest of the type section is in Aston Dingle (Figs. 3 & 4) which exposes some 100 m more of buff-weathering, blocky, bioturbated mudstones. They are calcareous and grey when fresh and have yielded a mixed shelly and graptolitic fauna (Figs. 5 & 6). The graptolites indicate the *G. nassa* Biozone (**Locs. a–d**), succeeded by the *M. ludensis* Biozone (**Locs. e–o**), with a characteristic association of *Monograptus ludensis* and less common examples of *Pristiograptus* cf. *jaegeri.* Thus the span of the Aston Mudstones Formation is precisely the *nassa* and *ludensis* biozones, making it the correlative of the Farley Member of the Coalbrookdale

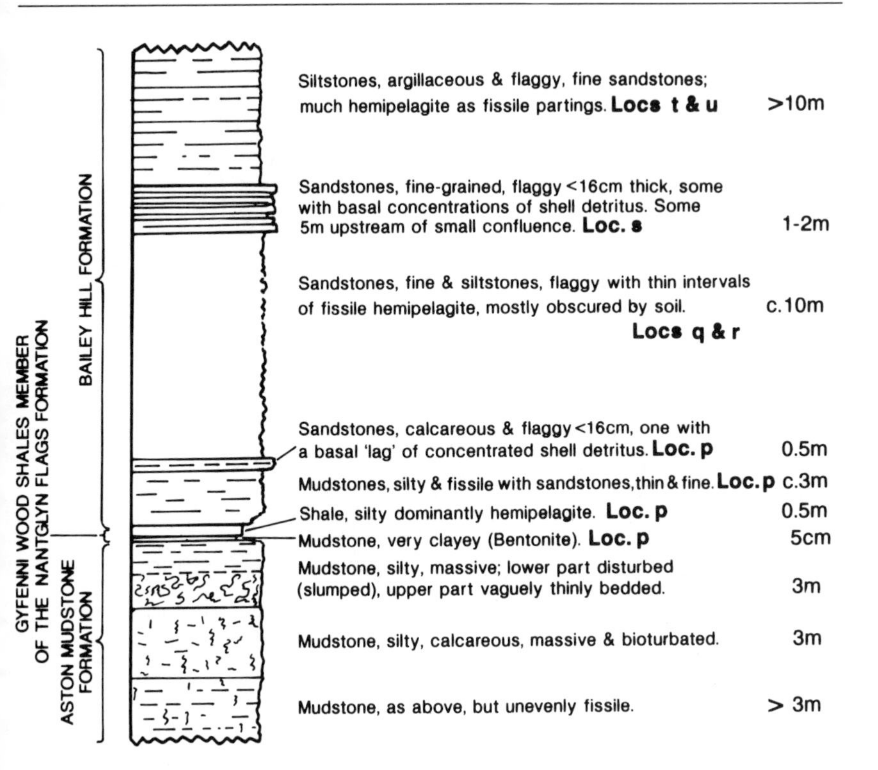

FIG. 4. Section of strata across the Wenlock–Ludlow boundary in upper Aston Dingle. (See also Fig. 5.)

Formation plus the Much Wenlock Limestone Formation of Wenlock Edge. The passage between these formations lies across the ground around the southern end of the Long Mynd in the form of an impure, rubbly limestone, the Edgton Limestone, the lower part of which extends as far west as Bishop's Castle (Fig. 1).

The top few metres of the Aston Mudstones Formation are described in Fig. 4 and are shown to be succeeded by some 50 cm of graptolitic silty shales (largely hemipelagite) (**Loc. p**), containing a graptolite fauna no more precisely indicative than of the *nilssoni-scanicus* biozones. These silty shales, which have a 5 cm bentonite at their base are thus referable to the Gyfenni Wood Shales (Fig. 1) and are probably the eastern extremity of a unit which is 100 m thick in places to the west. The 3.5 m of siltstones and sandstones which overlie this (also **Loc. p**), contain a diverse graptolite fauna indicative of the lower part of the *N. nilssoni* Biozone (Fig. 6, col. p).

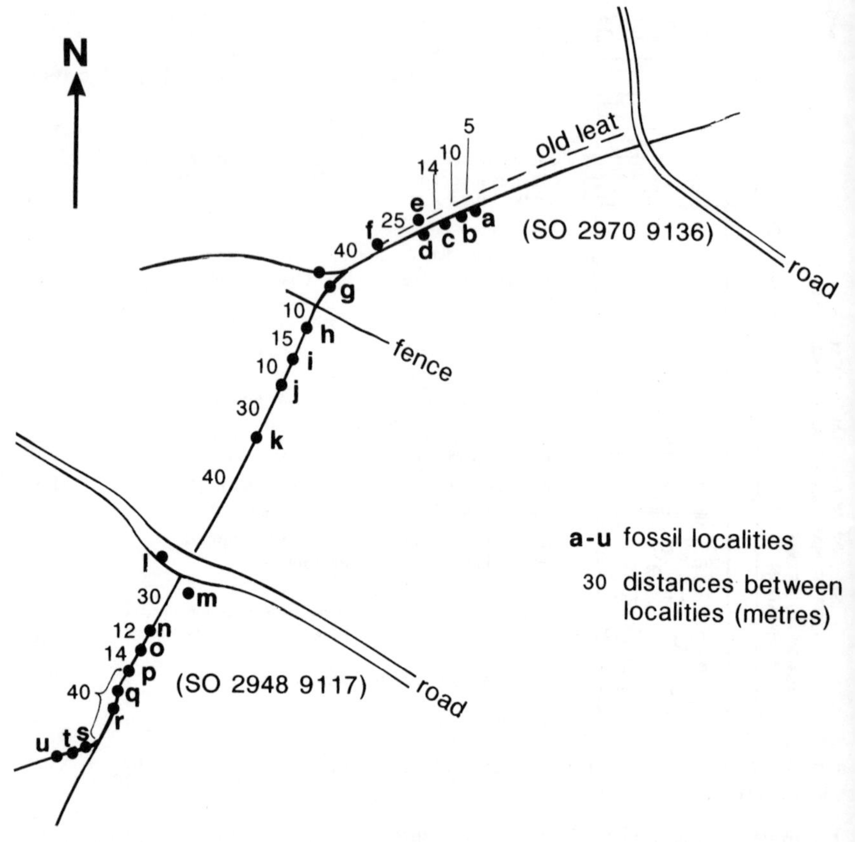

FIG. 5. Fossil localities in Aston Dingle.

The higher Bailey Hill Formation, with sandstones < 15 cm thick (**Locs. q–u**), contains saetograptids, proving a Ludlow age, but their zonal significance is uncertain. A small form is dominant, with features indicating affinities towards *Saetograptus leintwardinensis incipiens*.

From Aston Dingle **proceed westwards nearly to Pentre**, either along the B4385 or the lane past Upper Aston (Fig. 3).

3. Pentre (SO 2806 9158)

Locality 3 consists of excavations in the south bank of this lane, beneath the ancient 'Fort'. **Parking is available at the lane edge**.

ASTON DINGLE - BIOSTRATIGRAPHY

c =well represented
● =present
cf.=confer
? =uncertain

	a	b	c	d	e	f	g	h	i	j	k	l	m	n	o	p	q	r	s	t	u
Series	WENLOCK															LUDLOW					
Biozones	*G. nassa*				*M. ludensis*											*N. nilssoni* -?*S. scanicus*					
Localities	a	b	c	d	e	f	g	h	i	j	k	l	m	n	o	p	q	r	s	t	u
BRYOZOA																					
bryozoa -decalcified																		●			
BRACHIOPODA																					
Atrypa sp.																		●			
'*Clorinda*' *dormitzeri* (Barr.)														●		●					
Shagamella minor (Salter)																			●		
Strophochonetes sp.										c		●		●					●		
Visbyella trewerna Bassett								●							●	●					
CEPHALOPODA																					
'*Orthoceras*' aff. *recticinctum* Blake	●																				
'*O*' spp.							●					●	●		●	●					●
Parakionoceras originale (Barr.)															●	●					
aptychopsid plates													●								
BIVALVIA																					
Cardiola interrupta J. de C.Sow.																●					
C. spp.								●				●									
TRILOBITA																					
Dalmanites sp.								●													
cf. *Miraspis* sp.										●											
OSTRACODA																					
Beyrichia s.l.										●								●			
Bolbozoe sp.										●											
GRAPTOLITHINA																					
Gothograptus nassa (Holm)	c	c	c	●	●	●	●														
Monograptus deubeli Jaeger										cf.	●										
M. ludensis (Murchison)								cf.	cf.			c	c	c	c	●					
M. uncinatus orbatus Wood																●					
Neodiversograptus nilssoni (Barr.)																●					
Pristiograptus curtus Elles & Wood																●					
P. dubius (Suess) s.l.					●		●					●									
P. cf. *jaegeri* Holland, Rickards & Warren					●	●	●														
P. cf. *pseudodubius* (Bouček)		●		?																	
P.?	●									●				●					●	●	
Saetograptus chimaera (Barr.) s.l.																					●
S. colonus compactus (Wood)																c					
S. aff. *colonus* (Barr.) s.l.																	●				
S. aff. *incipiens* (Wood)																			c	c	
S. cf. *varians* (Wood)																●			●		
S. spp.																		●			●
Spinograptus spinosus (Wood)																●					
?algal or retiolitid strands				●	?																
	a	b	c	d	e	f	g	h	i	j	k	l	m	n	o	p	q	r	s	t	u

FIG. 6. Table of fossils from Aston Dingle – for locations see Fig. 5.

The laneside section consists of:

3. Shales, grey-brown, graptolitic. Thin layers of grey laminar hemipelagite alternate with thin layers of grey-buff homogeneous mudstone. *Cardiola interrupta, P.* cf. *jaegeri.* — seen 4.5 m

2. Mudstone buff to greenish, very blocky in rough beds < 8 cm thick. It is bioturbated and shows a scatter of brown blotches and 'pipes' of goethite (representing burrows). In the top few centimetres thin, graded layers with silt bases are present. *Slava?*, cf. *M. ludensis, P. jaegeri.*

1. Mudstones, buff, shaly to thinly flaggy and silty due to thin, parallel silt laminae and thin intervals of laminar carbonaceous hemipelagite. *Gothograptus* sp. (broad form) — seen 1.1 m

This exposure maps as part of the Aston Mudstones and its graptolites indicate a *ludensis* Biozone age. However, the 'bioturbated calcareous mudstone' characteristic of that formation is clearly divided by and partially replaced by dark-grey carbonaceous silty laminite (hemipelagite) and siltstone-mudstone rhythmite (possibly turbiditic) of an anoxic depositional environment. Such a lithofacies is characteristic of the Nantglyn Flags. One implication of these observations is that in *ludensis* times the sea bed here was at a depth where fluctuations in the surface of oxygen-depleted water were affecting sedimentation. It is thus reasonable to infer that the sea was deeper here than it was east of Aston Dingle. This is the second facies change within the *ludensis* Biozone to record westward-deepening marine conditions, the first being the passage from the Edgton Limestone to bioturbated calcareous mudstone east of Bishop's Castle.

From Pentre **follow the B4385 road**, turning left first to Bacheldre and thence to Hopton.

4. Hopton Dingle (SO 2298 9048)

A **right-angled bend in the lane approaching Hopton affords parking space** on the grass verge. A track opposite, to the south, leads past **a quarry** (SO 2289 9005). The track is rough but, **with the landowner's permission**, it will take cars.

The track closes with Hopton Dingle near to the outcrop of the base of the Bailey Hill Formation (SO 2298 9048). This occurs **at the foot of a 4 m waterfall** exposing a cobble conglomerate < 40 cm thick, consisting of round to sub-angular clasts of local rocks. Below it are some 60 m of shaly mudstones with at least two packets of slumped strata and much fissile graptolitic hemipelagite of the *nilssoni-scanicus* biozones, probably the latter in view of the presence of *Lobograptus scanicus* and *Saetograptus chimaera semispinosa*. The dingle is choked with dead trees and brambles and the same sequence is present at Locality 5.

Some 500 m to the south is a **trackside quarry** exposing 10 m of interbedded fine sandstones and argillaceous siltstones, or silty mudstones of the lower part of the Bailey Hill Formation. The thickness of the sandstones varies up to 30 cm. They are well bedded with plane bases and some waved surfaces. Internally there are cross, plane-parallel and convolute laminae. Some thick beds possess a basal lag deposit of coarse, shell detritus, up to 3 cm thick but lenticular. Sole marks are common and include bounce-casts, groove-casts, prod-marks and well formed flute-casts. From the last structures six separate bases indicated palaeocurrent directions towards 340°, 355°, 005°, 005°; 015° and 018° respectively (Fig. 12).

The top centimetre or so of beds with waved surfaces commonly consists of several parallel laminae approximately concordant with the surface, but which truncate the tops of convoluted laminae below. The former may have been deposited under the influence of marine waves. These beds belong to the lithofacies described earlier as calcareous, fine sandstone and argillaceous siltstone event-couplets and are considered to have been

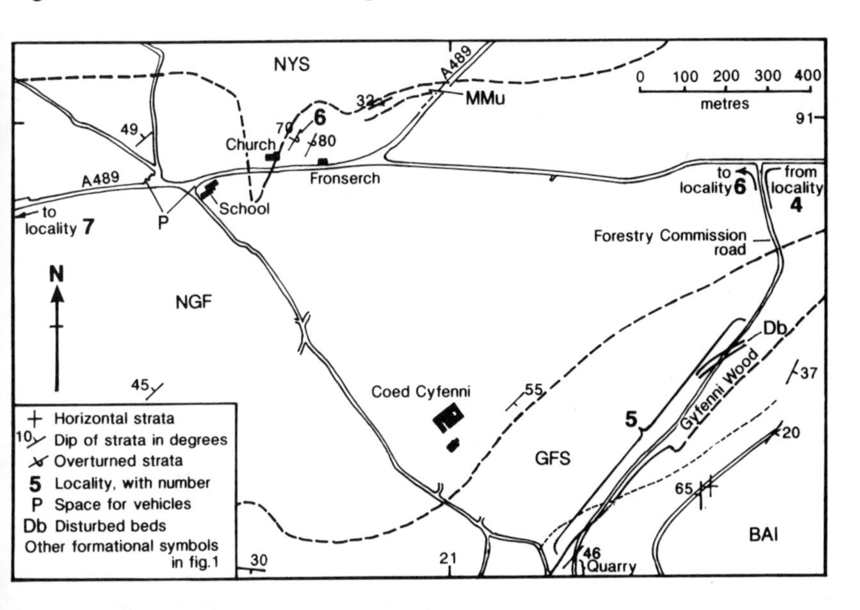

FIG. 7. Geological map of the Sarn area and Localities 5 and 6.

deposits from either storm-generated currents (Tyler & Woodcock 1987) or basinal turbidites (Bailey 1969). Fossils collected here include: chonetids, *Dayia navicula, Isorthis* sp., *Leptaena* sp., *Lingula* sp. and *Saetograptus incipiens?* The age is Ludlow, but the zone is uncertain. However, a trackside exposure (SO 2287 9024) several metres lower in the succession yields *Isorthis* sp., *Jonesea* sp., *Kirkidium* sp., *Pristiograptus* cf. *tumescens, P.* cf. *tumescens minor* and crinoid fragments, probably of the *tumescens* Biozone.

Continue westward on the lane from Hopton for about three-quarters of a mile where a **gated Forestry Commission road** (SO 2171 9090) leads to the left. The gate is locked and **permission will be required** to drive vehicles to the exposure, the end of which is 350 m beyond (Fig. 7).

5. Gyfenni Wood

This section is very similar to that in Hopton Dingle described by Allender (1958); details are given in Fig. 8.

Some 70 m of Gyfenni Wood Shales are exposed at the **edge of the new road** which has been dug below the level of the old road. They are fissile mudstones, buff and brown and non-bioturbated containing much carbonaceous silty laminite with graptolites. A slumped packet occurs in the lower part of the section.

The shales yield *Cardiola interrupta, 'Orthoceras' argus, 'O'. elongatocinctum, 'O'. recticinctum, 'O'. reticulum, Parakionoceras originale, Bohemograptus bohemicus bohemicus, Lobograptus progenitor* (common in lower parts), *L. scanicus, L. simplex, Monograptus* cf. *uncinatus orbatus, Monoclimacis micropoma*, cf. *Neodiversograptus nilssoni, Plectograptus macilentus, Saetograptus chimaera salweyi* (common at base), *S. c. semispinosus, S. colonus, S. varians* (common except at base) and *Spinograptus spinosus.*

This fauna indicates *N. nilssoni* Biozone, possibly rising into the *L. scanicus* Biozone at the very top. The contrast with Aston Dingle is thus great, for the contemporaneous rocks there formed part of the Bailey Hill Formation whereas here graptolites of the *Pristiograptus tumescens* Biozone are present in sandstones within 10 m of the base of the Bailey Hill Formation. Thus, whilst anoxic conditions of deposition prevailed in both areas during the *nilssoni* Zone, event-couplet sands reached the Sarn area only in *scanicus* Zone times.

The 3–4 m thick slumped mudstone (Fig. 8) has been placed in the Bailey Hill Formation rather than the underlying formation merely because it contains fragments of sandstone. The mass-flow conglomerate above it is probably very significant to the understanding of the depositional system

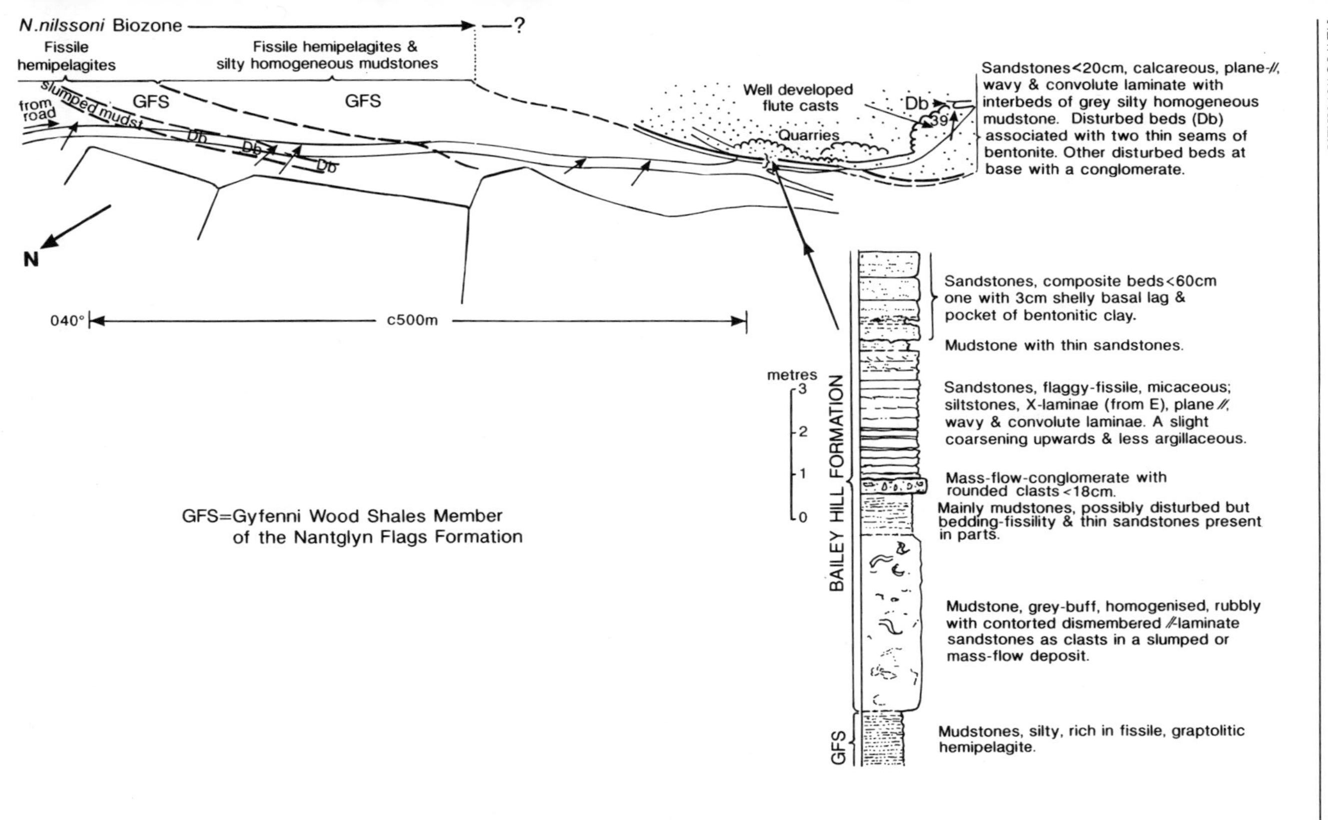

FIG. 8. Locality 5 (SO 2138 9015), Gyfenni Wood forestry road, Sarn.

of the early Ludlow and will be seen again at Loc. 8. The conglomerate and slump here are almost certainly within the *scanicus* Biozone. The evidence for this biozone at Loc. 5 is very scant, lending support to the view that the conglomerate might be associated with a short hiatus.

Continue to Sarn; **park near the cross-roads by the school**.

6. Sarn quarry (SO 2066 9097)
The quarry is situated on the **north side of a paddock behind Fronserch Cottage**, where the landowner lives (Fig. 7). It exposes some 50 m of beds, mostly overturned. They are thinly bedded, fine sandstone-mudstone rhythmites, and are interpreted as low density turbidites. Thin intervals of dark grey, carbonaceous, silty laminite (laminar hemipelagite) separate some of the rhythms and have yielded *Atrypta reticularis, P. ludensis*, and retiolitids, indicative of the *ludensis* Biozone. These beds fall comfortably within the litho-parameters of the Nantglyn Flags, but were deposited coevally with the Aston Formation of localities 2 and 3. The depositional environment at Sarn was anoxic, in even deeper water probably of the basin margin.

From Sarn **proceed westward along the A489 road** to the junction at Glanmule with the B4368. **Travel northwards and turn right** across the second bridge over the bed of the Kerry Railway in the Mule Gorge (SO 1676 9352).

7. Forestry Commission road, Abermule
The gated entrance to this forestry road is on the right over the bridge. The gate is locked, but **there is room here for vehicles**. Proceed on foot almost to the end of the road.

The reason for the locked gate is obvious. As soon as the road had been constructed the carriageway foundered in landslip. The valley side is steep, whilst well-bedded sandstones and argillaceous siltstones of the Bailey Hill Formation dip less steeply in the same direction. Excavation exposed the up-dip ends of the beds to direct ingress of drainage, while it redistributed the valley-side loading to create instability.

Towards the far end of the road – some 120 m before it joins the bed of the old railway – there is an abrupt base to the sequence with thick sandstones (Fig. 9). At this base the thickness of the sandstones reaches 30 cm and they form a thinning upwards packet which has yielded graptolites of the *scanicus* Biozone, but instead of resting upon a dominantly laminar hemipelagic facies with graptolites, there are at least 80 m of soft, homogenous, silty mudstones, closely similar to those in the Bailey Hill Formation above, but in which sandstones are very recessive. Isolated sandstones up to 20 cm do occur, especially in the upper portion, and there are graptolites, also of the *scanicus* Biozone, including *L. scanicus* and *S.* cf. *chimaera semispinosa* and the brachiopod *D. navicula*.

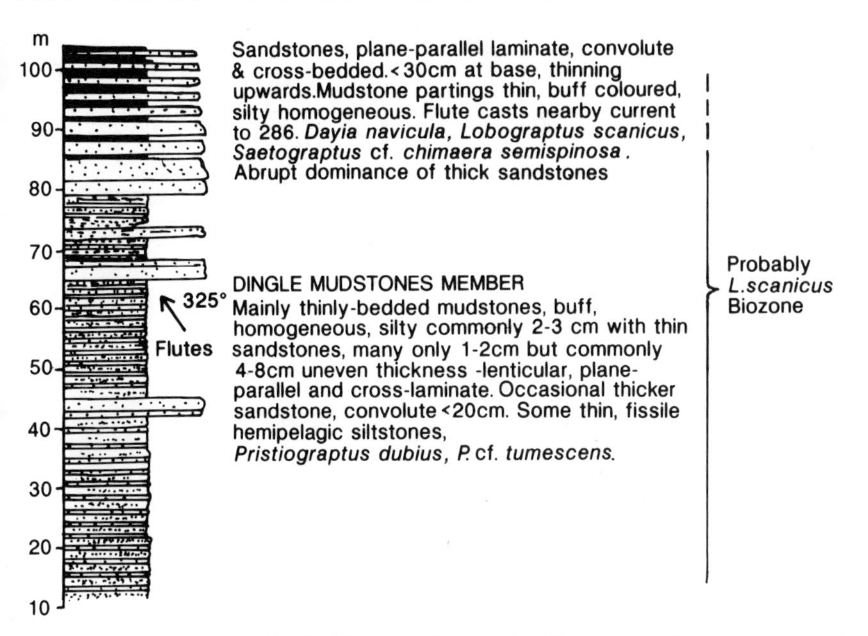

FIG. 9. Section (base 1652 9288) at southern end of forestry road in the basal part of the Bailey Hill Formation, Locality 6.

This new lithostratigraphical unit, at the base of the Bailey Hill Formation, occurs widely west of a line between here and Kerry and while clearly of the Bailey Hill facies (and age) it reveals that few of the event-couplets reached this area with any energy until late *scanicus* Biozone times. The few flute casts seen indicate palaeocurrents from the southeast-east. The unit is termed the Dingle Mudstone Member (Fig. 1), named from 'The Dingle' near Newtown.

Although the conglomerate present at many places at this stratigraphical level is not exposed here, debris revealed that it is present, probably sporadically, in a position at the base of the thicker sandstones.

Continue along the B4368 road to Abermule. Turn right (B4386), then left at the mini-roundabout, over the River Severn and right onto the A483 (T) road for 200 m where **a lane on the left leads west-north-west to New Mills**. There turn left (**unsuitable for coaches**), immediately before a small chapel, and proceed straight on for three-quarters of a mile where **there is parking space at the end of the lane.**

8. Bryn-rorin farm-track (1412 9535 to 1435 9553)

An exposed section lies along a track to the **north-west of the farm buildings**.

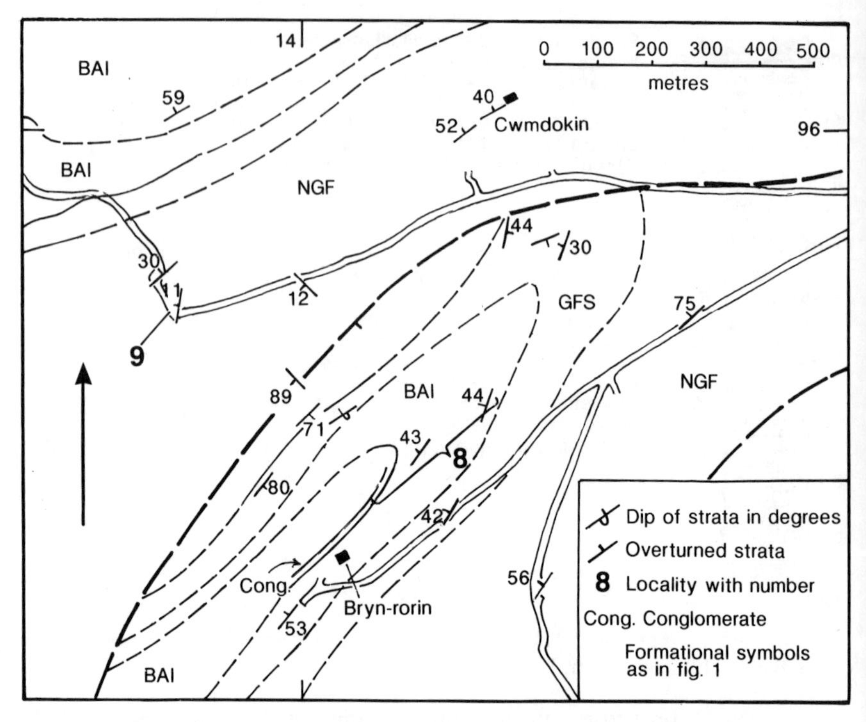

FIG. 10. Geological map of the Bryn-rorin area indicating Localities 8 and 9.

The exposure is part of a small, faulted outlier of the Bailey Hill Formation (Fig. 10). It is not the most 'distal' outcrop as there are two more further to the NNW, but this section offers the best exposure. The main details are given in Figure 11.

As at Locality 7, many metres of homogeneous mudstones with diminutive fine sandstones (or siltstones) and sparse, thin, graptolitic hemipelagites (Dingle Mudstones Member) separate the main Bailey Hill Formation, with its thick sandstones, from the Gyfenni Wood Shales dominated by fissile, graptolitic hemipelagites.

Several observations from this section appear to be important.

1. The Dingle Mudstones are contained entirely within the *scanicus* Biozone together with many metres of the Gyfenni Wood Shales here, revealing that the top of this lithofacies is still younging slightly north-westward.

Likewise, the evidence suggests that the basal part of the thicker sandstones of the Bailey Hill Formation could lie within the *tumescens* Biozone or possibly highest *scanicus* Biozone here.

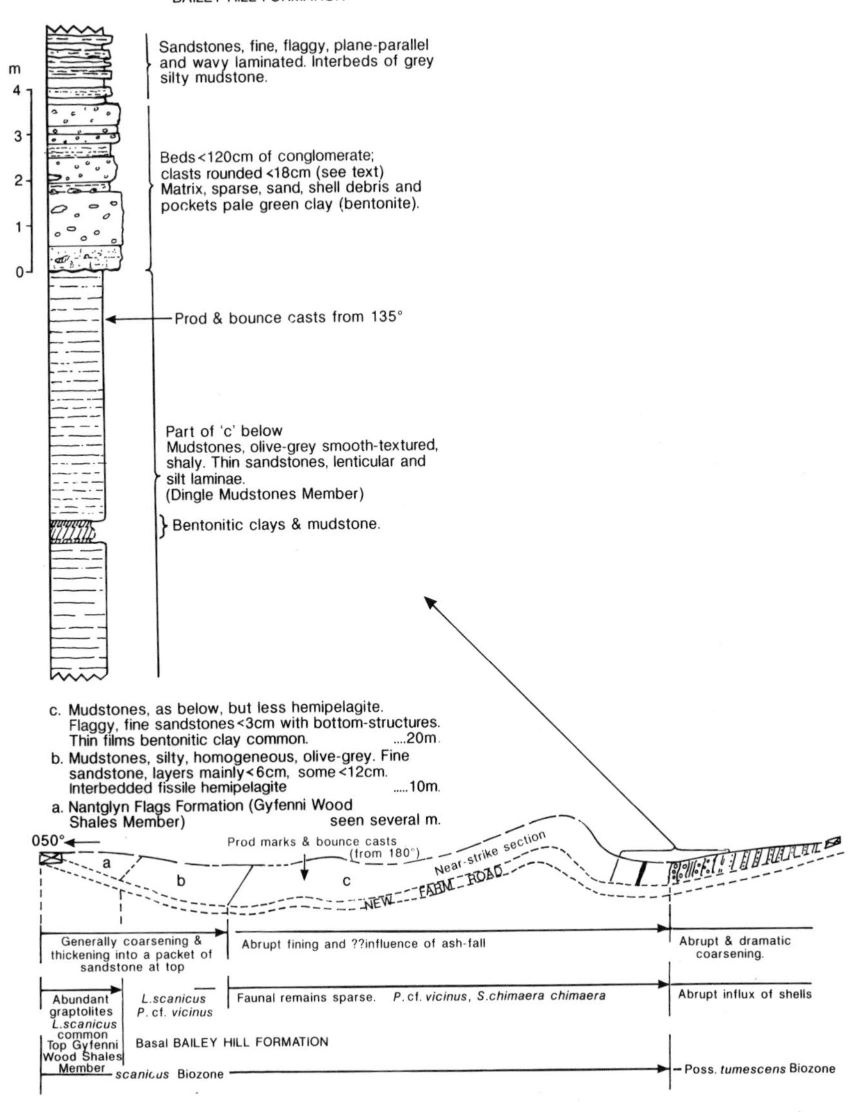
BAILEY HILL FORMATION
m
4
3
2
1
0
Sandstones, fine, flaggy, plane-parallel and wavy laminated. Interbeds of grey silty mudstone.
Beds<120cm of conglomerate; clasts rounded <18cm (see text) Matrix, sparse, sand, shell debris and pockets pale green clay (bentonite).
Prod & bounce casts from 135°
Part of 'c' below
Mudstones, olive-grey smooth-textured, shaly. Thin sandstones, lenticular and silt laminae.
(Dingle Mudstones Member)
Bentonitic clays & mudstone.
c. Mudstones, as below, but less hemipelagite. Flaggy, fine sandstones<3cm with bottom-structures. Thin films bentonitic clay common. ....20m.
b. Mudstones, silty, homogeneous, olive-grey. Fine sandstone, layers mainly<6cm, some<12cm. Interbedded fissile hemipelagite .....10m.
a. Nantglyn Flags Formation (Gyfenni Wood Shales Member) seen several m.
050°
Prod marks & bounce casts (from 180°)
Near-strike section
a
b
c
NEW FARM ROAD
Generally coarsening & thickening into a packet of sandstone at top
Abrupt fining and ??influence of ash-fall
Abrupt & dramatic coarsening.
Abundant graptolites L.scanicus common
L.scanicus P. cf. vicinus
Faunal remains sparse. P. cf. vicinus, S.chimaera chimaera
Abrupt influx of shells
Top Gyfenni Wood Shales Member
Basal BAILEY HILL FORMATION
scanicus Biozone
Poss. tumescens Biozone

FIG. 11. Section exposed in the bank of the farm track, Locality 8, at Bryn-rorin (SO 1406 9525).

2. Close examination reveals that, as well as the thicker bentonites near the top, the Dingle Mudstones possess innumerable thin films of bentonitic clay between its thin beds. It is possible that all such sections when fresh would show these; on the other hand it may be an associate specifically of the *scanicus* Biozone.

3. The conglomerate at the base of the sandstones and above the Dingle Mudstones is particularly well developed. In the road section it is composed of at least three separate beds. It was not, therefore, the product of a single event, yet **behind the barn nearby** (SO 1409 9530), there is a single bed thicker than these three combined, with no sign of division within it and with the clasts showing slight imbrication.

The generation of this conglomerate is not understood. Obviously there were strong erosive processes in order to produce the clasts, for in Hopton Dingle they include laminar hemipelagite and ovoid septarian concretions. Both of these would have required erosion of a compacted or cemented substrate down to several metres and in the case of Hopton Dingle they must have been derived very locally, for to the east there is no laminar hemipelagite of this level. At Bryn-rorin, higher beds of the conglomerate contain clasts of the basal bed, so the erosion too was not of one event.

The clasts are now closely packed, or in contact, having been concentrated into beds, albeit impersistently. The matrix contains abundant bryozoan and shell detritus including *A. reticularis, Leptaena* sp. and *Lingula* sp. and pockets of lemon-yellow bentonite are present in some places, squeezed between the clasts.

In one exposure, near Abermule, a bed composed of small, rounded, dark grey and brown clasts (possibly phosphatic), with abundant small shell detritus and sand possesses large-scale, steep cross-stratification, so that both mass flow and strong traction currents have been involved in transportation.

Brachiopod shells and bryozoan remains imply shallow, oxygenated water, but the beds above and below the conglomerate contain such remains only as detritus within traction current sandstones; anoxic graptolitic mud accumulated during quiescent intervals. Either the shelly material with the conglomerate was swept in from shallower areas and filled the voids between clasts, or coarse, cobble-gravel briefly provided a hospitable substrate for a shelly benthos. Concentrations of similar shells in the bases of some of the Bailey Hill sandstone beds above were undoubtedly transported, but the main evidence that they travelled far lies in the lack of bioturbation of the inter-sandstone silts and the low oxicity of the benthic environment.

Perhaps the two most significant factors when considering the origins of this conglomerate are that in this area it always underlies the Bailey Hill sandstones and it has not been recorded east of Hopton Dingle. It seems, therefore, that the conglomerate was formed in a phase of energetic, sediment-unsaturated currents which immediately preceded the arrival of thick event-generated sands as they prograded north-westwards from the shelf into the remnant of the Wenlock basin margin.

Precise correlation with inner shelf sequence like that at Tites Point, Gloucestershire, is not possible, but it is noteworthy that the limestone conglomerates of the Lower Blaisdon Beds (Cave & White 1971, 1978), though younger (basal Ludfordian), consist of storm-surge rip clasts from cemented, shallow, sea-floor sediments (hardgrounds; Cherns 1980) and presumably are associated with small hiati in the sequence. It is thus possible that the basal conglomerate of the Bailey Hill Formation is associated with a hiatus but of sub-biozonal span and thus difficult to prove.

Retrace the route as far as the chapel at New Mills. **Turn left in front of the chapel** and continue bearing left, along the lane, for one mile to a sharp right-hand bend where there is **a quarry and parking space (Fig. 10)**.

9. Quarry (SO 1376 9570) **near Bettws Cedewain**
This quarry exposes about 4 m of Nantglyn Flags dipping at 11° towards 280°. They are rhythmites, mainly of mudstone, considered to be turbiditic. The turbidite rhythms range in thickness up to 20 cm and are of the anoxic variety (Cave 1979). Here the rhythms are dominantly buff-grey homogeneous mudstone with spheroidal weathering properties. Fine sandstones/siltstones, normally forming the basal layer of the rhythm, are sparse and thin, up to 3 cm at maximum and lenticular with convex bases. On some of these bases are small scour casts, probably flute-casts, but none has been obtained intact as proof. It is interesting to note that apart from jointing with a strike of 309° and a dip *c.*87°SW, there is a weak, spaced cleavage striking 060°, dip 85°SE. It has imparted a rather pencil-like fracture to the beds.

The main interest in this exposure is its age. *Monograptus uncinatus orbatus* and *Pristiograptus* cf. *dubius* occur, indicating the *nilssoni* Biozone. Over the traverse from near Bishop's Castle, therefore, this biozone has passed from the Bailey Hill Formation (shelf edge) into the Gyfenni Wood Member (basin margin above the level of turbidity currents) and eventually into the Nantglyn Flags (turbidite basin) seen here.

Division of the Nantglyn Flags into Upper and Lower formations, as was done in Clwyd, is an impractical proposition here. The concept of the division is based in biostratigraphy (Warren *et al.* 1984).

SUPPLEMENTARY LOCALITIES

Two facies omitted from the main itinerary are a) the Nant-ysgollen Shales, which are the hemipelagic end-member of the passage westward from Bromleysmill Shales; and b) the slump facies. These facies can be examined at the supplementary localities outlined below and would add up to a day to the schedule.

Nant-ysgollon Shales Formation
Nant-ysgollon Shales are poorly exposed. They are soft mudstones dominated by dark grey silty laminate (anoxic hemipelagite) and crop out **in the scarp face just above Fron Church** (1785 9769, Garthmyl).

10. Llifior Brook
The least weathered exposure can be examined **just beyond the north end of the scarp in the right bank of Llifior Brook** upstream of the bridge (SO 1854

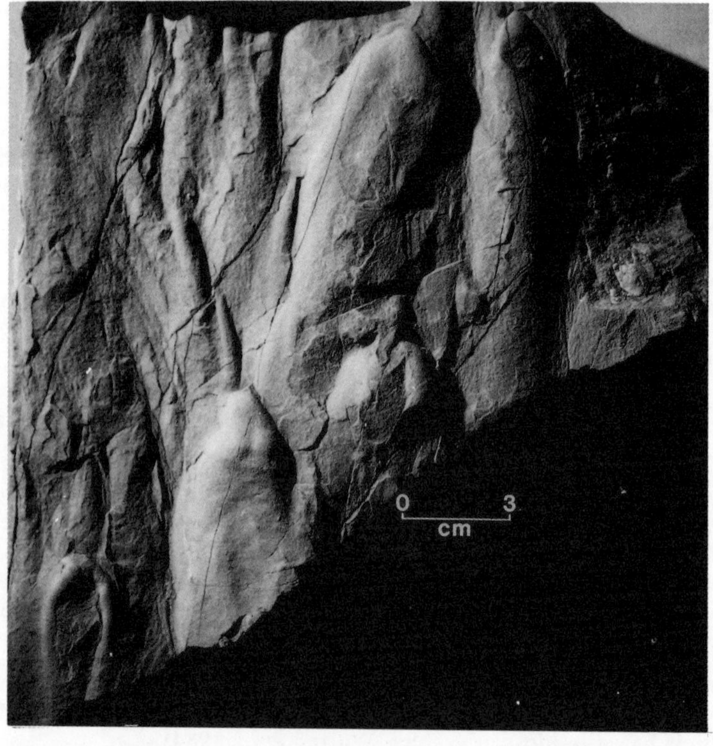

FIG. 12. Flute casts on the underside of a thin layer of sandstone in the Bailey Hill Formation at Locality 4.

9865) for a distance of 100 m. The shales have yielded *Pristiograptus ?lodenicensis, Monoclimacis flumendosae* cf. *kingi* and *Cyrtograptus* sp. within the range of the *ellesae-lundgreni* biozones. They are thus contemporaneous with the Bromleysmill Shales. These were the deposits of the basin margin on a sea bottom above the level of all but the weakest, low density, turbidity currents. The products of the latter are the fawn-brown weathered, thin layers of homogeneous mudstones which divide up the dark-grey hemipelagite.

Slump facies
Two different types of disturbed beds (slumps) occur in the Bailey Hill Formation. One consists largely of homogenized silty mudstone/argillaceous siltstone containing a scatter of rounded masses or lumps of fine sandstone/siltstone, commonly contorted internally, some of which are over 1 m across. This type occurs near the base of the formation, probably within the *scanicus* Biozone. Several metres are exposed without evidence of top or bottom at Localities 11 and 12.

11. Cutting (SO 1163 9071) **at Middle Brimmon Farm**
An old lane cut through a ridge of this slumped mudstone exposes some 10 m of rock in a face which now forms the field boundary 110 m south-east of Middle (or 'Great') Brimmon Farm. The farm is approached along a narrow lane from the bridge over the railway at Newtown Station and access to the exposure is by **permission from the farm** to proceed through an intervening field.

The slumped mudstone exposed here is part of a slump sheet at least 15 m thick. It has created a pronounced topographical ridge and thus, in places, it can be mapped.

12. Quarry (SO 0949 8954) **near Pen-y-banc**
A lane-side quarry 100 m from the Dolfor Road (A483 T) on the lane to Pen-y-Banc $1\frac{1}{2}$ miles south-south-west of Newtown exposes 25 m of steeply westward dipping slumped mudstone and siltstone. It is very similar to that exposed at Locality 11 and it is almost certainly part of the same slump sheet.

The second type of slump is characteristic of the *leintwardinensis* Biozone and forms Earp's (1940, p. 4) 'Main Contorted Group'. Bedding remains largely preserved but has been folded and distorted to varying degrees, 'dominantly of the intrafolial fold type' (Woodcock 1973, p. 65). Whereas the slumps of Type 1 could have translated some distance by mass flow, Type 2 has folded penecontemporaneously as a surficial package without complete rupture and detachment from the strata below, 'the lower contact of each slumped unit is invariably gradational' (Woodcock 1973).

13. The Ring Hole (SO 1205 8375)
A very good exposure of slumps of the second type occurs in the Ring Hole (SO 1205 8375) on the **north-east side of the B4355 road 4.75 miles south of Newtown Railway Station**. Locality 13 lies just beyond the southern edge of Figure 2, as indicated in the south-west corner.

The Ring Hole is a nearly circular excavation, 150 m across the top and 58 m deep. The sides are mainly of glacial Boulder Clay and it has formed as a result of erosion by the River Teme with consequential constant landslip of the bank into the river. At the bottom is a post-glacial gorge cut into several metres of solid rock.

These strata form part of the upper part of the Bailey Hill Formation where the 'sandstone' layers are thin and of even finer grain – probably of silt grade (Earp 1938; Woodcock 1973). They are organized into disturbed packets separated by evenly bedded packets. 'The average proportion of undisturbed units in the total sequence is about 25%' (Woodcock 1973) and they are exposed well on the northern side of the gorge with dips barely exceeding 10°. The disturbed packets range in thickness from 1 m to 12 m but in some the beds are only very gently disturbed and show mere discordance with undisturbed beds in places. The evenly bedded packets are usually thinner and, while packets thicken and thin and probably pass laterally one into another, a 2 m packet of even beds on a 4 m packet of disturbed beds can be traced for many metres in the walls of the gorge.

Apart from folds, other tectonic-like fabrics are present – for instance three sets of narrow kink bands in the bedding of some of the evenly bedded strata. These can be seen at stream level and probably reflect slump related stresses.

Woodcock related the slumping to a south-eastward basinal decline in a detailed study of fold style within the disturbed packets (Woodcock 1973, pp. 96 *et seq.*, 1976a).

Acknowledgements. Of great help to the authors have been the discussions held with members of the Ludlow Research Group when they visited the area in 1988 and also with Dr T. J. Palmer on the origins of intraformational conglomerate. We are also grateful to Mr S. P. Tunnicliff and Dr J. A. Zalasiewicz for their assistance in the field and with fossil identification. The itinerary is published with the permission of the Director, British Geological Survey and is based upon fieldwork conducted for BGS, Aberystwyth while the authors were members of staff.

4. WENLOCK TURBIDITES BETWEEN RADNOR FOREST AND THE NEWTOWN AREA

by A. J. DIMBERLINE *and* N. H. WOODCOCK

Maps *Topographical:* 1:50 000 Sheets 136 Newtown and Llanidloes, 147 Elan Valley and Builth Wells, 148 Presteigne and Hay-on-Wye

1:25 000 Sheets 887 (SJ 0010) Llanfair Caereinion, 908 (SO 0919) Newtown, 949 (SO 0117) Llanbister and Abbeycwmhir, 970 (SO 0616) Llandrindod Wells

Geological: 1:250 000 Mid-Wales and Marches

1:50 000 Sheet 179 Rhayader

THE purpose of this itinerary (Fig. 1) is to examine the turbidites that dominate the basinal part of the Wenlock Series in Powys. The focus is a sand-dominated turbidite system, the Penstrowed Grits Formation, supplied from the south-west of the basin and confined by the basin margin at its south-eastern edge. The sandy turbidity currents are shown to have coexisted with two finer-grained depositional systems; low-volume mud turbidites and a background hemipelagic fallout. These Wenlock systems provide the opportunity to contrast a range of turbidite depositional facies.

GEOLOGICAL SETTING AND STRATIGRAPHY

The Wenlock sand-turbidites of Powys crop out in a belt from near Llanbister north-north-westwards past Newtown and Llanfair Caereinion towards Llangadfan and Lake Efyrnwy (Fig. 1; Cummins 1957, 1959*a*, 1963). At its south-eastern end the sand turbidite system abuts against NNE-striking faults on the north-west flank of the Pontesford Lineament (Dimberline & Woodcock 1987). South-east of here the Wenlock comprises first mud turbidites and hemipelagic mudstones, then, on the Midland Platform, bioturbated mudstones and limestones.

Traced to the north-west, the Wenlock turbidites are affected by the folds and faults of the Tywi Lineament, which crosses their outcrop in the region of Localities 4 to 6 (Fig. 1). This line also marks the south-eastern limit of

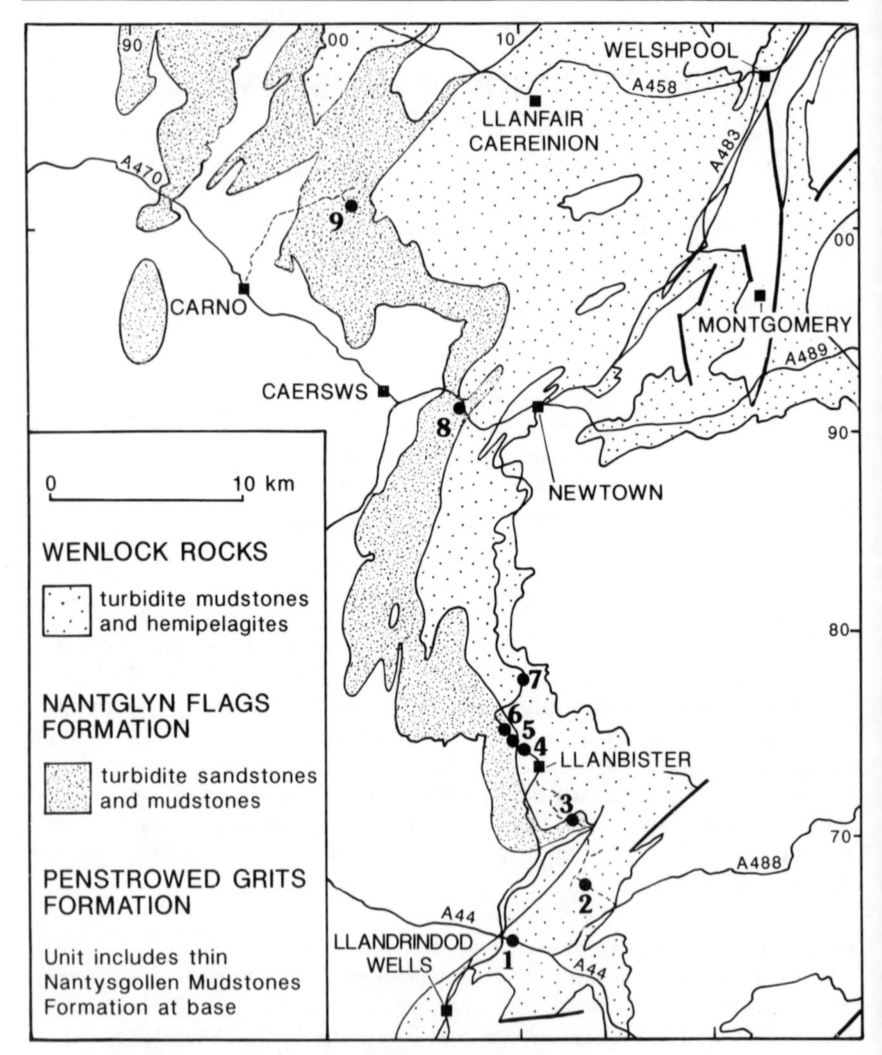

FIG. 1. Outcrop of Wenlock rocks in central Powys, with localities for the itinerary and routes between them (dotted away from the main roads).

strong cleavage. However, throughout the outcrop of the turbidites, metamorphic grade remains in the anchizone or diagenetic zone, and cleavage is sometimes weak enough to appear absent in outcrop. North-west of Newtown, the Wenlock rocks are deformed by major south-east-verging folds, containing in particular large tracts of steeply dipping south-east-younging strata (e.g. Tan Y Foel Quarry, Locality 9). These large folds probably nucleated on basement fractures that had earlier influenced sedimentation.

Three stratigraphical units are relevant to this itinerary (Fig. 1). They have been referred to by a variety of lithostratigraphical names; the usage here is approximately that on the Rhayader map sheet (British Geological Survey 1991).

The *Penstrowed Grits Formation* is a unit dominated by sand-turbidites, named after Penstrowed Quarry (Locality 8) near Newtown (Institute of Geological Sciences 1972). Synonyms are the Castle Vale Formation at the southern end of the outcrop (Dimberline & Woodcock 1987) and the Fynyddog Grits (Wood 1906) in the Tarannon outlier, probably both deposited in continuity with the type area. The original continuity with the Denbigh Grits Group of North Wales (Warren *et al.* 1984) is less well established (Cummins 1957), although this group is the approximate time equivalent of the Penstrowed Grits.

The *Nant-ysgollon Mudstones Formation* (Wood 1906) is a thin unit (<100 m) of interbedded hemipelagic and turbiditic mudstones below the Penstrowed Grits. It corresponds to the Benarth Flags of North Wales.

The *Nantglyn Flags Formation* is a thicker unit (>300 m) of similar mudstones above the Penstrowed Grits. It is the equivalent of the Lower Nantglyn Flags Group in North Wales, and of the Llanbadarn Formation in the south (Locality 7, Dimberline & Woodcock 1987).

The Penstrowed Grits probably span at most the *riccartonensis* and *rigidus* biozones of the lower Wenlock (Dimberline 1987). Their thickness and age span decrease south-eastward towards their feather edge at the basin margin, probably within the *riccartonensis* Biozone.

FACIES INTERPRETATION

The Wenlock sediments can be described in terms of six facies (Dimberline 1987), designated B to G in conformity with the scheme of Pickering *et al.* (1989).

Facies B comprises medium to very thick beds (10 to >100 cm) of granule to medium sand grade (Fig. 2a). Some beds show no internal organization. More typically they show an upward sequence of an inversely graded granule to medium sand base, a structureless coarse to medium sand division, and parallel-laminated then cross-laminated divisions of medium to fine sandstone. Beds of this facies are the deposits of a high density turbidity current decaying to a low density turbidity current. Facies B forms a minor component of the Wenlock turbidites, but good examples occur at Penstrowed and Tan Y Foel Quarries (Localities 8 and 9).

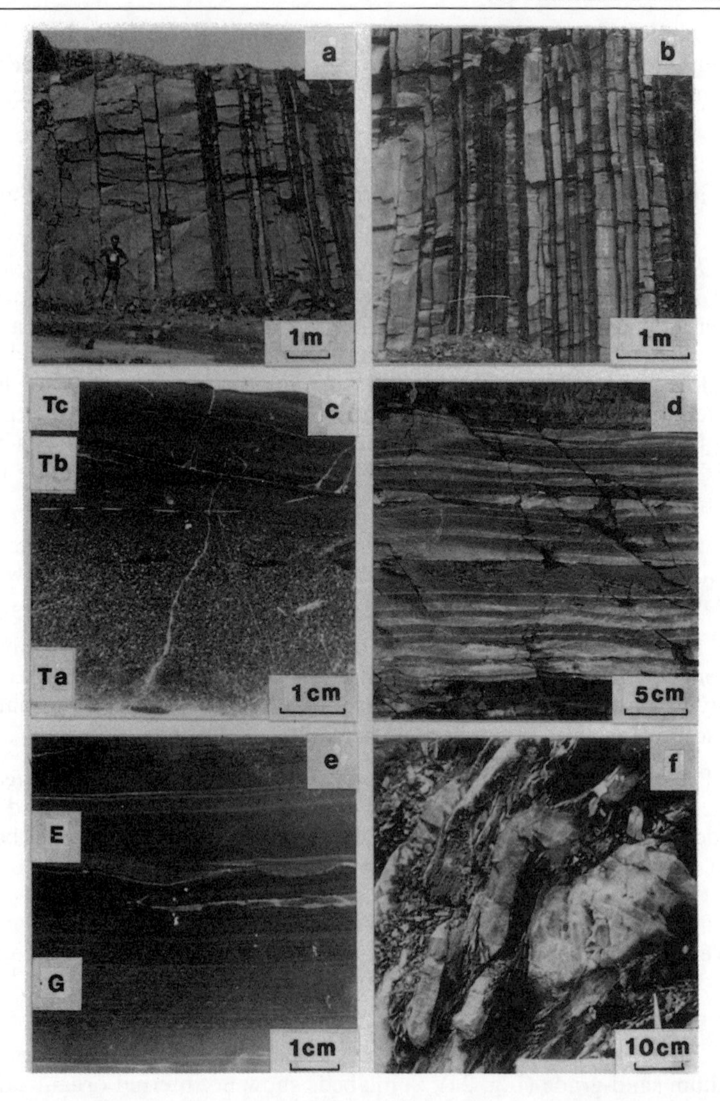

FIG. 2. The main basinal facies in the Wenlock. a) Facies B, NW end of main face, Tan Y Foel Quarry (Loc. 9, Fig. 6); b) Facies C, main face, Tan Y Foel (Loc. 9, Fig. 6); c) Facies C sandstone showing T_{abc} sequence, Llananno west (Loc. 5), d) Facies D, stacked silt-mud turbidites, main face, Tan Y Foel Quarry (Loc. 9), e) Facies E and G, Castle Vale (Loc. 6), f) Facies F, hinge of slump fold, Castle Vale (Loc. 6).

Facies C units are typically thin to medium-bedded (3–30 cm) sand-mud couplets (Fig. 2b). The beds usually have well developed normal grading, sometimes stepped rather than continuous. They display the upward Bouma sequence (Fig. 2c) of structureless graded sand (Ta), parallel-laminated sand (Tb), cross- and convolute-laminated sand (Tc), parallel-laminated fine sand and mud (Td), and structureless mud (Te). The Ta and the Tde divisions may be missing in finer and coarser beds respectively. Facies C beds are the product of traction sedimentation from a low density turbidity current, preceded for coarser beds by a phase of suspension sedimentation from a high density turbidity current. This facies dominates the Wenlock sand turbidite system, and is seen at Localities 3, 5, 6, 8 and 9.

Facies D comprises thinly to thickly laminated (1–10 mm), graded silt-muds displaying a variety of parallel-, cross- and convolute- lamination. They often form graded-laminated units in which successive silt laminae become finer upwards over about 2–10 cm intervals (Fig. 2d). Deposition occurred from low-concentration, slow-moving, fine-grained turbidity currents. It is not always easy to tell the repeated products of successive single turbidity flows from the silt-mud lamination produced by sorting processes within one flow, except where hemipelagite (Facies G) intervenes between flows. Facies D, together with E, makes up much of the Nantglyn Flags (Localities 4, 7) and forms finer-grained packets within the Penstrowed Grits (Localities 3, 5, 6, 8, 9).

Facies E is graded mud in very thin to thin beds (1–10 cm), often with a very thin silt lamina at the base (Fig. 2e). These beds were deposited from muddy turbidity currents. They are associated with Facies D as a component of the Nantglyn Flags (Localities 4, 7).

Facies F comprises strata which have been folded or disaggregated by soft-sediment deformation (Fig. 2f). Original bed fragments may be preserved in a muddy matrix, or may be disrupted more pervasively into an intimate mix of sediment types. The soft-sediment deformation in the Wenlock occurs near the basin-margin faults at Localities 2, 3 and 6.

Facies G consists mainly of very thinly laminated siltstone (Fig. 2e). The lamination is defined by silt laminae alternating with organic carbon-rich laminae on a scale of 3 to 4 couplets per millimetre. Graptolites are common on lamination surfaces. The lamination is occasionally disrupted by bioturbation. Facies G is interpreted as hemipelagic fallout (Dimberline & Woodcock 1987; Dimberline *et al.* 1990). Each couplet may represent a near-annual cycle of phytoplankton blooms alternating with enhanced silt input into the basin. In this case it can be used to estimate the time interval between turbidite events. The lamination was preserved because the bottom water conditions were anoxic, prohibiting benthic activity. Bioturbated hemipelagites record brief intervals of more oxic bottom

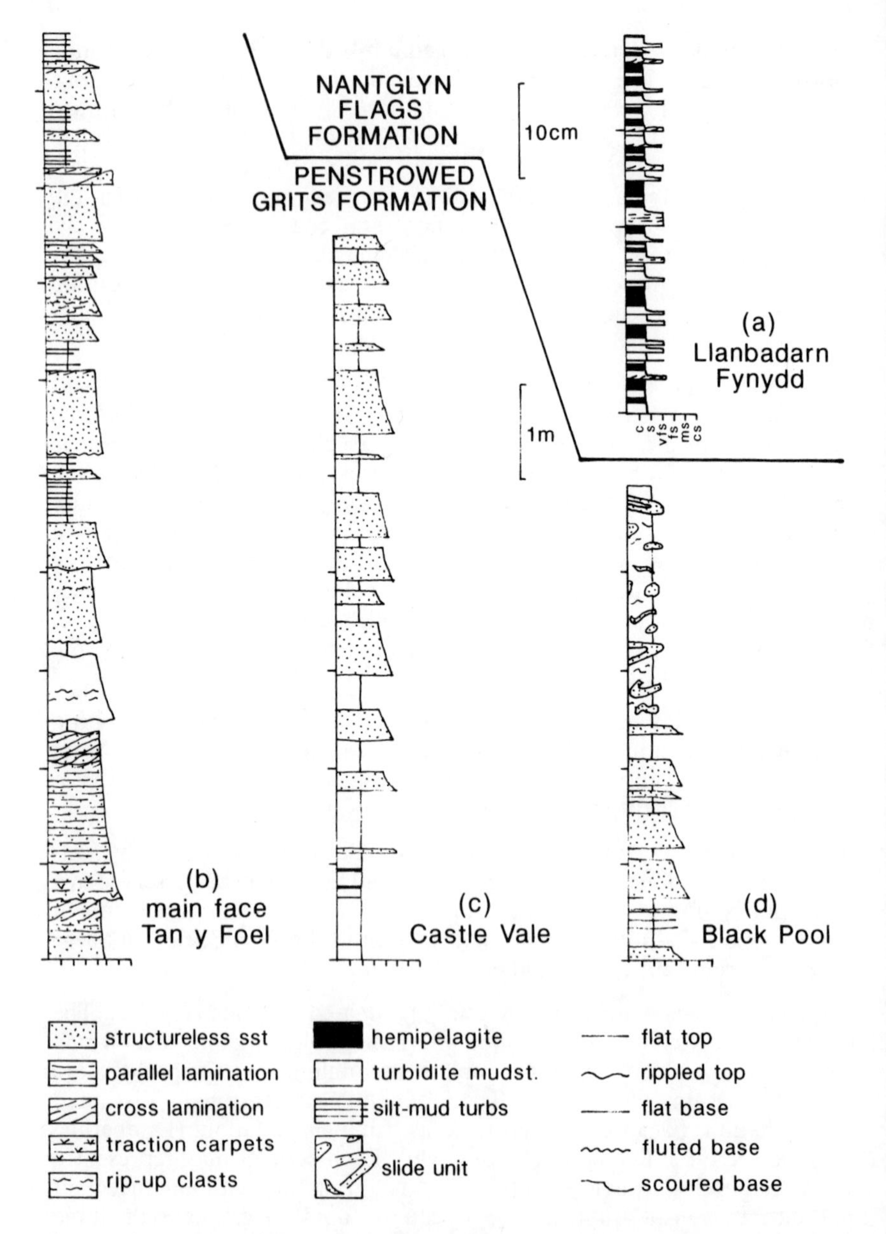

FIG. 3. Representative segments of lithological logs showing lateral and vertical facies variations. a) Llanbadarn Fynydd (Loc. 7), b) base of main face, Tan Y Foel Quarry (Loc. 9), c) top of Castle Vale section (Loc. 6), d) Black Pool Quarry (Loc. 3).

water. Facies G is present at all localities, but is best preserved in the Nantglyn Flags (Localities 4, 7) where it is least diluted by turbidite sediment.

Each Wenlock unit in this itinerary is characterized by an association of the facies above. The Nantglyn Flags Formation comprises mud-turbidites of Facies D and E intercalated with intervals of hemipelagic mudstones of Facies G (Fig. 3a). The Penstrowed Grits Formation is a sand-turbidite association dominated by Facies C, with beds becoming thicker and coarser basinward, and being augmented by beds of Facies B (Fig. 3b, c, d). Intervals of the finer-grained Facies D, E, G are variably intercalated. Facies F results from slumping of both associations near the basin margin.

BASIN MODEL

Four sediment distribution systems were operating in this part of the Welsh Basin during Wenlock time (Fig. 4).

a) *Sand-bearing turbidity currents* were supplied from the south-west end of the basin. Their considerable distance of travel suggests that they were fed from a source, such as slumps on a delta front, with a high mud content in addition to the sand. Such turbidity currents build leveed channels, allowing sediment to be transported far out into the basin. Palaeoflow estimates from sole structures on sandstones are NNE-directed at the southern end of the Penstrowed Grits outcrop, swinging to N-directed to the north of Newtown. The turbidity flows were constrained on their south-eastern and eastern flank by the basin slope underlain and controlled by the Welsh Borderland Fault System.

b) *Muddy turbidity currents* were probably supplied more locally from the adjacent platform and slope. Their greater frequency and smaller volume suggest a different trigger to the sand-bearing flows. Mean repeat times of about 120 years for those in the Penstrowed Grits Formation, and seventy-five years for those in the Nantglyn Flags Formation, are compatible with mobilization of sediment during strong storms (Dimberline & Woodcock 1987).

c) *Hemipelagic fallout* occurred as background sedimentation throughout the basin. It was fed by settling of pelagic organisms, their faecal pellets, and terrigenous silt washed off the adjacent platform.

d) *Submarine slumping* redistributed sediment locally down the basin slope, probably triggered by intermittent seismic activity of the basin-bounding faults.

Finally, an important factor in preserving the Wenlock basinal sequences was the prevailing anoxia of the bottom waters. Bioturbation is largely absent, allowing original detail of sedimentary lamination to be seen, even in the finer grained sediments.

ITINERARY

The itinerary is a transect across the sand turbidite system, oblique to palaeoflow (Fig. 1). Localities are described basinwards and down-flow, but a reverse itinerary is equally satisfactory. The whole itinerary can be completed in one day, although more time is needed to study Localities 6 and 9 in detail. Localities 2 and 3 are **not easily accessible to a party with a large coach.** Localities 4, 5, 6 and 7 provide a satisfactory short itinerary, without the need for obtaining prior permissions for access. Some supplementary localities are identified at the end of the itinerary, for those wanting to see further examples of the sand turbidites.

The first locality is approached on the A44 from Kington in the east or from Rhayader in the west.

1. Cross Gates road cut (SO 0940 6480) is on the north side of the A44 just west of the the railway bridge 0.5 km east of Cross Gates roundabout. Parking for one or two small vehicles is available in the lay-by or the lane

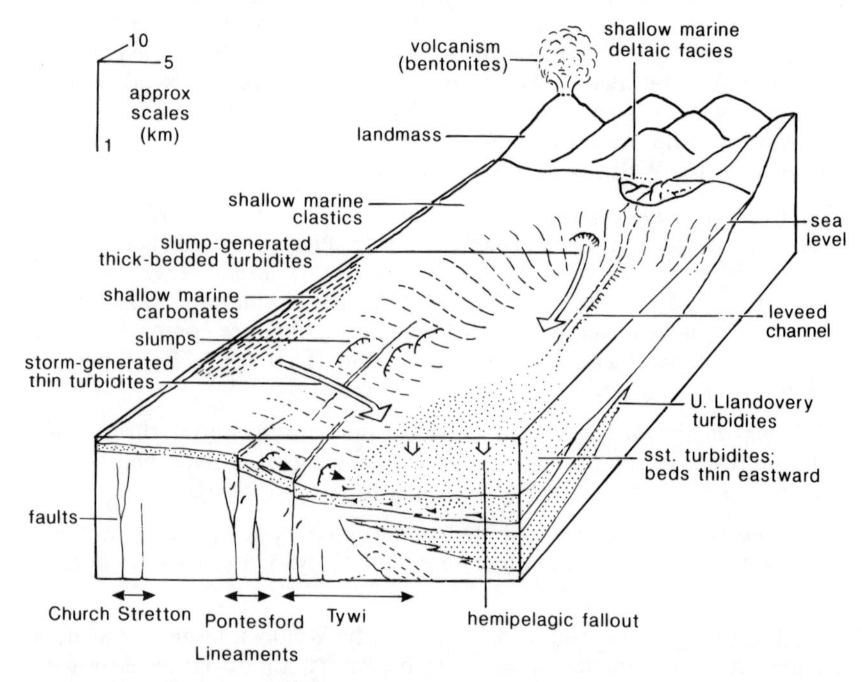

FIG. 4. Cartoon of depositional processes in the Welsh basin during the early Wenlock (after Dimberline & Woodcock 1987).

end opposite the exposure. **With permission,** there is also parking available in the yard of the community centre opposite the exposure. The road cut exposes late Wenlock (*lundgreni* Biozone) laminated hemipelagic mudstones (Facies G) interbedded with thin-bedded turbidite mudstones (Facies E) in a gently dipping sequence. Sedimentation took place in a series of small-volume events. The carbonaceous laminae that define the hemipelagite lamination may be close to annual. However, the sequence contains some zones of truncated lamination on a scale of tens of centimetres. These might be interpreted as current scours, but are more likely to be the sediment-draped scars from which slump sheets have detached themselves. The outcrop also preserves some post-depositional carbonate nodules.

The locality lies south-east of the presumed site of deposition of the Wenlock sand turbidites. The sandy turbidity currents, supplied from the south-west (Fig. 4), were probably confined to a topographical low by the fault-founded north-west-facing slope to the Welsh Basin. Slope sedimentation comprised vertical hemipelagic fallout augmented by muddy turbidity currents, probably sourced from the adjacent platform. The slump scars suggest a slope that was intermittently unstable. A site of accumulation of slumped slope muds can be examined at Locality 2.

Follow the A44 eastwards for about 2 km and in Penybont turn left on to the A488. Turn left again in 4.5 km (at SO 1488 6656), 250 m south-west of Llanfihangel Rhydithon. Proceed 1.6 km to the north-west, crossing the railway and the River Aran.

2. Abermain Quarry (SO 1347 6747) is a disused quarry on the north-east side of the road. Vehicles can be parked on the roadside by the quarry. **Permission for access should be obtained from Abermain Farm,** adjoining the quarry to the south-east.

The quarry exposes calcareous, yellowing-weathering mudstones in a strongly deformed or homogenized state. A moderately north-dipping fault defines the north side of the quarry, and part of this deformation may therefore be tectonic. However, the chaotic folding and the breakdown of lamination is characteristic of soft-sediment deformation soon after deposition. Outcrops with preserved lamination occur on the south-west side of the **road opposite the quarry**, revealing hemipelagites and turbidite mudstones (Facies G and D) similar to those at Locality 1. These mudstones are of uncertain Wenlock or earliest Ludlow age.

Cross Gates road cut (Locality 1) represented the primary slope facies, from which slump sheets had been detached. By contrast Abermain Quarry represents a site lower on the basin slope where slump sheets accumulated. The two localities together are typical of Wenlock facies preserved at the basin margin along the Pontesford Lineament. They were deposited on a fault-controlled slope which constrained the sandy turbidity currents to the basin further north-west.

From Abermain Quarry drive NW for 1.1 km. Take the right turn at SO 1280 6810. There is a good view north-eastward from this road of the ridge and valley topography caused by erosion into the folds and faults of the Pontesford Lineament. Bear left towards Llanbister after 1.2 km and continue for a further 1.8 km. Park on the verge at SO 1248 7062.

3. Black Pool Quarry (SO 1248 7076) is a small, disused quarry 200 m north of the road. Facies types C, D, F and G are present, in a sequence dipping at about 40° NE. Graptolites indicate an early Wenlock (*riccartonensis* Biozone) age. The base of the 10 m thick exposed sequence (Fig. 3d) consists of sandstones with a maximum bed thickness of 40 cm and grain size of coarse sand. Minor scours are preserved on bed bases, indicating palaeoflow towards the NNE. The most significant feature at this locality is the slide sheet which makes up the top 2.7 m of the section (Fig. 3d). Folded and balled sandstones are well displayed. On the assumption that slump fold axes parallel the strike of the palaeoslope, this slide sheet moved downslope to the west.

Black Pool Quarry is the south-easternmost locality exposing sand turbidites in the whole Wenlock turbidite system. It records a depositional site a little way up the fault-founded south-eastern slope to the Welsh Basin (Fig. 4). Occasional large volume, sand-bearing turbidity currents sourced in the south-west could just encroach this high on the basin slope, although most were confined to the axial part of the basin further north-west. The sand turbidites on the slope became interbedded with mud turbidites and hemipelagic fallout. The resulting unconsolidated sequence was prone to downslope sliding, perhaps initiated by seismic activity on the underlying basin-bounding faults. Later localities demonstrate a decrease in slide sheet frequency away from the basin margin.

Drive north-west from Black Pool for 4.2 km to join the A483 at SO 1058 7296. Head north on the A483 for 3.5 km. **Park in the lay-bys at SO 0938 7465**, which are convenient for Localities 4, 5 & 6, all road cuts on the A483 (Fig. 5).

4. Llananno east (SO 0998 7407 to SO 0979 7422) is a **small road cut** about 300 m south-east of Llananno church (Fig. 5). It displays Nantglyn Flags in a weakly deformed east-dipping section. Poorly preserved graptolites include *Monograptus flemingii elegans* and ?*Pristiograptus dubius* indicative of a probable *lundgreni* Biozone (Wenlock) age. Turbidites of Facies D and E are intercalated with hemipelagite divisions of Facies G. These relationships are displayed more clearly at Locality 7.

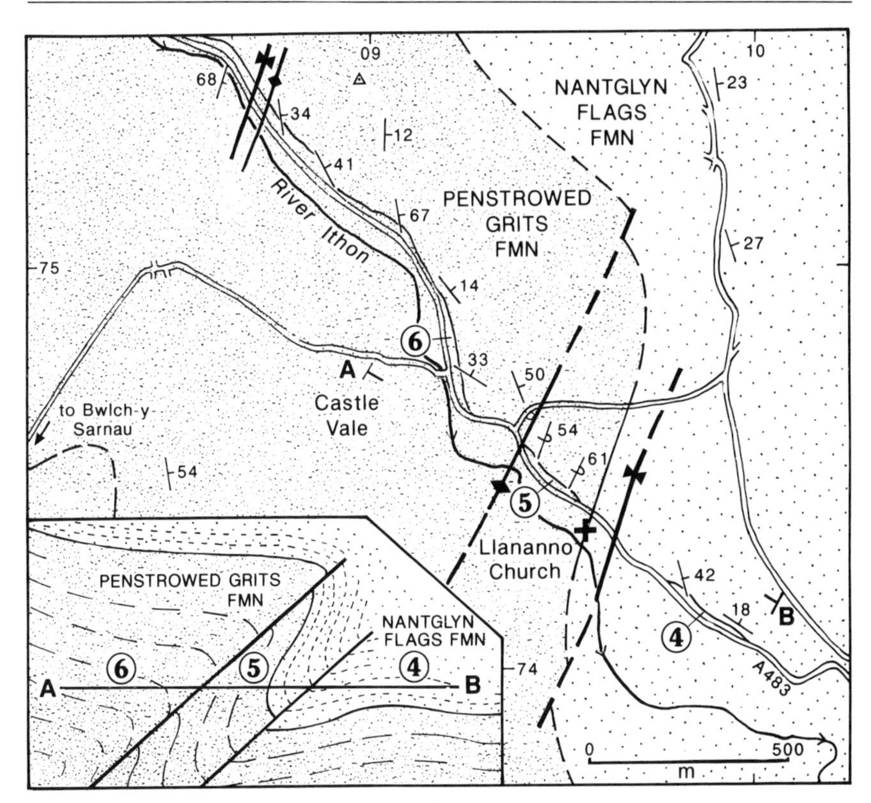

FIG. 5. Map and sketch section of the Llananno area (Locs. 4, 5, 6). Note that the folds plunge moderately north-eastward, so that the vertical section does not give a true profile view.

5. Llananno west (SO 0953 7442 to SO 0941 7456) is a **road cut** starting just north-west of Llananno church (Fig. 5). It exposes a 40 m thick section of inverted sandstones and mudstones, dipping 40° to the north-west. Cleavage in the mudstones dips more shallowly in the same direction. Facies types C, D, E and G are present. Graptolites suggest an age in the late *riccartonensis* to late *rigidus* biozones of the lower–mid Wenlock. Sandstone beds reach a thickness of 55 cm and grain size of granule grade. Due to the inverted bedding the section provides good exposures of flute casts on the base of facies C sandstones. When rotated back to correct for the overturning, these flutes show palaeoflow to the NNE.

The facies at this locality are interpreted in conjunction with those at the next locality.

6. Castle Vale (SO 0928 7465 to SO 0920 7490) is a road-cut starting 500 m north-west of Llananno church, **opposite the road junction for Bwlch-y-sarnau** (Fig. 5). The section is one of the most informative in the Wenlock turbidite system. Over 50 m of turbidite and hemipelagic sediments are exposed, dated somewhere in the interval from the *rigidus* to *lundgreni* biozones, and all assigned to the Penstrowed Grits Formation. Facies C, D, E, F and G are represented, but the sequence is dominated by alternations of the fine-grained facies D and G. On the assumption that the lamination in the hemipelagic sediments of Facies G is annual, the repeat time of the silt-mud turbidites (Facies D) has a mean of 123 years (standard deviation 163 years).

Sandstones are thickest towards the top of the exposed section, reaching a maximum bed thickness of 65 cm and a grain size of very coarse sand (Fig. 3c). Scour surfaces and the amalgamation of sandstones occur in the top sandstone packet. This packet has a crude thinning- and fining-up trend in successive beds. Such patterns are rare in the Wenlock turbidite system, with statistical tests revealing random bed thickness sequences (Dimberline 1987).

An interval of intense pre-cleavage deformation occurs at the south-east end of the section. Its base is not exposed. Disaggregated and folded sandstone bed fragments and balled mudstones are contained in a fine-grained matrix. This structure indicates some cohesion of beds at the time of sliding. The original lithology of the slide sheet matches that in the overlying undisturbed sequence, suggesting a local intrabasinal slope failure event. Cleavage clearly transects several of the slump folds.

The sequence dips at about 40° NE, with the weak cleavage dipping west at about 65°. A pair of steep NNE-striking faults with oblique-slip slickensides occurs near the south-east end of the outcrop.

This and the previous locality record sedimentation basinwards of Black Pool (Locality 2), but still on the lowest part of the north-west dipping basin slope. This slope site is indicated by the presence of the slide sheet, but the slope must have been insignificant further north-west where such sheets are absent. Localities 4, 5 and 6 are in three separate limbs of a major asymmetrical fold pair (Fig. 5), which plunges gently to the north-east and is overturned to the south-east, with faulted north-west dipping axial surfaces. The fold lies directly along strike from the core of the Tywi Lineament further south-west (Excursions 6 and 7). It probably developed in the Acadian deformation by reversal of movement on an underlying basin-margin fault, formerly with basin-down normal displacement. Reverse displacements in the basement formed a step, over which the softer Silurian cover was forced to drape itself. Such forced folds with south-eastward asymmetry are common along the basin margin and above major intrabasinal faults.

From Llananno take the A483 northward for 5 km, and park in a lay-by on the east of the road (SO 0992 7781) in the village of Llanbadarn Fynydd.

7. Llanbadarn Fynydd (SO 0997 7793). The exposure is an **old quarry and a track section** on the east side of the A483. Bedding dips only gently and cleavage is very weak. A 15 m thick sequence of Nantglyn Flags is well exposed, comprising the fine-grained facies D, E and G (Fig. 3a). Graptolites suggest an age somewhere in the *rigidus* to *ludensis* range, but most probably in the *lundgreni* Biozone.

The coarsest grain size at this locality is very fine sand, often incorporated in clearly cross-laminated units at the bases of mud turbidites. The laminated hemipelagic facies (G) is very well displayed. Its lamination can be interpreted as annual, allowing the calculation of repeat times for the turbidite events (Dimberline & Woodcock 1987, Dimberline *et al.* 1990). One centimetre of hemipelagite is equivalent to about thirty years, giving repeat times for the silt-mud turbidites in the range 10–150 years. This is roughly the frequency with which major storms would have occurred on the adjacent shelf, suggesting these as a possible trigger for the mud turbidite events.

Drive northwards on the A483 for 19 km to Newtown, and turn west onto the A489. In a further 4 km (at SO 0713 9095) take the track leading south-west from the A489, by the sign for RMC (Wales) Ltd.

8. Penstrowed Quarry (SO 0680 9095) is a working quarry approached past the concrete plant. Small vehicles can be parked in the quarry car park by arrangement. **Permission for access should be obtained** on site or from the operators, Goetre Ltd, Station Yard, Abermule, Newtown (tel. 0686 86667).

The quarry is the type locality for the Penstrowed Grits. A sequence approximately 90 m thick is exposed, dipping at about 50° NW and with a steep NW-dipping cleavage. Facies types B, C, D, E, and G are represented. The locality is of most interest for the thick-bedded Facies B sandstones showing well-developed Bouma sequences. Single event sandstones are up to 50 cm thick and up to granule grain size. Amalgamation of sandstones is common. Flute marks are particularly well developed on the large quarried blocks of Facies C sandstones. Corkscrew and nested flutes can be observed. A thin (3 cm) bentonite is present towards the base of the main quarry face.

The presence of Facies B sandstones at this locality is consistent with a more axial position in the depositional basin than previous localities further to the south-east. These sandstones were probably deposited from turbidity currents with a high density of grains in water. Such currents were constrained to follow the axial trough of the basin, and encroached

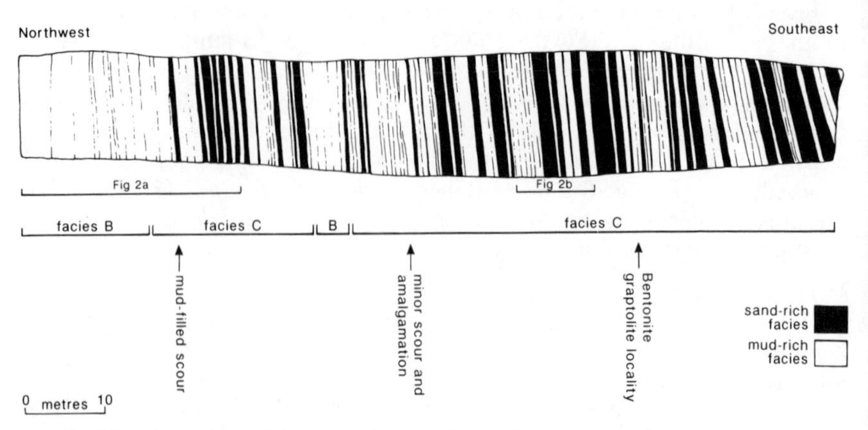

FIG. 6. Sketch section of the main face of Tan Y Foel Quarry (Loc. 9).

less onto its bounding slopes than did the lower density currents that produced Facies C. Facies B turbidites are present throughout the more axial regions of the basin to the north-west (e.g. Locality 9).

From Penstrowed take the A489 westward for 4.5 km. Turn north-west onto the A470 and travel 11 km to Carno. Turn right at the far end of the town (SN 9575 9730) and follow the minor road for 8 km towards Llanllugan.

9. Tan Y Foel Quarry (SJ 0130 0140) is a working quarry 5 km west of Llanllugan. **Permission for access should be obtained** from H. V. Bowen & Sons, Tan Y Foel Quarry, New Mills, Newtown, Powys. The quarry provides the most extensive exposures of the Wenlock turbidite system in Powys, and indeed in the whole of Wales. At least 300 m of the Penstrowed Grits Formation are exposed in a south-east younging, steeply dipping sequence. Graptolites from both the base and the top of the sequence indicate a *rigidus* Biozone age. Sediments of facies types B, C, D, E and G are well displayed. The most convenient exposures occur in the main working face (Fig. 6), which forms the top 130 m of the exposed sequence, but comparable exposure occurs in smaller faces down-sequence to the north-west.

Tan Y Foel Quarry exposes the thickest bedded and coarsest sandstones in the turbidite system. The percentage of sand-grade sediment is about 52 per cent, the maximum sandstone bed thickness is 300 cm and the maximum grain size is granule grade. Amalgamated sandstones occur, with a maximum amalgamated bed thickness of 480 cm.

Facies C sandstones are numerically most important and the cross-laminated Bouma Tc divisions are exceptionally well exposed on the large bedding surfaces. Flute casts are well displayed on the sandstones of the main face. Sandstone beds are laterally continuous over the maximum exposed length of 30 m, but some minor scouring occurs (Fig. 6). The thickest bedded, coarsest sandstones, assigned to Facies B, occur at the north-west end of the main face. Statistical tests show that the sandstones occur in random bed thickness sequences.

When logged in detail in 1986 (Dimberline 1987), the 320 m thick sequence at Tan Y Foel revealed 953 sandstones turbidites and 1,567 silt-mud turbidites. If these were deposited within one graptolite biozone of about 0.67 Ma duration, the sand turbidites were deposited at least every 700 years and the silt-mud turbidites at least every 425 years. The actual repeat intervals were probably shorter than this, and the interval for silt-mud turbidites was therefore probably of the same order as the 100-year interval determined from counting hemipelagite laminae at Localities 6 and 7.

SUPPLEMENTARY LOCALITIES

The localities described above are chosen to provide a convenient transect from the margin to the centre of the early Wenlock basin in Powys. However, there are a number of other good exposures that will repay a visit from anyone wishing to explore the Wenlock turbidite system in more detail. The best of those in the Penstrowed Grits Formation are listed below from south to north, with their grid references, constituent facies and any previous description. All are described in detail by Dimberline (1987).

10. Cefn Bronllys: road cut (SO 100 709): Facies C, D, E, G: Dimberline and Woodcock (1987), Davies *et al.* (1978).

11. Bwlch-y-sarnau: disused quarry (SO 037 751): Facies C, D, E, G: Dimberline and Woodcock (1987), Davies *et al.* (1978).

12. Maesyrhelem: road cut (SO 085 756); Facies C, D, E, G.

13. Bryn Yr Oerfa: forestry road section (SN 943 903): Facies B, C, D, E, G: Wood (1906).

14. Ffrwd Wen: disused quarry (SO 046 971): Facies B, C, D, E.

15. Garreg Hir: crags (SN 999 978): Facies C, fine facies unexposed.

16. Nant Carfan: stream section (SH 896 077 to 901 079): Facies A, B, C, D, E, G: D. A. Bassett (1955).

17. Moel Bentyrch: crag (SJ 057 095): Facies A, B, C, D, E.

18. Gro: crags and small quarry (SJ 066 112): Facies A, C, E: Cummins (1963).

5. THE MACHYNLLETH AND LLANIDLOES AREAS

by M. J. LENG *and* R. CAVE

Maps	*Topographical*:	1:25 000 Sheets SH70, SN69, SN79, SN98
		1:50 000 Sheets 135 (Aberystwyth) & 136 (Newtown & Llanidloes)
	Geological:	1:100 000 Central Wales Mining Field
		1:50 000 Sheet 163 (Aberystwyth) & Sheet 149 (Cadair Idris in preparation)
		1:63 360 Sheets 59SE & 60SW (old series)
	Aerial photographs:	RAF Sortie No. CPE UK 2531 Prints 3102-3104 incl. 4101-4103 incl.

Powys extends as far west as the Dovey estuary. In this northerly part of the county (Montgomeryshire) five Ordovician inliers occur within the Silurian outcrop, of which three are just west of Van, near Llanidloes, whilst the other two are found near Plynlimon and Machynlleth (Fig. 1). The Van Inlier itself and the Machynlleth Inlier (Fig. 1) provide the itinerary, with exposures of Ordovician (Ashgill) and Silurian rocks (Llandovery), respectively.

In Ordovician and Silurian times the area lay within the deeper parts of a submarine basin formed on continental crust. The shallow water margins of this basin have been identified in eastern Wales, the Welsh Borderland and South Wales, so that a submarine slope from the shallow areas to the deeper parts lay somewhere east of Llanidloes. The Ordovician and Silurian rocks (sandstones and mudstones) of the area were sediments transported down the slope from the shallow water shelf. The dominant processes were turbidity currents, hemipelagic fallout, and mass flow. Turbidity currents were engendered by the slope and spread material at its base and across the basin-plain as turbidites to form submarine fans of sand and mud. Vertical fallout of suspended matter (mud and organic debris) in the basin resulted in hemipelagic deposits, usually deposited on the basin floor between turbidites. Mass-flow deposits were initially turbidites or hemipelagites deposited on the slope, but instabilities there remobilized them to form down-slope slurries and slides.

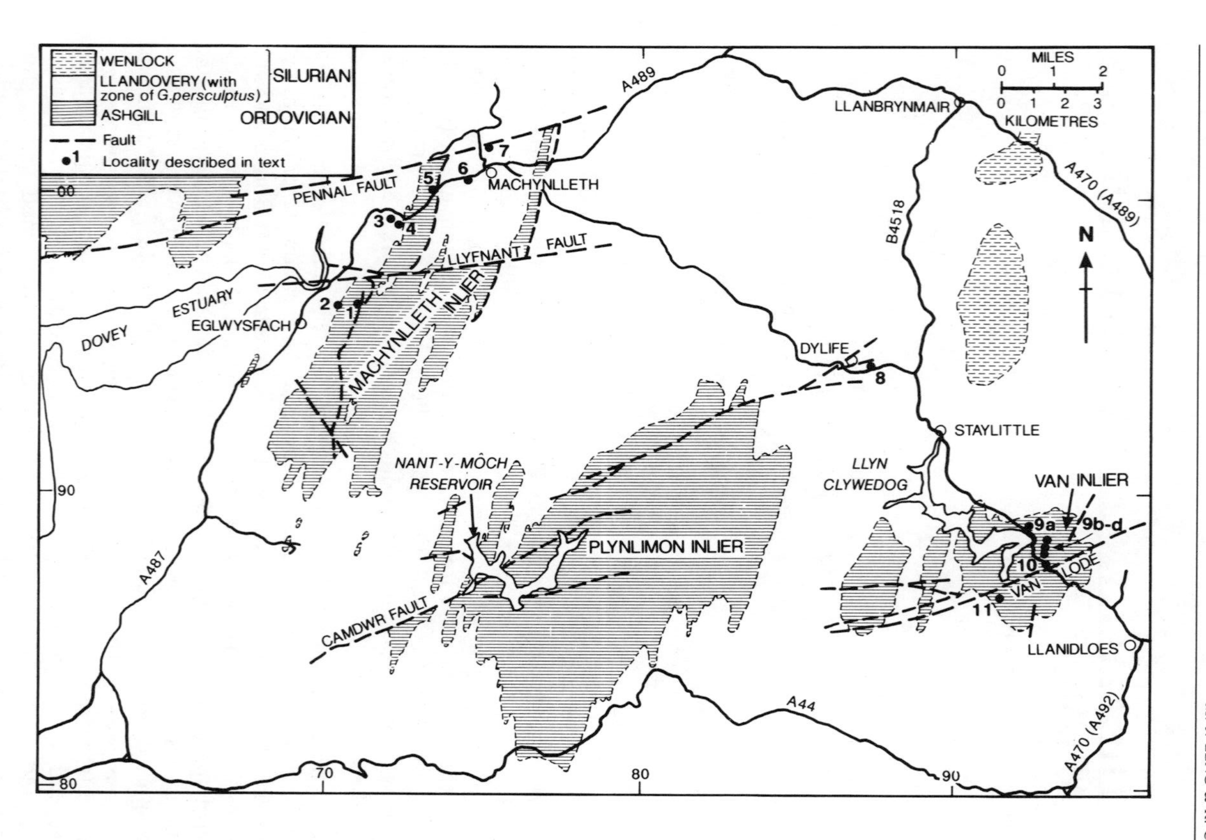

FIG. 1. General map of the area and localities.

STRATIGRAPHY

Ordovician

Only rocks of the uppermost Ashgill Series are present in the inliers of the area. Elsewhere in mid Wales the Ashgill Series is fully exposed and those sequences reveal that the overall basinal succession of litho-formations possesses a distinct pattern. This pattern forms the basis for a suggested correlation of formations (Fig. 2). Fossils are sparse and thus the correlation relies on a supposition that the pattern of lithofacies reflects events which affected sedimentation synchronously over the whole basin, and was not the product of progressive evolutionary infill of the basin from its margins.

Lower Ashgill. Most basinal sequences commence with a rhythmically layered, mud-dominated formation, much of which is intensely bioturbated. Each layer is thought to be a low-density (mud) turbidite. Distal from the source, in areas such as Corris and Llangranog, the resulting rock is a colour-banded silty mudstone with each medium grey homogenous layer of turbidite silty mudstones being succeeded by a paler grey layer of mottled silty mudstones. This mottling is the trace fossil *Chondrites*, the result of burrowing by an unknown organism. A layer of small dark grey phosphate concretions commonly occurs below the paler band but in the portions of the formation most intensely bioturbated all layering has been obliterated and even the concretions were disturbed. Such is the state of the Broad Vein Formation near Corris (Leng 1990) on the northern borders of Powys. In other areas, more proximal to source, where the turbidite rhythms possess sandy bases, burrowing failed to homogenize the sediment, e.g. the Craig-y-Dullfan subfacies of the Nant-y-Moch Formation near Plynlimon, on the western edge of the county (Cave & Hains 1986).

The high degree of bioturbation and the pale grey colouration indicate a general marine oxygenation of the early Ashgill basin, in sharp contrast to the immediately preceding Caradoc which produced dark grey graptolitic mudstones deposited under anoxic conditions.

Towards the top of these lower Ashgill formations are several dark grey mudstone layers that yield graptolites of the *Dicellograptus anceps* Biozone. The presence of these rocks at this level represents basin-wide events that recreated reducing conditions in the basin and thus allowed the preservation of layers of graptolitic hemipelagic mudstones between the turbiditic mudstones.

Middle Ashgill. In most places in the basin the silty mudstones of the lower Ashgill are succeeded by a thick sequence of rocks which are very obtrusive in the succession. Their unusual litho-characteristics, thickness, and basin-wide distribution can be explained only by invoking some extraordinary event.

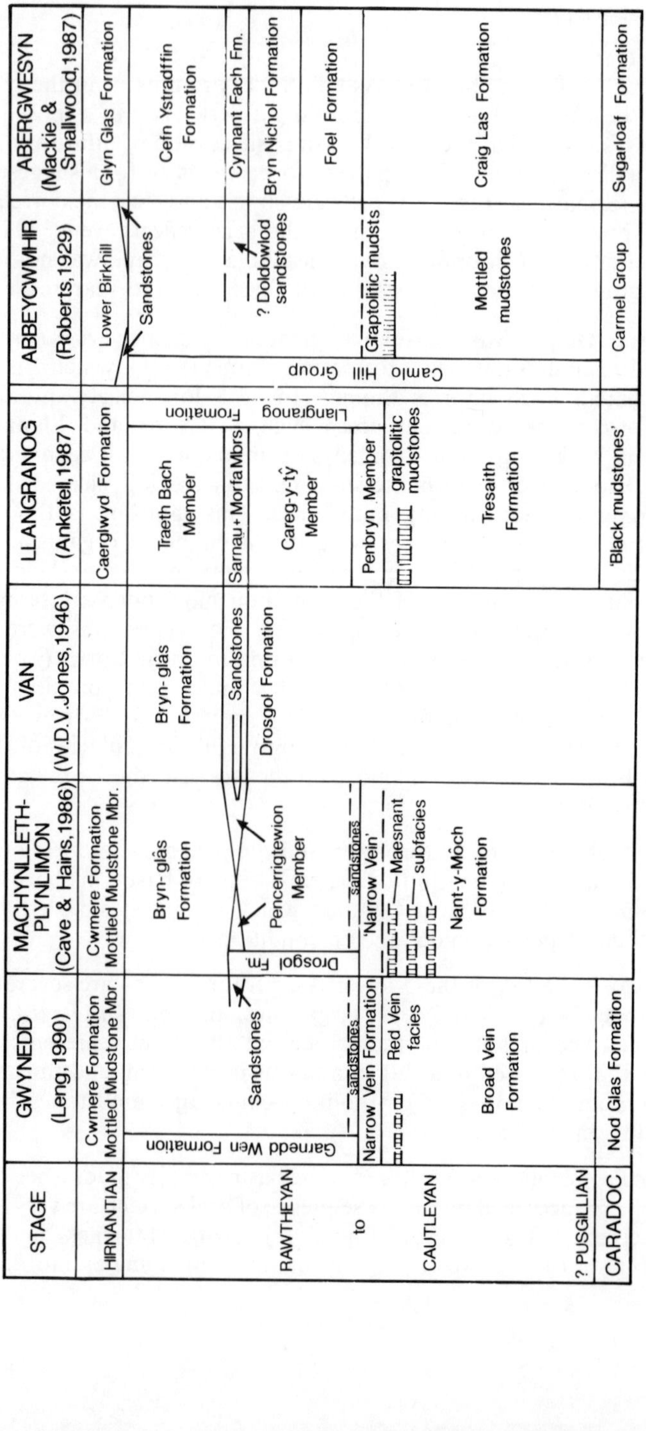

STAGE: HIRNANTIAN; RAWTHEYAN to CAUTLEYAN; ? PUSGILLIAN; CARADOC

GWYNEDD (Leng, 1990): Cwmere Formation Mottled Mudstone Mbr.; Garnedd Wen Formation; Sandstones; sandstones; Narrow Vein Formation; Red Vein facies; Broad Vein Formation; Nod Glas Formation

MACHYNLLETH-PLYNLIMON (Cave & Hains, 1986): Cwmere Formation Mottled Mudstone Mbr.; Bryn-glâs Formation; Drosgol Fm.; Pencerrigtewion Member; sandstones; 'Narrow Vein'; Maesnant subfacies; Nant-y-Môch Formation

VAN (W.D.V. Jones, 1946): Bryn-glâs Formation; Sandstones; Drosgol Formation

LLANGRANOG (Anketell, 1987): Caerglwyd Formation; Traeth Bach Member; Sarnau + Morfa Mbrs; Careg-y-tŷ Member; Llangranog Formation; Penbryn Member; graptolitic mudstones; Tresaith Formation; 'Black mudstones'

ABBEYCWMHIR (Roberts, 1929): Lower Birkhill; Sandstones; ? Doldowlod sandstones; Camlo Hill Group; Graptolitic mudsts; Mottled mudstones; Carmel Group

ABERGWESYN (Mackie & Smallwood, 1987): Glyn Glas Formation; Cefn Ystradffin Formation; Cynnant Fach Fm.; Bryn Nichol Formation; Foel Formation; Craig Las Formation; Sugarloaf Formation

FIG. 2. Suggested correlation of mid Wales basinal Ashgill sequence.

One possibility is the rapid growth of an ice-cap, which is known to have depressed sea-level abruptly at the time (Brenchley & Newall 1980); another could be basin margin tectonism. These rocks form the Garnedd-wen Formation around Corris; the Drosgol and Bryn-glâs Formations of Machynlleth, Plynlimon and Van; the Llangranog Formation (with the possible exception of the basal few metres) further south; and the upper part of the Camlo Hill Group near Abbeycwmhir (Fig. 2). The rocks at this level are dominantly silty mudstones and siltstones, medium to dark grey in colour. They commonly lack a normal bedding fabric although it may be present in a highly disturbed and distorted state. In places, largely in the lower parts of the formations, the silty mudstones are pebbly. These pebbles are either concentrated in lenses or scattered within certain layers. Some conglomeratic masses are clast-supported and form bodies that either have gradational or sharp contacts with the enclosing sediment. Similarly rounded bodies of sandstone occur, usually as quartz arenites, with diameters ranging from centimetres to metres. These sandstones have a contorted internal structure, but large raft-like bodies also occur in which stratification is preserved relatively undisturbed. The margins of such rafts are usually sharp.

At the base of the mid Ashgill formations, evenly bedded sandstones form a packet between 0–25 m thick. In places, flute-casts on the bases of these sandstones show that turbidity currents came from the east and south-east. These sandstones have been included in the Garnedd-wen Formation, Drosgol Formation and Llangranog Formation (as part of the Penbryn Member), although in places they seem to be channelled by the overlying silty mudstones (Fig. 6). Nevertheless they clearly represent a brief period heralding much more energetic deposition of the disturbed silty mudstones above. This high energy regime is considered to have been responsible also for the coarse-grained and conglomeratic Foel and Bryn Nicol formations near Abergwesyn to the south-east (Mackie & Smallwood 1987) (Fig. 2).

Within the mid Ashgill group of disturbed deposits there is commonly a thick packet of sandstones, e.g. the Pencerrigtewion Member (Fig. 2), often well bedded and quartzose, forming layers up to 4 m thick. This packet is uneven in thickness and is absent in places. It is also divided in places by silty mudstone intervals and, although there is no proof of correlation of this packet in the various areas where Ashgill rocks are exposed, it seems likely that the sandstone packet represents an event which affected the whole basin.

Upper Ashgill. The disturbed silty mudstones are succeeded abruptly by evenly bedded mudstones – the Mottled Mudstone Member of the Cwmere Formation in the Corris and Plynlimon areas. The Mottled Mudstone (Fig. 2) is a sequence of rhythmically bedded, low-density mud turbidites within the *Glyptograptus persculptus* Biozone and thus belonging to the Hirnantian Stage. A layer of dark grey graptolitic mudstone, several centimetres thick and occurring approximately 1 m above the base of the member, is present in many areas in mid Wales. Most of the rest of the member comprises colour-banded mudstones similar to the mudstones that comprise the lower Ashgill succession. It would seem that conditions of deposition returned to those of the lower Ashgill. Phosphate concretions and bands of pale grey mottled mudstones are especially well developed and record a degree of oxygenation of the basin which prevented the fossilization of carbonaceous pelagic deposits. However, towards the top of the biozone (i.e. the Mottled Member) the pale grey bands are replaced by bands of dark grey laminated mudstones (hemipelagite) of similar thickness. These mudstones contain graptolites (*G. persculptus*), indicating that there was a return to anoxic conditions in the basin before the close of that biozone.

Silurian.
The Silurian Period thus commenced with the Welsh Basin in an anoxic condition, and the remainder of the Cwmere Formation (Rhuddanian Stage of the Llandovery) consists mainly of rhythmically and thinly interbedded layers of medium–dark, grey, homogeneous turbiditic mudstones and dark grey laminated hemipelagic mudstones. Such rocks reveal that the basin was starved of coarse terrigenous detritus, but received abundant organic debris and nepheloid silt.

All these factors were the result of the relative rise in sea-level that occurred during the early Llandovery. Once the concomitant marine transgression of the marginal areas of lowland had ceased, the shallow shelf-sea was again able to process terrigenous detritus and deliver it to the basin as turbidites. So, during the succeeding Aeronian Stage, marine conditions again became mainly oxygenated and thin rhythmic deposits of turbidite mud accumulated. Short periods of anoxicity recurred, however, punctuating the sequence with dark grey graptolitic hemipelagic bands (Jones & Pugh 1916).

Such rhythmic turbidites became the common denominator of late Llandovery (Telychian) times, with the special element being influxes of coarse, high-energy turbidites from the south. One of these influxes resulted in the deposition of the Talerddig Grits of the *Monoclimacis griestoniensis* Biozone, which reached the Machynlleth–Llanidloes region of Powys and can be seen in the famous railway cutting at Talerddig (D.A. Bassett 1955).

One of the widest outcrops of the Llandovery 'background' rocks is that of the Telychian Devil's Bridge Formation in the *Monograptus turriculatus* Biozone, and the itinerary includes visits to exposures of these rocks, mainly to examine folds.

In western areas of Ceredigion the higher parts of the Devil's Bridge Formation are replaced laterally by the Borth Mudstones Formation (Cave & Hains 1986), but east of the N–S Glandyfi Tract, which lies along the A487 trunk road (Fig. 1), the Devil's Bridge Formation rises into the mid part of the *turriculatus* Biozone (Fig. 3) where it is overlain by dark grey turbiditic mudstones with black anoxic hemipelagite layers and fewer thin sandstones. These constitute the Crewi Formation (Scott & Cave 1990b), which are well exposed in and adjacent to the Afon Crewi 2 km ESE of Penegoes and have yielded many graptolites, e.g. farm track (7940 0000) and (7938 0003). The graptolites prove the presence of the *utilis* Subzone of the mid to upper *turriculatus* Biozone (Fig. 3) (Loydell 1991). They also reveal that the Crewi Formation is of the same age as the Blaen Myherin Mudstones of the south side of the Plynlimon Inlier, near Devil's Bridge (Cave & Hains 1986). The Crewi Formation was, in fact, the product of yet another period of basinal anoxicity. From having been predominantly well oxygenated throughout the deposition of the Devil's Bridge Formation (and Borth Mudstones on the west) the bottom waters of the basin became starved of oxygen in high *utilis* Subzone times, permitting the preservation of graptolites in abundance.

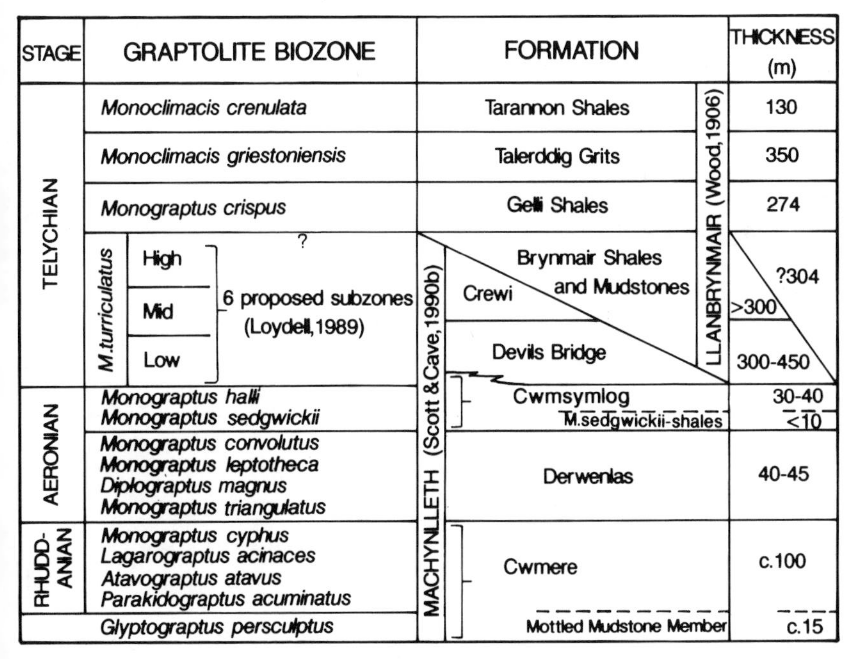

FIG. 3. Table of strata of the Llandovery Series and the *persculptus* Biozone near Machynlleth.

ITINERARY

The localities are described from west to east. They could form a western group near Machynlleth and an eastern group near Van to occupy two separate days. The road to Localities 1 to 3 will accommodate **nothing larger than minibuses**. All others can be reached with coaches (Fig. 4).

Llyfnant Valley area.

1. Craig Caerhedyn. An examination of the Drosgol Formation (Locality 1) entails a hill walk of about 2 km (Fig. 4). The area is approached from the A487 at Eglwys Fach along a narrow lane to a parking space (SN 7075 9635) near Brwyno. Here the road ends and a track takes a sharp bend to the south over bare rock. Many metres of evenly and rhythmically bedded thin turbidites of the Nant-y-Moch Formation are visible east of the track

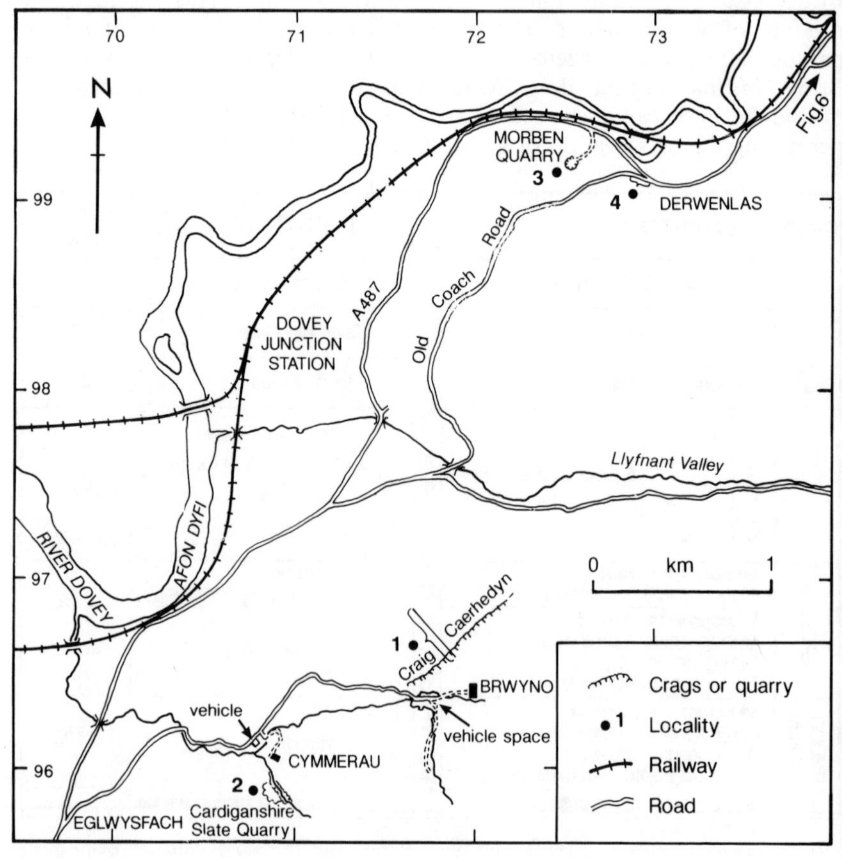

FIG. 4. Sketch map showing positions of Localities 1–4.

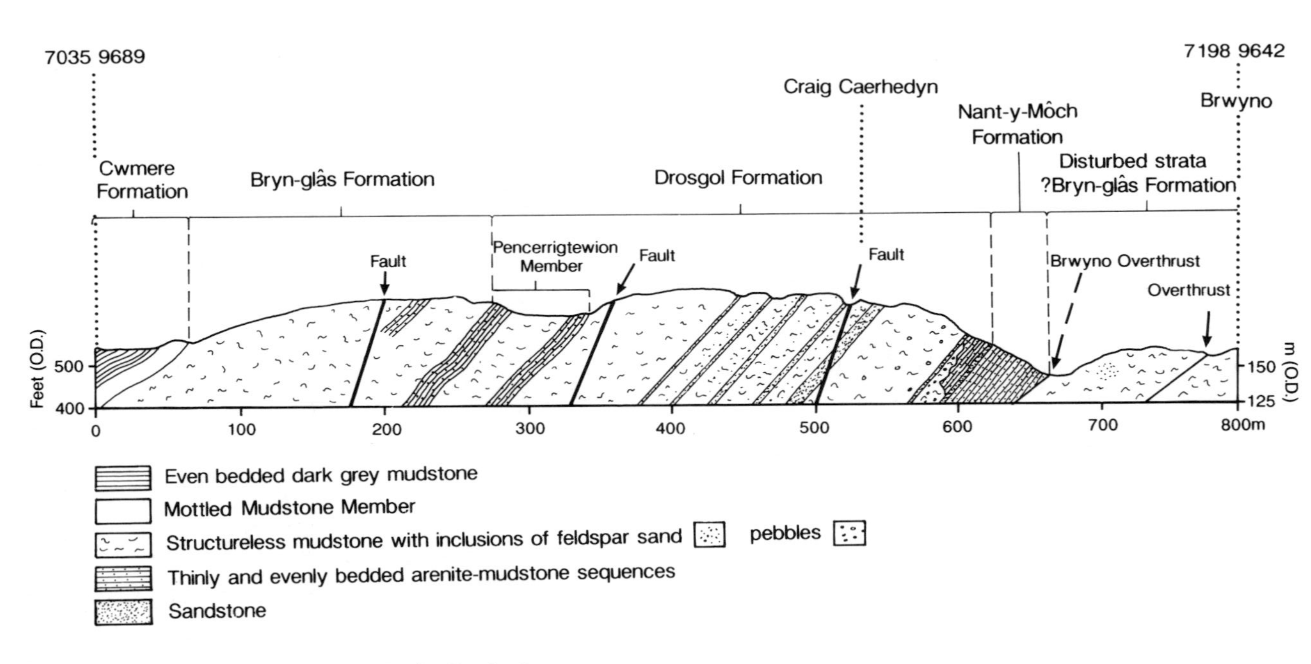

FIG. 5. Locality 1, section across Craig Caerhedyn.

and they continue across it to the west. There they are overlain by massive sandstones up to 2.5 m thick. These sandstones have been taken to mark the base of the Drosgol Formation (Cave & Hains 1986), whilst a second similar set of sandstones, some 20 m higher, is exposed further west. These two sets of sandstone are laterally persistent and they vary little. The beds between them are, however, variable and can be examined in the crags 280 m to the north-north-east at the start of the traverse through the Drosgol Formation (Fig. 5).

This variable sequence consists in part of evenly-bedded mudstones and thin sandstones which are replaced laterally across erosive contacts by interbedded pebbly mudstones or mudstones with sandstone intraclasts (Fig. 6). These mudstones, which become the main component of the rest of the formation, appear to scour into the well-bedded parts, yet they have not been seen channelling into, or below, the bottom sandstone. The sandstones were probably delivered into the basin by an agency independent of that which delivered the unlayered channel-fill of pebbly mudstones. The main significance of this portion of the lithostratigraphy is, however, that a fundamental change in the depositional system of the

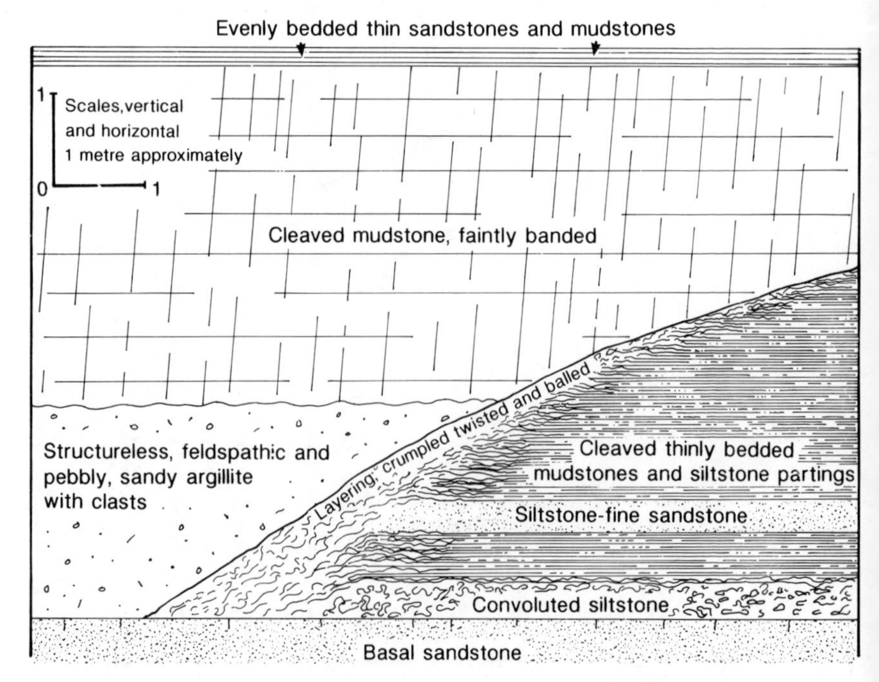

FIG. 6. Locality 1, channel margin at the base of the Drosgol Formation, Craig Caerhedyn.

basin occurred. The change probably commenced with the deposition of the bottom sandstones, but was over when the second set of sandstones was deposited. A new source of terrigenous detritus had therefore arisen near the basin margin to the east and a very energetic system of delivery into the basin developed coincidentally. This has been attributed to a drop in sea-level connected with the Saharan late Ordovician glaciation (Brenchley & Newall 1980).

From the basal part, the rest of the Drosgol Formation is traversed by proceeding **westward, up-slope**. The main part of the hill is composed of mudstones and siltstones which are medium to dark grey in colour, without a bedding fabric, and with an uneven phacoidal cleavage. Inclusions of paler grey, fine-grained sandstones as isolated lumps and wisps are common. These sandstones are distributed randomly and any internal bedding fabric that remains is intensely contorted. Feldspathic sandstones are important locally. The sandstone inclusions are attributed to soft-sediment disturbance caused either by reverse density gradients between the muds and layers of sand (Cave & Hains 1986) or by the down-slope translation (slumping) of sediment (Leng 1990). The latter probably played the more important role.

Walking across the hill, five parallel, furrow-like, trenches can be observed for long distances along strike (NNE–SSW). Looking southwards, the same trenches are visible above Bwlch Einion and on the slopes of Foel Goch about 4 km away. Inspection reveals that these trenches are occupied by exposures of evenly bedded, often massive, sandstones and bedded mudstones. Obviously the deposition of disturbed silty mudstone was punctuated by at least five periods when sandy turbidites entered the basin, forming sandstone lobes. It is worth noting that outcrops of Ashgill sandstones in mid Wales often produce topographic hollows and not the normal ridge features expected from such 'hard' rocks. This was the result of the Pleistocene ice sheets plucking out the well-jointed sandstones.

The top of the Drosgol Formation is marked by a broad hollow occupied by the Pencerrigtewion Member, but in this area massive sandstones are absent and exposure is poor.

The road can be rejoined by **walking south** down the hollow, and rather than returning to the point of departure it is easier to **walk west** directly to Cymmerau Farm and Locality 2.

2. Cardiganshire Slate Quarry. This quarry (SN 6991 9595) (Fig. 4) belongs to the adjacent farm (SN 6990 9605), where **permission to visit and to park vehicles** can be obtained. Great care should be taken at this locality. The route to the south-east side of the quarry is via the bank of an incised stream and the descent into the quarry is difficult; loose, often wet rocks occur and the north-west end of the quarry has a deep pool. The succession is shown in Fig. 7. Details of the lower part of the section are tabulated below:

BED		THICKNESS (CM)
10.	Mudstone, grey-brown, tough, cleaved. Top surface grooved NW–SE by sliding of the superincumbent bed	22.0
9.	*G. persculptus* Band. Mudstone, grey-brown, soft, rubbly. Abundant small limonitic flecks. *G. persculptus* present. Top part of the bed is softer than the bottom part and laminated.	30.5
8.	Mudstone, grey, cleaved, rather tough	10.2
7.	Mudstone, dark grey, shaly, with limonitic specks and pale grey weathered spots	5.3
6.	Mudstone, medium grey, tough, mainly structureless appearance though bedding laminations faintly visible on weathered surfaces together with layers of weathered pyrite particles	30.5
5.	As bed 3.	8.3
4.	Mudstone, medium grey, cleaved and shaly in parts	10.2
3.	Shale, grey with limonitic (weathered) laminae, graptolitic remains.	10.2
2.	Mudstone, medium grey, structureless, evenly cleaved	17.8
1.	Plane surface, weathered out on mudstone, rather dark grey, structureless, silty and tough, unevenly cleaved. Contains a scatter of silty wisps and clots up to 2 cm in cross-section	61.0

The Bryn-glâs Formation is exposed in the incised cascade near the south-east corner of the quarry (Bed 1, Fig. 7) and beds numbered 1–10 (above) crop out in the north-west wall of the cutting, with the *persculptus* Band (Jones & Pugh 1916) near the top (Bed 9). (Any graptolites found in Bed 3 are particularly worthy of submission for specialist advice.)

The *persculptus* Band is also exposed at the top of the adjacent dip surface down into the quarry, where *G. persculptus* is present in abundance and preserved in relief. The south-west wall exposes the characteristically banded mudstones with medium grey layers of turbiditic mudstones alternating with bioturbated paler bands that display the trace fossil *Chondrites*. At the bottom of the quarry, in the NW-facing wall, the sequence rises through the top of this oxic facies into dark grey, rusty-

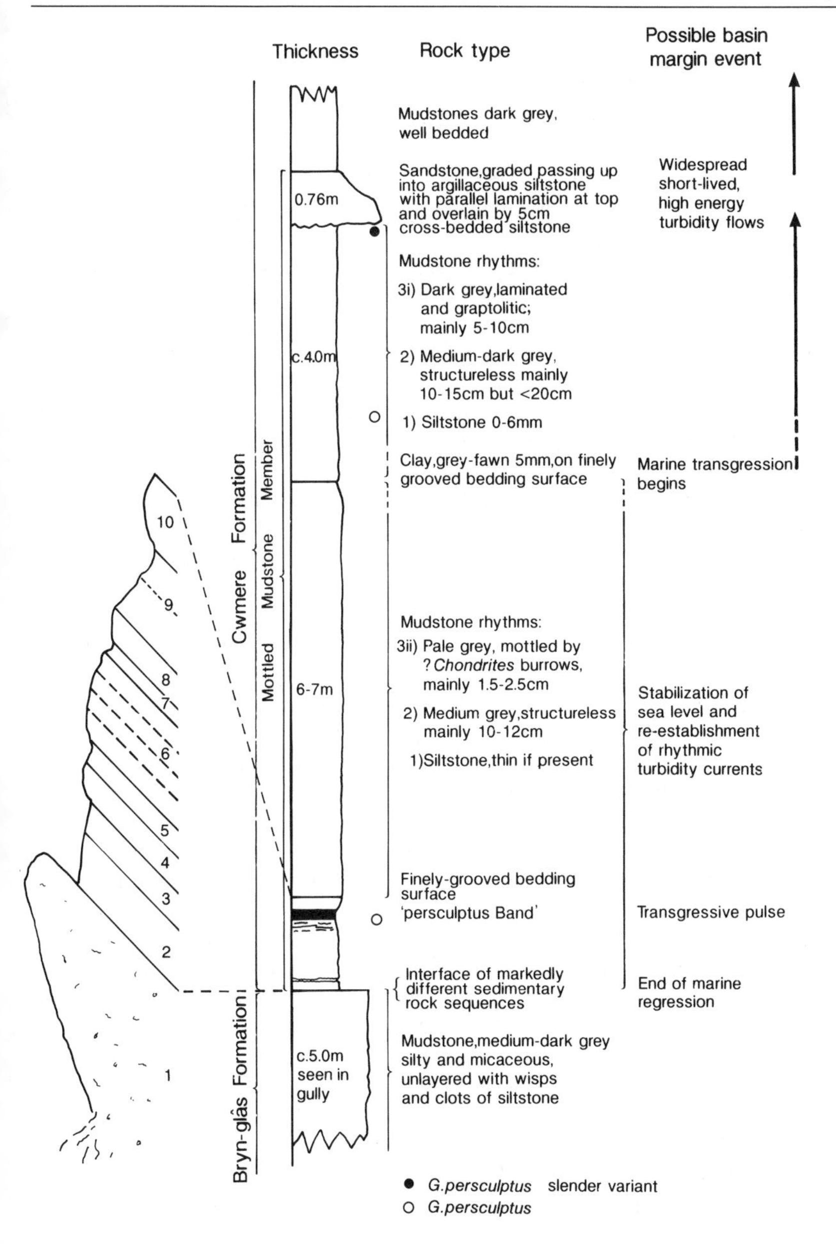

FIG. 7. Locality 2, the succession in Cardiganshire Slate Quarry, Cymmerau Farm. See text for description of numbered beds.

weathering mudstones of the succeeding anoxic facies. The change in the nature of the banding can be examined and compared with the generalized scheme tabulated in Fig. 7. The slate waste yields blocks with good examples of this banding and can be examined in waste tips on the return to the farm.

Near the top of the north-west wall of the quarry is a thick (0.76 m), feldspathic sandstone which can be examined best in the fallen blocks. Sandstones at this level persist widely in the area and they were chosen to represent the top of the Mottled Mudstone Member for the practical reason that they form a mappable lithostratigraphical feature.

The Mottled Mudstone Member signified the end of the period of mass flow into the basin, just beneath the Ordovician–Silurian boundary. The base of the Member is sharp, indicating abrupt reversion to low-density turbidite deposition and the colour of the banding reveals that basin conditions were generally oxygenated. As in the early Ashgill mudstones, phosphate (apatite enriched) nodules occur at the bases of the pale grey bioturbated bands, and some of these were burrowed (not bored), indicating that the enriched sediment was not initially cemented.

The *persculptus* Band represents a brief interlude of anoxicity of wide extent in the basin. This was almost certainly the product of a regional, or wider, event such as a brief marine transgression. The band illustrates the synchronicity of the cessation of the mass flows which formed the underlying Bryn-glâs Formation, because about 1 m of bioturbated grey silty mudstones consistently form the interval between the *persculptus* Band and the disturbed formations in mid Wales.

Both the top and bottom of the mass-flow formations below the *persculptus* Band are thus sharply defined by synchronous boundaries, yet their precise biostratigraphical ages are not known. Although the slumped formations record a major change in the environment of the basin as a whole, it was ephemeral and did not fundamentally alter the nature of the basin. The Mottled Mudstone Member reveals that deposition regained the pattern of the Nant-y-Moch and equivalent formations (Fig. 2). The change might well have been climatic and connected with the late Ordovician glaciation of the Saharan area. It is thought that this event might have lowered sea-level by up to 100 m (Brenchley & Newall 1980). The sandstones at the top of the quarry might also be basin-wide. They are coarse grained and in parts feldspathic. Similar thin layers of turbiditic sandstones are of wide occurrence at this level in the stratigraphy. These sandstones correlate with massive quartzose sandstones at Rhayader, called the Cerig Gwinion Grits (Chapter **6**), which were affected by syn-sedimentational sliding.

Localities 6 and **7** near Machynlleth are approached north-eastward on the A487. *En route* there are three sections that could provide extra stops, time permitting. These localities are described here as Localities **3, 4,** and **5**.

3. Morben Quarry (SN 7152 9918) is owned by the Garth Owen Estate, and exploited slate slabs from the Cwmsymlog Formation (top Aeronian Stage and bottom Telychian Stage, Llandovery (Cave & Hains 1986, p. 89)) (Fig. 4). It is approached (on foot) by a track from a gate on the east side of the A487.

There is **limited parking space by the gate. A coach should park in Derwenlas**.

The quarry was entered on two levels, of which the upper working was an opencast trench and was abandoned when it became too deep. The sides of this trench comprise steeply overturned, dark grey, rusty weathering mudstones of the '*Monograptus sedgwickii* shales'. The lower level is a tunnel, the entrance of which is now blocked. Access into the quarry can be gained from its north-west corner. In the bottom of the quarry is an upstanding whale-back of rock left by the quarrymen because of its poor quality. This rock is in fact the '*Monograptus sedgwickii* shales' occupying the core of an asymmetrical anticline with an axial plane dipping 68°W and striking 025°. The east face of the quarry is in the steep limb and comprises very thinly bedded, fine-grained turbiditic sandstones and mudstones from the upper parts of the Cwmsymlog Formation. The limb is overturned with a dip of 75°W and strike of 017°. The overturned sandstone bases display some excellent flute-casts and other bed form structures indicating palaeocurrents from the south-east. The thicker turbiditic sandstones seen at the base of the overlying Devil's Bridge Formation are well exposed at the top of the western side of the quarry at the entrance. Again **care is essential** in descending and ascending the quarry. This quarry would have much of interest and educational value if it were to be made safe.

4. Derwenlas. Close by (SN 7190 9910) is the **old coach road section** documented by Jones & Pugh (1916, p. 353–8) (Fig. 4). They described this section bed-by-bed and a plan of their localities was produced by Cave & Hains (1986, p. 80). The section is the type for the Derwenlas Formation, comprising some 37 m of tough, medium-grey, turbiditic mudstones in thin rhythmic layers, divided by several thin layers of dark grey, graptolitic, hemipelagic-rich mudstones. Each layer was given the specific name of its characteristic graptolite by Jones & Pugh and they constitute their classic Graptolite Bands.

5. Gelli-goch. One mile north-east of Derwenlas a **road cutting on the A487** (SH 7325 0015) exposes the Gelli-goch Thrust (Fig. 8). Quartz veining marks the thrust plane and separates tough, unbedded medium to dark grey silty mudstone of the Bryn-glâs Formation on the west from well cleaved, medium grey, thinly bedded mudstones of the Devil's Bridge Formation.

6. South-west Machynlleth. A further half a mile north-eastwards along the A487 (SH 7380 0037)–(SH 7388 0041) is the first of **two cuttings near Machynlleth** (Fig. 8). These display examples of mesoscopic folds in the Devil's Bridge Formation (Telychian, Fig. 3).

The formation in this area comprises siltstone–mudstone couplets, which are between 10–45 cm thick (average 15–20 cm), with a laminated fine-grained sandstone or siltstone base, <3 cm thick, which grades up into mudstones. The mudstone portions each consist of <10 cm of turbiditic mudstone, <1 cm of phosphatic concretion, and <4–5 cm of pale grey (oxic) mudstone (partly hemipelagic). The pale grey layers are of uniform texture and colour apart from occasional dark grey *Chondrites* burrows.

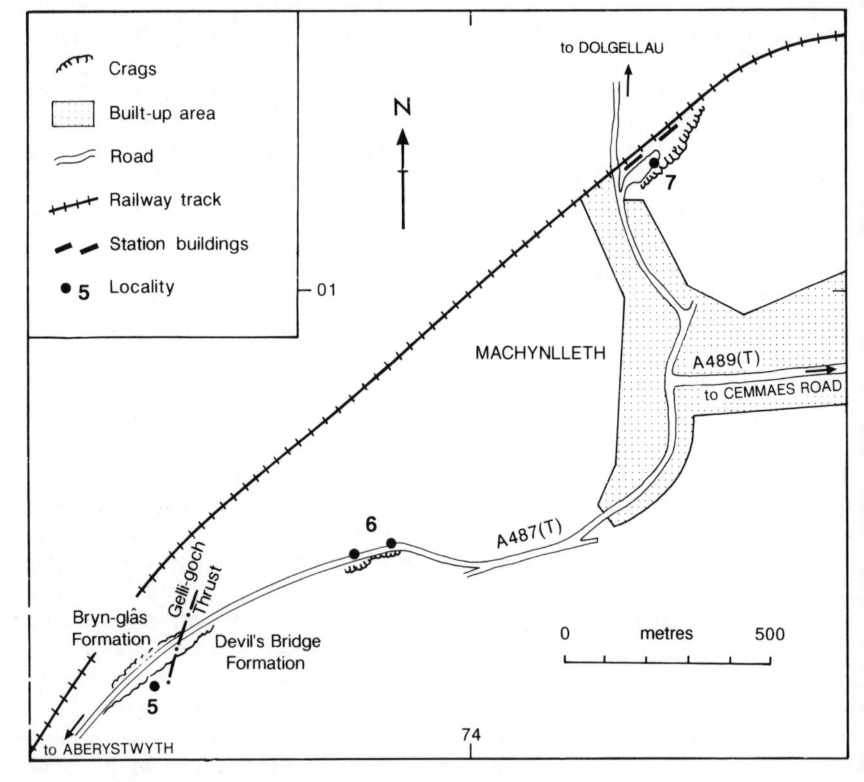

FIG. 8. Sketch map showing Localities 5, 6 and 7.

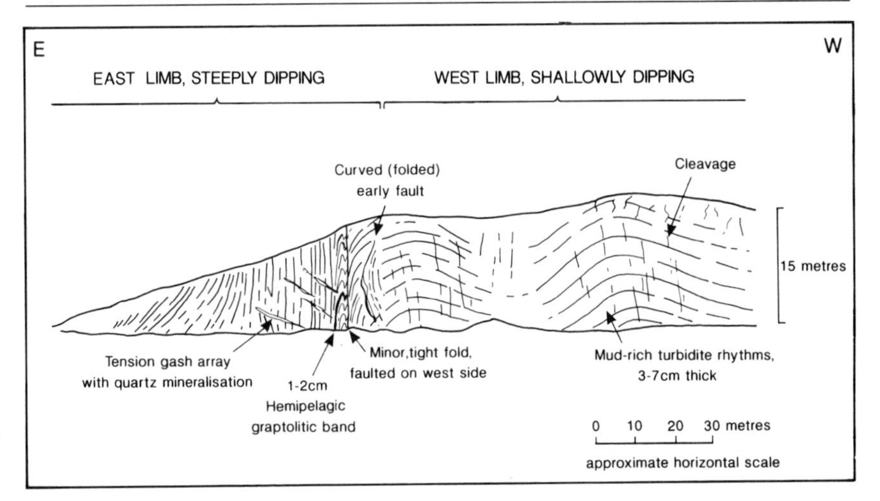

FIG. 9. Locality 6, large asymmetrical fold, roadside section near Machynlleth.

They are low density, mud-rich turbidites that accumulated in a mostly oxygenated marine basin. Thin layers of dark grey mudstone with graptolites do occur sparsely, and these layers of hemipelagite indicate that anoxic conditions prevailed briefly at times. One such layer is present at Locality 6. Basal sandstones occasionally show current-ripple laminae. Fine (<0.5 mm) laminations are common in both the basal sandstones and siltstones. Discoid cone-in-cone nodules containing silica and iron rich carbonate occur, <10 cm in diameter. The nodules are resistant and eventually weather out of the enclosing rock (Scott & Cave 1990a).

A large asymmetrical fold is exposed at Locality **6**, with a shallowly dipping west limb showing mesoscale open folds and a vertical east limb. The east limb shows two faults and an array of quartz-filled tension gashes (Fig. 9).

The folds of Localities **6** and **7** are typical of the multilayered mudstones and sandstones of the area. The main fold at Locality **6** was tightened as much as the rocks would allow until they failed in the form of an array of tension gashes in the vertical limb. These are now visible as quartz veins (Fig. 9). The next stage in the development of these folds might have been the complete failure in the form of a high angle thrust plane overriding the tension-gash array, such as seen in the thrust fold in the Ystrad Meurig Grits of the Hendre Quarry, Pontrhydygroes. Perhaps the ultimate product of the compressive processes is large-scale thrusting of the type seen in the Gelli-Goch road cutting (Locality 5).

7. Machynlleth Railway Station. The second cutting lies **behind the sidings at Machynlleth railway station** (SH 7455 0129) to the north of the town (Fig. 8). The rock face north-east of the fold has collapsed twice in recent years and must be avoided. The Devil's Bridge Formation is similar here to that described at Locality 6. The rocks are in the form of a north-east plunging anticline (Fig. 10). Bedding-cleavage intersections trend and plunge at 016°/21°NE. The axial plane of this fold is transected by the cleavage at a clockwise angle of approximately 16°. This transection can be calculated by plotting poles to bedding, poles to cleavage, and the bedding-cleavage intersection lineations on a stereogram (Fig. 10) (Duncan 1984).

8. Dylife. The second part of the itinerary is devoted to the centre of the Van Inlier near the Clywedog dams. Access is via the minor Machynlleth to Llanidloes road, which passes through the old mining village of Dylife. Here conservation work has rid the area of old mine buildings and waste tips. Even the church has been demolished so that little remains to reflect a once important mining community (O. T. Jones 1922, p. 154–5).

Approximately **1 km east of Dylife** on the Forge road is an example of river capture (Fig. 11). An original eastwards-flowing stream has been diverted to flow northwards. Viewed northwards from the lay-by (SN 8730 9394) two valleys intersect, resulting in a rectangular bend of the stream. The N–S trending V-shaped valley of the River Twymyn has been eroded headwards (southwards) and intersected the W–E flowing stream which occupied the broad U-shaped valley. The capture point is at the waterfall, as the S–N flowing stream has been incised to a lower level (Locality **8**, Fig. 11).

The broad U-shaped E–W valley is filled with Boulder Clay, whilst the upper reaches of the N–S valley contain only scree and river alluvium. This suggests that the W–E valley was pre-glacial, whilst the southern end of the N–S valley was eroded post-glacially.

9, 10, 11. Clywedog Dam area. This area has a two-fold value in the interpretation of the geology: first, as an exercise in aerial photograph interpretation, using the RAF 1:10,560 (approx.) aerial photographs; and secondly, as a study of structure and bedform of the Pencerrigtewion Member of the Drosgol Formation.

This member is well developed here as a packet of thick and massive beds of sandstones which has suffered large to medium scale folding and subsequent faulting. The cover of superficial deposits is sparse and so, using the sandstones as a major marker band visible on the aerial photographs, a geological interpretation map can be produced. As well as the folds, several major faults can also be interpreted. A ground check of the map can follow, for most of the area is easily accessible provided

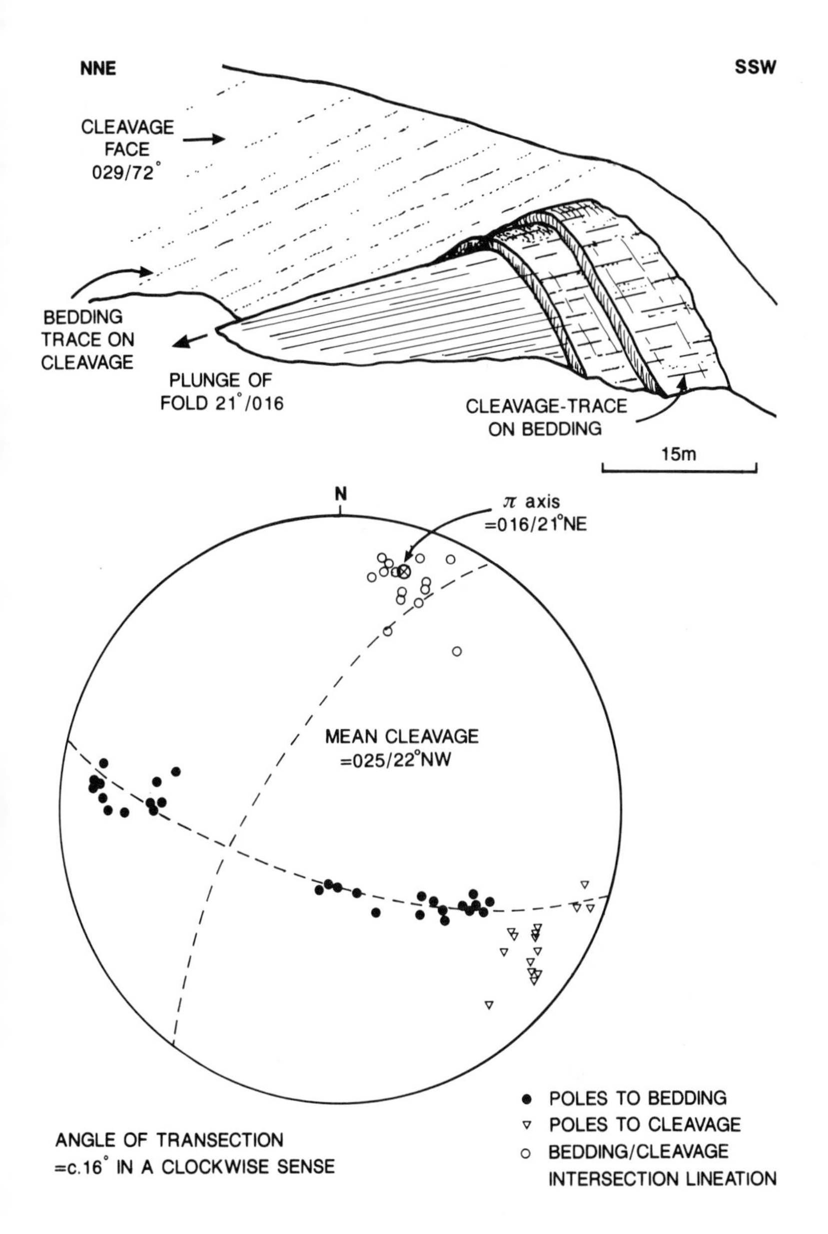

FIG. 10. Locality 7, block diagram and a lower hemisphere equal-area projection of a plunging anticline, Machynlleth railway sidings.

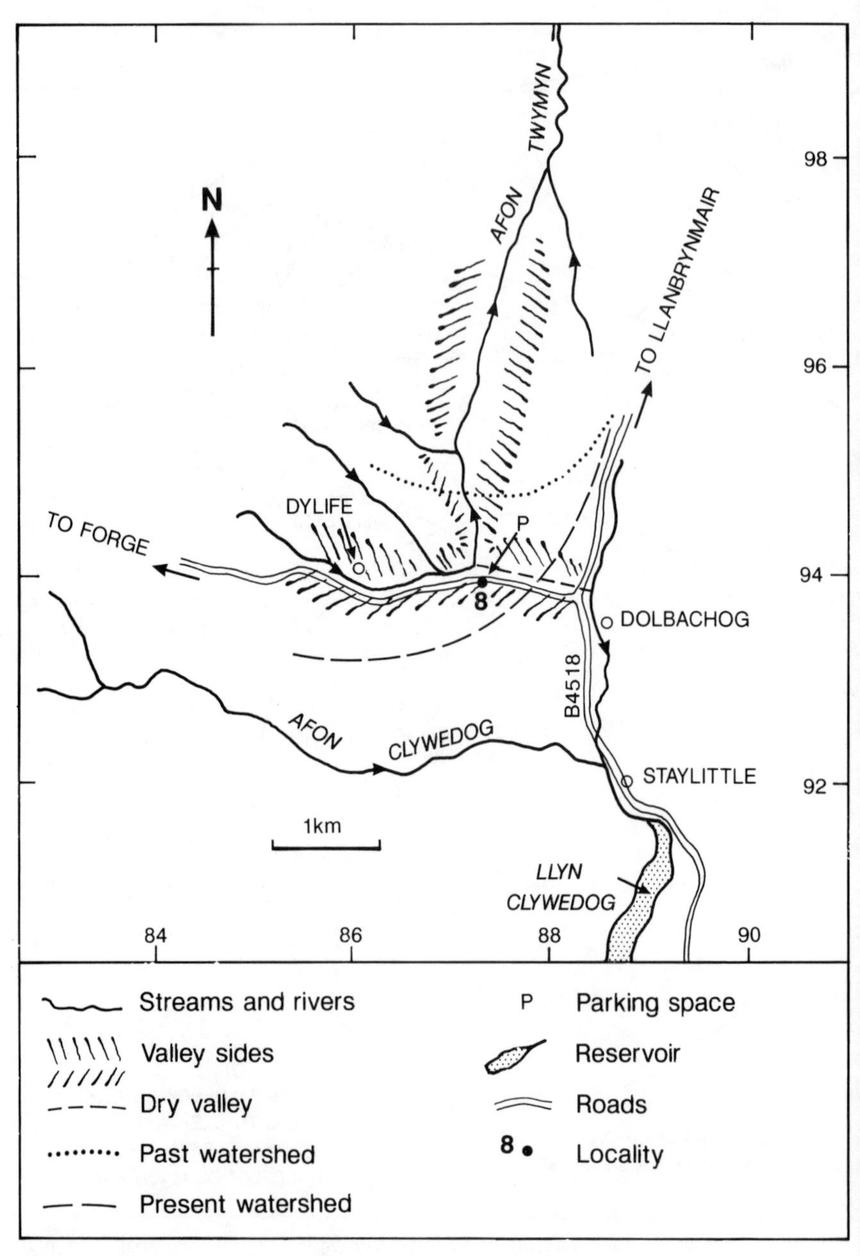

FIG. 11. Locality 8, river capture near Dylife.

permission from farmers and the Severn and Trent River Authority is obtained. It should be noted that the massive homogeneous mudstones of the Bryn-glâs and Drosgol Formations are stratified into thick sheets, which also produce very clear features in places. Other attributes of the area are good roadside exposure and quarries in both the Bryn-glâs Formation and the Pencerrigtewion Member of the Drosgol Formation.

An aerial photographic interpretation of the geology in this area is given as Figure 12, which includes the following localities.

9A. Waun y Gadair. The Bryn-glas Formation is exposed in the **road cutting** at (SN 9170 8870) (Fig. 12). There is **parking space** available both at the top (SN 9145 8875) and the bottom (SN 9215 8820) of this steep incline. Many metres of medium–dark grey, splintery mudstones are visible which display the diagnostic uneven, phacoidal cleavage. No undisturbed bedding fabric can be seen in these silty mudstones. However, a large 'raft' of thinly bedded sandstones and mudstones occurs near the top of the incline while unevenly rounded inclusions of sandstone are present at several places lower down.

9B,C. On the NW-facing slope of Bryn y Fan (SN 9242 8823) sandstones of the Pencerrigtewion Member of the Drosgol Formation are exposed. The bedding surfaces of these sandstones (i.e. 043°/32°NW) are approximately parallel with the slope (ie. a dip-slope) (Fig. 12). Glacial striations on rock surfaces are visible here trending 100°. **Walking southwards along the slope and over the shoulder** of the hill, lower beds of sandstones are encountered, and a section through these can be seen in natural exposures. The sandstones comprise a series of layers (<0.5 m thick), are grey/brown, and medium grained. The majority of the beds are massive, although parallel lamination, cross-lamination, and convolute lamination occur. Where thin mudstone partings are present between beds of sandstone the latter reveal reverse-density load-structures in the form of well developed pseudonodules.

A weakly developed spaced cleavage (026°/76°E) can be seen in these sandstones, which is accompanied occasionally by fine (<1 cm) quartz veins developed along the cleavage.

The sandstones are interpreted as turbidites comprising the basal Bouma divisions Ta, Tb and Tc. The Ta–c sequences form cycles with rare developments of the Te division separating successive beds. Although the primary, depositional, fabric is present, as listed above (i.e. parallel- and cross-lamination etc.) it has suffered much modification during pre-lithification deformation (e.g. loading effects which have produced loaded-ripples, pseudonodules etc.).

These sandstones are comparable to, and probably the stratigraphical equivalents of, sandstones exposed around Plynlimon (SN 7895 8695)

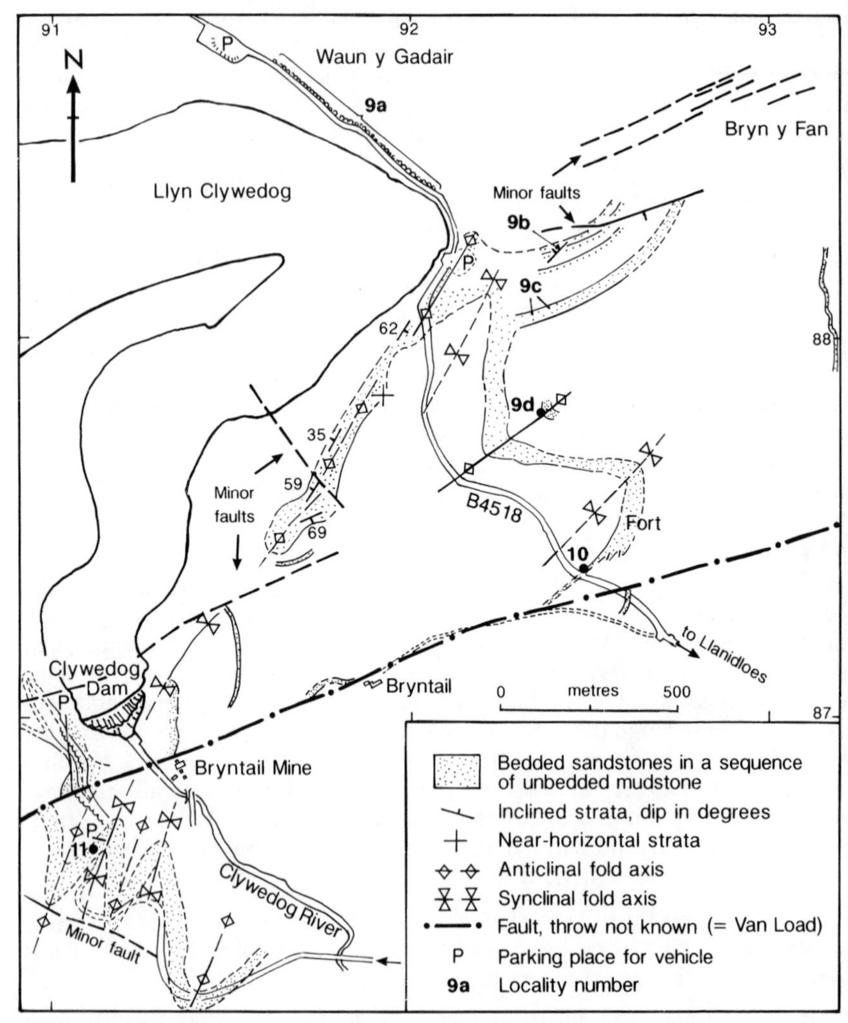

FIG. 12. Sketch map of sandstones in the Drosgol Formation of the Llyn Clywedog area, based largely on aerial photographs.

(Pencerrigtewion Member of the Drosgol Formation (Cave & Hains 1986)) and between Tywyn (SH 5870 0070) and Corris (SH 7499 0786) (sandstones within the Garnedd-wen Formation (Leng 1990)) (Fig. 2). They have been interpreted as outer-fan sandstone lobe deposits carried into the depositional basin through channels that were incised into the west-facing basin-margin slope.

9D. Pen y Clun Hill. The Pencerrigtewion sandstones exposed on Bryn y Fan can be traced from (SN 9236 8804) across the valley to the **NNE-facing valley side** of Pen y Clun Hill (SN 9239 8773) (Fig. 12). An excellent vantage point from which to view them is Locality **9C**, looking to the south-south-west.

Two packets of sandstones form an asymmetrical anticline here; the lower packet occurs below the Pencerrigtewion Member in the core of the fold at (SN 9239 8773) and comprises *c*.12 m of interlayered (<1 m thick) sandstones and high matrix (muddy) sandstones. The high matrix sandstones can be identified by the well-defined, but uneven, spaced cleavage that they possess. A near-vertical fault displaces the sandstones by a few centimetres, approximately along the axial plane of the fold. This anticline is underlain at a distance of several metres by horizontally bedded sandstones. It is difficult to accommodate the two exposures structurally without invoking a decollement between them. Between these sandstones and the stratigraphical equivalents of the Pencerrigtewion Member (of the Drosgol Formation) above, are structureless silty mudstones which display the regional cleavage.

10. The Fort. (SN 9260 8750). An ancient earthwork occurs on a steeply west-dipping outcrop of massive sandstone, medium to pale grey and fine-grained, equivalent to the Pencerrigtewion Member. On line of strike of these beds is a **road cutting** (SN 9250 8737) where the upper part of the sandstones can be examined (Fig. 12). The lowest beds have become disturbed and disrupted into balls and pillows enveloped by medium–dark grey splintery mudstones, while one portion of massive sandstone is curled into a partially unrolled 'swiss-roll' structure. This condition is very similar to that at Carn Owen, west of Plynlimon (Cave & Hains 1986, p. 26–9). The beds have been disturbed penecontemporaneously by slumping accentuated by diapiric fluid escape process. One sandstone bed shows internal, bedding parallel laminae which form curved, concave upward cusps. This is a dish-structure, a fabric caused by pore water escape during rapid deposition of sand (Allen 1984).

11. Roadside quarry near the Clywedog dam. Up to 20 m of thick-bedded sandstones occur in a **linear quarry face** (SN 9112 8665) near the main Clywedog Dam (Locality **11**, Fig. 12). The quarry extends for over 100 m south-west from the road. The sequence was inspected initially by J. H. Davies *et al.* (1978, item 29) and their log is reproduced here with minor changes (Fig. 13). The bedding is steeply inclined and, near the road, dips 77°SE and strikes 060°; at the south-west end of the quarry the bedding dips 42°SSE and strikes 078°. The sandstones are medium to coarse grained and mostly massive, forming beds <2 m thick. Mudstone interbeds are

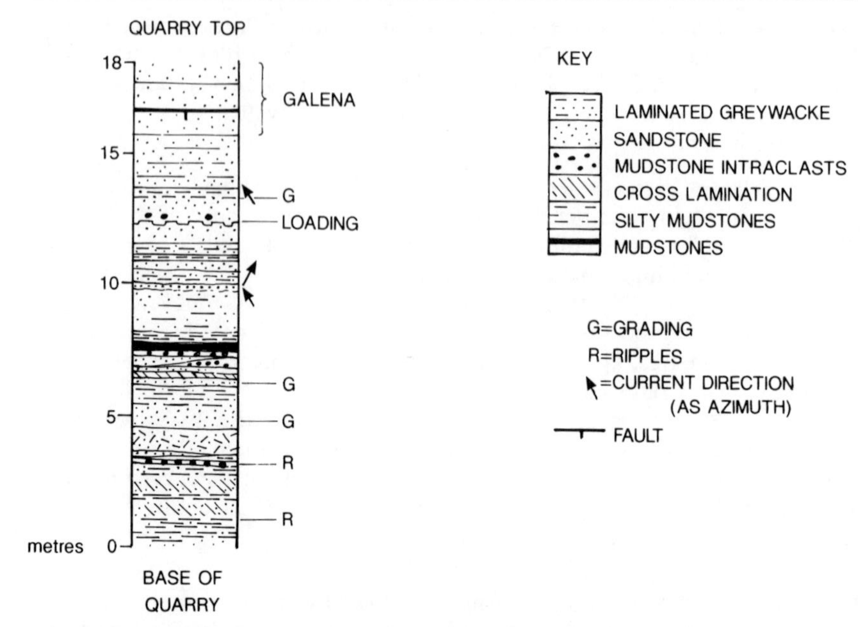

FIG. 13. Locality 11, section through part of the Pencerrigtewion Member in a quarry (SN 9112 8665) near Clywedog Dam (after J. H. Davies *et al.* 1978).

common, forming layers <15 cm thick. Two rippled sandstone surfaces crop out immediately next to the road; the first occurs about 75 cm above the base of this sequence, the second is a little higher. These sandstones represent Bouma Ta, Tb, and Tc intervals. At the extreme south-west part of the quarry (SN 9105 8656) a thin vein of galena (PbS, lead sulphide) with quartz occurs. This vein trends 350° and has a westward hade. In the surrounding sandstones, galena has also impregnated the pore spaces between sand grains, whilst an adjacent massive sandstone contains sigmoidal tension gashes <10 cm long and <2 cm wide, infilled with galena but no associated quartz. This mode of mineralization of the galena is reminiscent of that in the nearby Van Mine (SN 940 880) where galena was worked from 'flats' in the same sandstones (O. T. Jones 1922, pp. 159–60). Other minerals present are chalcopyrite and pyrite. The weathered galena has been altered to white powdery minerals, probably a mixture of anglesite and pyromorphite.

Nearby (SN 9136 8685) are the preserved remains of the Bryntail Mine. This mine exploited galena, witherite ($BaCO_3$, barium carbonate) and barytes ($BaSO_4$, barium sulphate) from a lode which extends for more than 1 km from the River Clywedog to east of Bryntail Farm. This is the line of a major fault which crosses the river just east of the dam and continues into the Van Mine to the east-north-east.

Acknowledgements. The authors wish to acknowledge the use of facilities provided by the Institute of Earth Studies, University College of Wales, Aberystwyth and helpful discussion with Dr David Loydell.

6. LLANDOVERY BASINAL AND SLOPE SEQUENCES OF THE RHAYADER DISTRICT

by R. A. WATERS, J. R. DAVIES, C. J. N. FLETCHER, D. WILSON, J. A. ZALASIEWICZ, *and* R. CAVE

Maps	*Topographical*:	1:50 000 Sheet 147 Elan Valley and Builth Wells
		1:25 000 Sheets SN 86, SN 87, SN 96, SN 97, SO 06, SO 07
	Geological:	1:100 000 Central Wales Mining Field
		1:50 000 179 Rhayader

THE Llandovery strata of the Rhayader district flank the Tywi Lineament (Figs. 1 and 10). Although this comprises a broadly anticlinal structure in Ordovician rocks, closing to the north-east near Abbeycwmhir, in detail it represents a complex zone of folding and faulting (see Excursion 7). The Llandovery sequence lying to the west of the Tywi Lineament consists of a *c*.4 km thick, graptolitic succession predominantly of turbiditic mudstones and sandstones, deposited in a rapidly subsiding, deep-water environment at the eastern margin of the Early Palaeozoic Welsh Basin. This basinal succession contrasts markedly with the thinner (locally as thin as 15 m), commonly incomplete, shelly Llandovery sequences deposited in a shallow shelf environment to the east of the Tywi Lineament and exposed around Garth and Builth (see Excursion 12). Transitional slope deposits, comprising an incomplete but mainly Telychian graptolitic sequence, are exposed across the Tywi Lineament where the Ordovician rocks plunge out to the north-east near Abbeycwmhir. The junction between slope and shelf sequences is taken at the Garth Fault, which runs along the south-east flank of the Tywi Lineament (Fig. 1).

The basinal and slope sequences were folded during the Acadian (early to mid Devonian) orogeny into a series of SE-facing, asymmetric, open to tight and commonly periclinal folds, of varying wavelength. In addition to the Tywi Lineament, major folds within the district include the Rhiwnant and Tylwch anticlines and the Central Wales and Waun Marteg synclines (Fig. 3). All the mudstones exhibit a well-developed cleavage.

Previous stratigraphical studies in the area include those of Lapworth (1900), K. A. Davies & Platt (1933), K. A. Davies (1928), Roberts (1929)

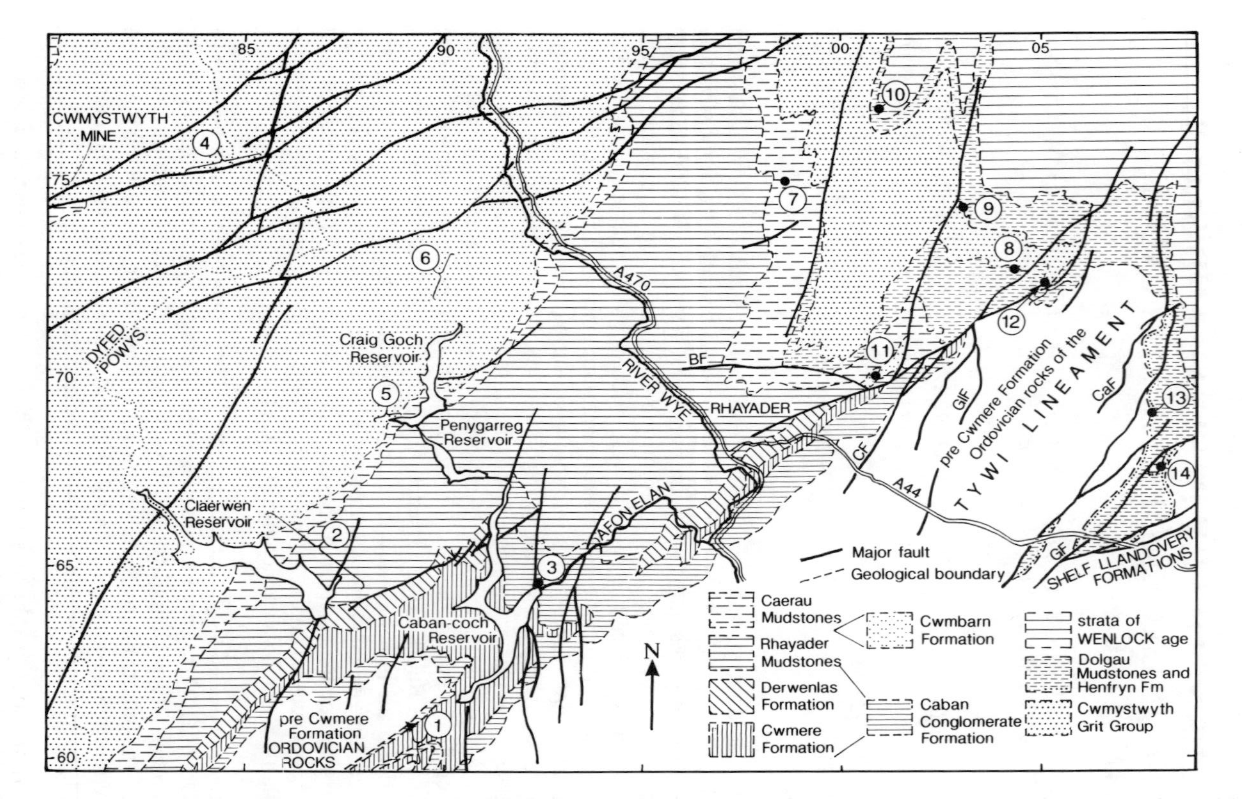

FIG. 1. Geological map of the Llandovery basinal and slope sequences of the Rhayader district showing positions of localities 1–14. BF – Brynscolfa Fault, CaF – Carmel Fault, CF – Cwmysgawen Fault, GF – Garth Fault, GlF – Glanalders fault.

and W. D. V. Jones (1945), but the lack of modern geological maps and a detailed regional stratigraphical framework has largely precluded modern sedimentological studies of the basinal and slope Llandovery in the district. Exceptions include the work of Kelling & Woollands (1969) on proximal turbidites around Rhayader and Smith's (1987*a*) study of turbidite facies within the *griestoniensis* Biozone. Recent mapping by the British Geological Survey (BGS) of the 1:50 000 Sheet 179 (Rhayader) has now established the framework for future studies and this itinerary is largely based on the results of that survey.

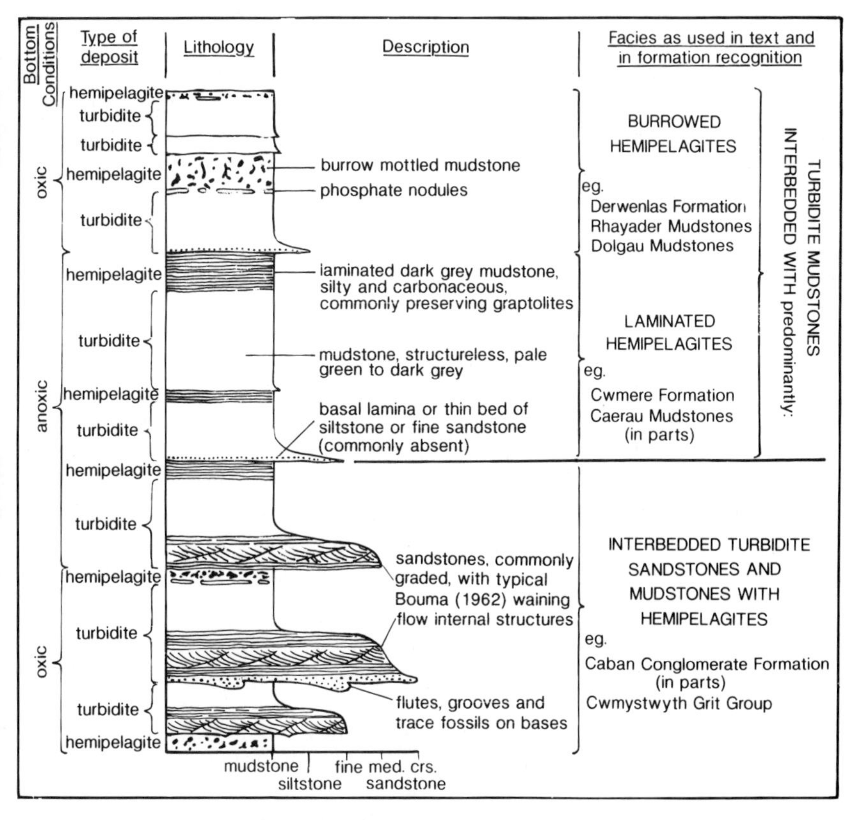

FIG. 2. Turbiditic and associated lithologies seen in the Rhayader district: their nomenclature and interpretation, and use in formational recognition. Not to scale. The type of hemipelagite present is important for the recognition of mudstone-dominated formations but is not used to subdivide sandstone-rich divisions. Note that in the latter, turbidite mudstone may still comprise as much as 80 per cent of an individual turbidite unit and of a formation.

GEOLOGICAL SEQUENCE

The Ordovician–Silurian boundary (base *acuminatus* Biozone) falls within formations which are predominantly Llandovery in age. In the Rhayader district and throughout the Welsh Basin, major lithological changes occur within the late Ashgill *persculptus* Biozone at the base of these formations, reflecting the transgressive rise in sea-level which followed the late Ordovician glacio-eustatic low stand (Brenchley 1988). These latest Ashgill strata are therefore associated sedimentologically with the early Llandovery deposits and are included in this itinerary. Additional localities from the Tywi Lineament are described in Excursion 7.

The Llandovery formations of the district predominantly record the introduction of terrigenous sediment into the basin by mass gravity transport processes. Thick-bedded, resedimented conglomerates, pebbly sandstones and mudstones are developed locally (Figs. 3 and 4), but the more commonly encountered thinner-bedded turbiditic and associated background lithologies are summarised in Fig. 2. Abundant classical sandstone turbidites exhibiting the Bouma (1962) sequence of internal structures are largely confined to the Caban Conglomerate Formation and Cwmystwyth Grits Group. Remaining Llandovery formations are largely composed of interbedded, mudstone-dominated turbidites and hemipelagites. Individual turbidite–hemipelagite couplets range from 2 to 30 cm in thickness. Laminae or very thin beds of siltstone or fine sandstone are commonly developed at the base of each turbidite unit (Fig. 2) and the presence of such thin arenaceous layers, locally or throughout mudstone-dominated formations should be assumed. The overlying and thicker turbidite mudstones are structureless, ranging from dark grey to pale green in colour. These allochthonous and periodic mass-flow sediments are capped by thin layers (<2 cm) of background hemipelagite, recording the slow but constant rain of fine terrigenous particles and planktonic organisms which accumulated on the sea floor between turbidity flows.

Two types of hemipelagite, laminated and burrowed, can be recognized (Fig. 2). The former comprise dark grey, laminated, carbonaceous and pyritic mudstones, the weathering of which commonly results in rusty staining of adjacent strata. Their recognition in the field is important for it is within these lithologies that graptolites are predominantly preserved. The lack of burrowing, pyritic character and preservation of organic material, suggest that they record accumulation under anoxic bottom conditions (Cave 1979). Burrowed hemipelagites comprise mottled pale grey or pale green-grey mudstones. The mottles represent the darker mudstone fills of burrow systems predominantly of *Chondrites* type. At the interface between these pale hemipelagites and underlying darker turbidite mudstone, small lenticular phosphatic concretions are commonly

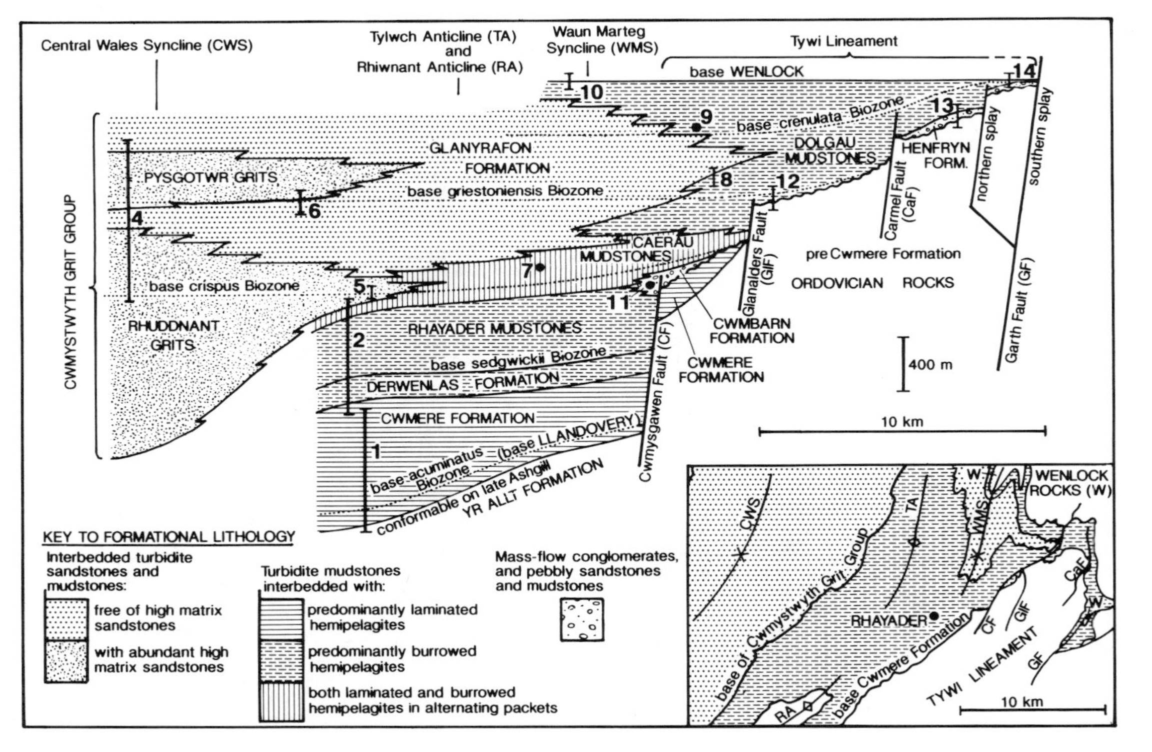

FIG. 3. Schematic NW–SE section across the Llandovery basinal and slope sequences of the Rhayader district, showing the distribution of formation and facies in relation to major structural elements. The laterally derived Caban Conglomerate Formation is not shown for reasons of clarity (see Fig. 4).

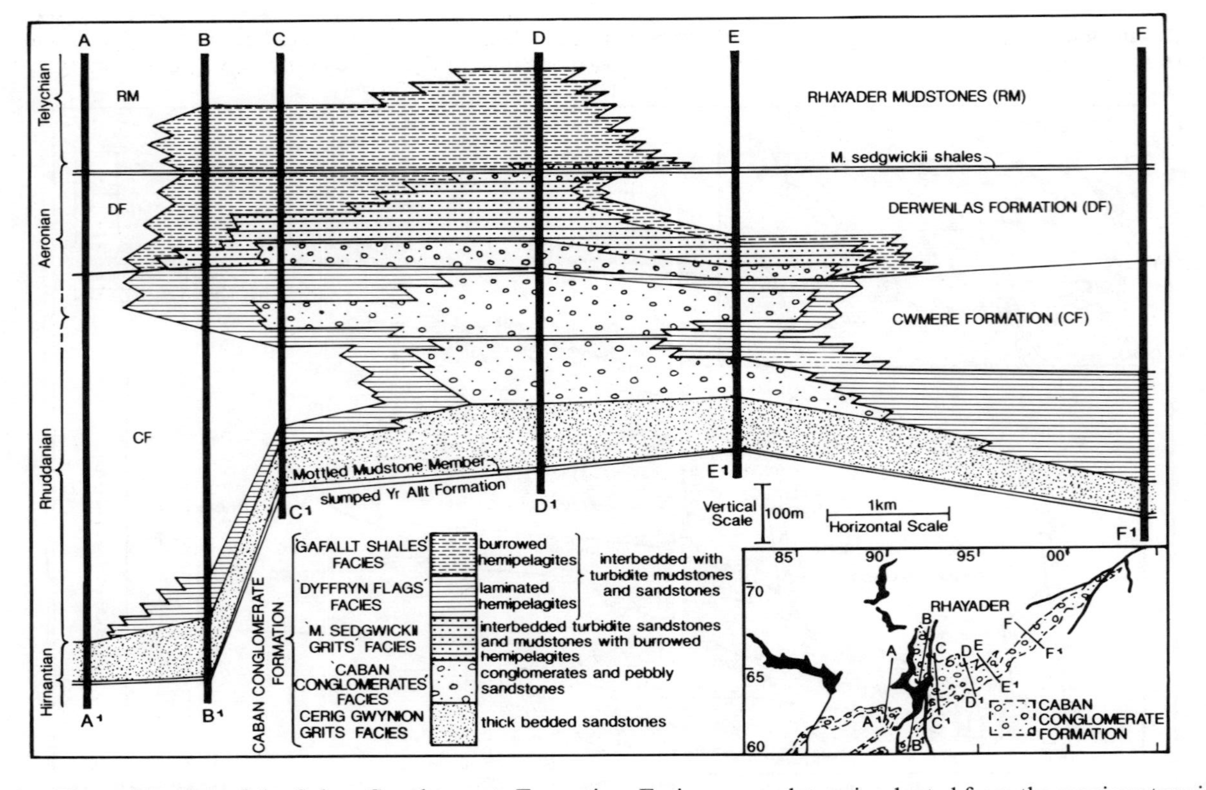

FIG. 4. The architecture of the Caban Conglomerate Formation. Facies nomenclature is adapted from the previous terminology of Lapworth (1900) and Kelling & Woollands (1969).

developed. Black when fresh, these weather to form conspicuous white lenses, sub-parallel to the bedding. The presence of burrowing and lack of graptolites and other organic matter suggest deposition under oxic bottom conditions (Cave 1979).

The distribution of Llandovery facies and formations within the Rhayader area is depicted on Fig. 3. The lithostratigraphical nomenclature used follows that of the BGS 1:50 000 Sheet 179 (Rhayader). The Llandovery succession of the district is subdivided by the Cwmysgawen Fault (Fig. 1), which broadly coincides with the basin–slope transition. West of this structure, a full Llandovery basinal succession conformably overlies the Ashgill rocks on the western side of the Tywi Lineament and is conformably overlain by basinal Wenlock strata. In contrast, between this structure and the Garth Fault, an area spanning the Tywi Lineament, a Llandovery slope sequence is represented by rocks of Telychian age, resting with marked disconformity on early Rhuddanian or Ashgill strata (Fig. 10).

Basinal sequence west of the Cwmysgawen Fault (Fig. 3)
The Cwmere Formation, the basal division of the basinal Llandovery sequence, comprises thinly interbedded grey turbidite mudstones and laminated hemipelagites. At the base, the Mottled Mudstone Member, a unit of thinly interbedded grey turbidite mudstones and pale grey burrowed hemipelagites, up to 10 m thick, sharply overlies the late Ashgill Yr Allt Formation. The Mottled Mudstone Member yields *persculptus* Biozone graptolites from a thin laminated hemipelagite, about a metre above the base. The base of the *acuminatus* Biozone and therefore the Silurian System occurs *c.*90 m above the base of the formation. Deposited under predominantly anoxic bottom conditions, the formation reflects the transgressive establishment of deeper water conditions across the Welsh Basin (Temple 1988) following the late Ordovician glacio-eustatic low stand. The top of the formation falls within the Aeronian *magnus* Biozone.

The mid to late Aeronian Derwenlas Formation, comprising thinly interbedded grey and green-grey turbidite mudstones and burrowed hemipelagites, sharply overlies the Cwmere Formation. Punctuating the formation are several units, up to a few metres thick, in which laminated hemipelagites predominate.

The overlying Rhayader Mudstones are very similar to the Derwenlas Formation. Packets with thin turbidite sandstones are locally developed in the upper part of the formation. A few thin units with laminated hemipelagites are present and the base of the formation is taken at the most conspicuous of these, the '*Monograptus sedgwickii* shales' (Jones & Pugh 1916). These yield a *sedgwickii* Biozone graptolite fauna. This unit records the introduction of deeper water, anoxic bottom conditions in response to the late Aeronian transgression of the adjacent shelf (Ziegler *et al.* 1968). The remainder of the formation falls within the *turriculatus* Biozone.

Near Rhayader the latest Ashgill to early Telychian sequence described above is punctuated by a system of mass-flow conglomerates and sandstones, collectively termed the Caban Conglomerate Formation. The lithostratigraphical and facies nomenclature of the formation and its spatial and temporal equivalence to enveloping mudstone formations are shown in Figure 4. It represents a major contemporaneous channel complex, infilled by proximal turbidites, derived from the basin margin to the south-east (Kelling & Woollands 1969). Following a gradual decline in coarse clastic supply, the Caban channel system was finally abandoned in the early Telychian, coincident with the transgression of easterly source areas (Ziegler *et al.* 1968).

The Rhayader Mudstones are overlain by the Caerau Mudstones, a sequence of thick bedded, grey turbidite mudstones in which alternating packets are dominated by either laminated or burrowed hemipelagites. The top of the formation falls within the early *crispus* Biozone.

Some 4 km north-east of Rhayader, a lenticular fining upwards sequence of pebbly mudstones, turbidite conglomerates, sandstones and mudstones, comprises the Cwmbarn Formation. Confined between the Brynscolfa and Cwmysgawen faults, these strata yield mid to high *turriculatus* Biozone graptolites and are overlain by the Caerau Mudstones.

Over much of the district the Caerau Mudstones are succeeded by the varying facies of the Cwmystwyth Grits Group. Thinly interbedded turbidite sandstones and mudstones with both burrowed and laminated hemipelagites form a ubiquitous background lithofacies, but also present in places are thick beds of high-matrix sandstone, the deposits of high-concentration turbidity currents (see Locality **4**). The group is divided into three formations. The Glanyrafon Formation, lacking high-matrix sandstones, is the sole representative of the group east of the Tylwch Anticline. However, to the west of the anticline, a lower tongue of this formation separates the Rhuddnant Grits from the younger Pysgotwr Grits (Fig. 3), both characterized by the presence of abundant high-matrix sandstones.

The Cwmystwyth Grits Group, marking a significant change in basinal sedimentation, records the introduction of southwesterly-derived sandstone turbidites into the district. The group as a whole thins noticeably from over 2,300 m in the west to a feather edge in the east. Much of this thinning reflects the pronounced diachroneity of the base of the group, which rises eastwards, from within the *turriculatus* to within the *griestoniensis* Biozone and records progressive easterly sidelap against the marginal slope of the Welsh Basin (Cave & Hains 1986; Smith 1987a) (Fig. 3).

The Cwmystwyth Grits Group is succeeded on the eastern limb of the Tylwch Anticline by the Dolgau Mudstones, a sequence of pale green turbidite mudstones with predominantly burrowed hemipelagites, of *crenulata* Biozone age. However, in response to easterly sidelap, grey and green turbidite mudstones, assigned to the Dolgau Mudstones, also emerge from beneath the Cwmystwyth Grits Group around the southerly closure of the Waun Marteg Syncline (Figs. 1 and 3). This lower development of the formation displays a greater proportion of thin laminated hemipelagites than the upper tongue. Commensurate with the continued thinning of the Cwmystwyth Grits Group, this lower unit of Dolgau Mudstones thickens eastwards, its top rising to within the *griestoniensis* Biozone (Fig. 3).

The Dolgau Mudstones are overlain by the basal Wenlock Nant-ysgollen Mudstones comprising grey turbidite mudstones and laminated hemipelagites. Although the base of the Wenlock Series has not been delineated, it is thought to approximate to the base of the Nant-ysgollen Mudstones.

Slope sequence east of the Cwmysgawen Fault (Figs. 3 and 10)
Across the Tywi Lineament the degree of sub-Llandovery (Telychian) overstep and the development of coarse basal clastics vary markedly across major NE-trending faults.

Between the Cwmysgawen and Glanalders faults, at Lan Goch, conglomerates of the Cwmbarn Formation and the overlying Caerau Mudstones are preserved resting on remnant Rhuddanian rocks ('Dyffryn Flags' facies and Cwmere Formation). However, east of the junction of these structures, Llandovery strata older than the uppermost *crispus* Biozone are absent. Failure of the Cwmystwyth Grits Group allows the lower and upper tongues of the Dolgau Mudstones to merge, and it is these strata which rest eastwards on varying levels of the Ashgill sequence of the Tywi Lineament (Fig. 10). Thin laminated hemipelagites are more common in the lower parts of these Llandovery mudstones, as in the west, whereas pale green colouration and burrowed hemipelagites characterize the upper part (the 'Tarannon Pale Shales' of previous authors). Graptolites of the *griestoniensis* Biozone have been proven *c.*50 m above the plane of disconformity (Locality **11**), but the formation probably ranges from high within the *crispus* Biozone into the *crenulata* Biozone. It is overlain, as in the west, by the basal Wenlock Nant-ysgollen Mudstones.

Between the Carmel Fault and the northernmost splay of the Garth Fault, a heterolithic unit of pebbly mudstones, conglomerates and sandstones, the

Henfryn Formation, rests disconformably on low levels in the Yr Allt Formation. Both the Henfryn Formation and overlying Dolgau Mudstones contain *griestoniensis* Biozone graptolites. Further south, within splays of the Garth Fault, a similar but much thinner succession rests disconformably on the Ashgill Pentre Formation (see Excursion 7).

The sub-Telychian disconformity has been attributed both to uplift and erosion, subsequent to intra-Llandovery folding (Roberts 1929; George 1963) or to overstep onto a Llandovery submarine 'non-depositional slope' (James 1983). Since shelf regions to the east had been submerged by earlier Aeronian and Telychian transgressions (Ziegler *et al.* 1968; Fortey & Cocks 1988), it seems likely that the sub-Telychian disconformity across the Tywi Lineament does not record uplift and subaerial exposure, but an episode of substantial submarine erosion. That the region occupied a setting on the basin margin or slope, is suggested by the hemipelagite-bearing mudstone sequence, sidelap and failure of the Cwmystwyth Grits Group and presence of resedimented coarse basal clastics (Cwm Barn and Henfryn formations). In such a setting, submarine mass wasting appears the most likely mechanism to have generated the disconformity (Saxov & Nieuwenhuis 1982; Rad & Wissmann 1982), with the dramatic changes in the degree of overstep across faults pointing to an underlying tectonic control.

ITINERARY

The itinerary describes three thematic traverses: a) the mudstone-dominated Rhuddanian to early Telychian sequence below the Cwmystwyth Grits Group, punctuated by the easterly derived proximal turbidites of the Caban Conglomerate Formation (Localities **1–3**); b) easterly sidelap of the southerly derived mid to late Telychian Cwmystwyth Grits Group against the basin margin (Localities **4–10**); c) the sub-Telychian disconformity and associated onlap of Telychian slope deposits onto Ordovician strata across the Tywi Lineament around Abbeycwmhir (Localities **11–14**). A full day should be allowed for each traverse. However, for a general one-day excursion to examine the basinal succession, only Localities **2**, **3** and **4** need be visited. Apart from Localities **1** and **2**, all the roads are passable in vehicles up to the size of a small coach. However, **some roads are narrow and cars or minibuses should be used if possible**. Although Localities **1**, **2**, **3**, **5** and **6** are owned by Welsh Water, **permission should be sought from tenant farmers where indicated**.

1. Nant Paradwys. From Rhayader take the B4518 along the Elan Valley. Turn left over the bridge at (SN 9111 6400) following the minor road to the south-west end of the Caban-coch Reservoir. Here, turn left by the phone box (SN 9007 6162), down the track into Llannerch-y-cawr, where **permission** should be obtained from the tenant farmer. Turn sharp right onto the road leading west from the farm. After 170 m, take the left fork along the

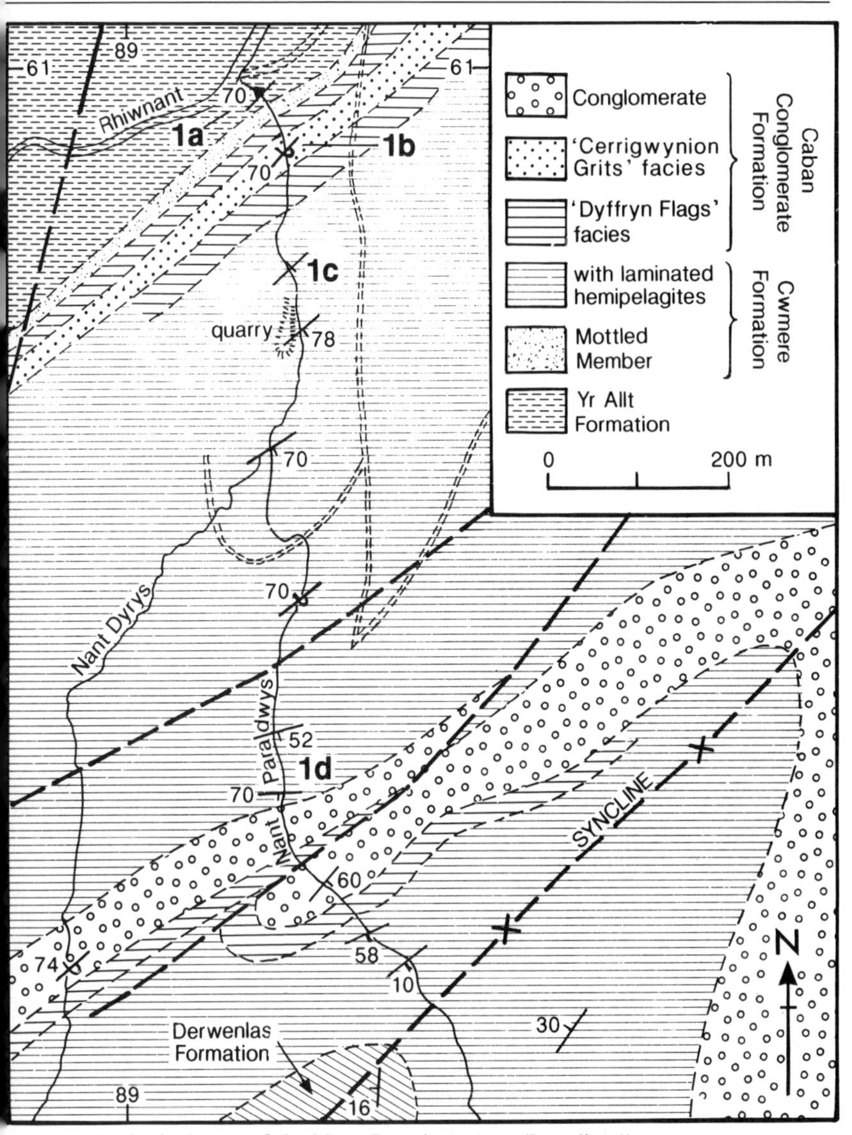

FIG. 5. Geological map of the Nant Paradwys area (Locality 1).

unmetalled track that follows the eastern side of the Rhiwnant Valley. **Cars can be parked** at (SN 8931 6117) where the track widens. From here take the poorly defined path leading south-east to the confluence of the Afon Rhiwnant and the Nant Paradwys (SN 8918 6098).

Nant Paradwys exposes most of the Cwmere Formation, together with two intercalations of Caban Conglomerate Formation (Fig. 5). The sequence is situated on the eastern limb of the Rhiwnant Anticline, dipping steeply south-east and locally inverted. About 30 m upstream of the confluence (Locality **1a**), tough, weakly laminated, dark grey silty mudstones of the Yr Allt Formation are separated by a small exposure gap from the overlying Cwmere Formation. The latter, seen under the first tree, comprises *c.*10 m of pale grey, burrow-mottled mudstones of the Mottled Mudstone Member. They are overlain by the lowest intercalation of Caban Conglomerate Formation. Interbedded turbidite sandstones and mudstones represent the 'Dyffryn Flags' facies of the formation (Fig. 4). Intercalated hemipelagites are burrow-mottled in the lower part but give way upwards to the laminated variety. Overlying thick-bedded, medium to coarse, turbidite sandstones, that form the first waterfall (SN 8917 6092) (Locality **1b**), comprise the 'Cerig Gwynion Grits' facies. The soles of the inverted sandstone beds exhibit large grooves and crescentic flute casts. The orientation of these sole marks demonstrates transport towards west-south-west. Current data from sole marks obtained by Kelling & Woollands (1969) from the facies as a whole suggested transport mainly towards the north-north-east but also towards the south-west. They suggested that the facies represented mid-fan turbidites deposited by currents flowing parallel to the palaeoslope but fed by north-west directed canyons. Smith (1987b) has invoked a tectonically-induced slope to the north-west to explain the axial deflection of such laterally derived currents. An overlying sequence of 'Dyffryn Flags' facies is exposed in the stream where the quarry spoil starts.

From this point upstream to where the track crosses at (SN 8917 6050), Nant Paradwys provides virtually continuous exposure in Cwmere Formation mudstones. The typical banded lithology of interbedded grey turbidite mudstone and generally thinner, dark laminated hemipelagites is well displayed on the water-washed surfaces e.g. (SN 8918 6080) (Locality **1c**). Packets with laminae or thin beds of turbidite sandstone, up to a centimetre thick, are scattered throughout. Climacograptids and diplograptids are relatively common in the hemipelagites. The *persculptus* Biozone has not been proved in the section but faunas of the *acuminatus* Biozone have been recovered from the northern (downstream) end of the ridge (SN 8919 6072) between the old quarry and the stream.

Proceed from the track upstream to the next waterfall (SN 8917 6022), (Locality **1d**) where, immediately downstream, largely sand-free Cwmere Formation mudstones yield a graptolite fauna indicative of the *acinaces* to *cyphus* Biozone interval. The waterfall is formed by the second intercalation of Caban Conglomerate Formation. Thick-bedded, coarse conglomerates exhibit a sharp erosive base and both normal and inverted grading is present. Upstream from the waterfall there is a rapid transition into overlying interbedded sandstones and mudstones of 'Dyffryn Flags' facies. At the base of the next waterfall (SN 8920 6016) is a fault that repeats the conglomerate and overlying 'Dyffryn Flags' facies. Beyond the waterfall the uppermost part of the Cwmere Formation can be examined in scattered exposures up to the core of the syncline (SN 8932 6001). Graptolites of probable *triangulatus* Biozone age have been obtained at (SN 8926 6009), 40 m stratigraphically above the top of the conglomerate. The top of the formation is not exposed.

2. Claerwen Reservoir. Exposures around the Claerwen Reservoir provide a semi-continuous section from the top of the Cwmere Formation to the base of the Rhuddnant Grits (Fig. 6).

From Nant Paradwys return to Llanerch-y-cawr and turn left onto the road leading to the Claerwen Reservoir. At (SN 8770 6302) take the left fork to the **car park (SN 8710 6335) below the dam**. Cross the bridge over the Afon Claerwen and walk south-south-west across the fields to the wooden bridge over the Afon Arban (SN 8686 6329) (Locality **2a**), where the contact between the Cwmere and Derwenlas formations is well displayed. **Immediately downstream of the bridge**, thinly interbedded turbidite mudstones and laminated hemipelagites, inclined north-west, pass upwards into thinly interbedded turbidite mudstones and pale grey, burrow-mottled hemipelagites.

Return to the car park and drive to the **car park at the top of the dam** (SN 8707 6364). Walk back southwards down the road where a major line of north-west dipping crags reaches the road (Locality **2b**) (SN 8722 6340). The base of the crags approximates to the Cwmere/Derwenlas Formation junction, although the contact at road level is hidden by a wall. Walking back towards the dam the lower part of the Derwenlas Formation may be examined. Medium grey turbidite mudstones and paler, burrow-mottled hemipelagites are thinly interbedded and give rise to a prominent colour banding on a 1–3 cm scale. Thin, white weathering phosphate lenses are common. Scattered rusty weathering packets, with thin laminated hemipelagites occur in the lowest part of the formation but die out rapidly upwards. Immediately north-west of the public toilets (SN 8711 6352), siltstone laminae make their first appearance at the bases of individual turbidites. Lithologies characteristic of the higher part of the formation are well exposed at the parking area by the dam (SN 8706 6365). They comprise diffusely colour banded, pale green-grey turbidite mudstones and homogeneous hemipelagites, in which burrowing cannot be generally observed. Siltstone laminae remain abundant. Proceed along the track beside the reservoir, noting thin rusty weathering units in which laminated hemipelagites are dominant, as at (SN 8706 6386). (The track, although passable by cars and minibuses, is rough going in places.)

Where a small stream passes under the track, climb down to the reservoir shore (SN 8708 6405) (Locality **2c**) to examine water-washed surfaces in turbidite mudstones and laminated hemipelagites of the '*M. sedgwickii* shales'. Graptolites are sparse. The contact with the Derwenlas Formation can be seen here but the upper contact is faulted.

Continue to Locality **2d** (SN 8705 6442) beyond the first inlet, where the lower part of the Rhayader Mudstones is well displayed. These are

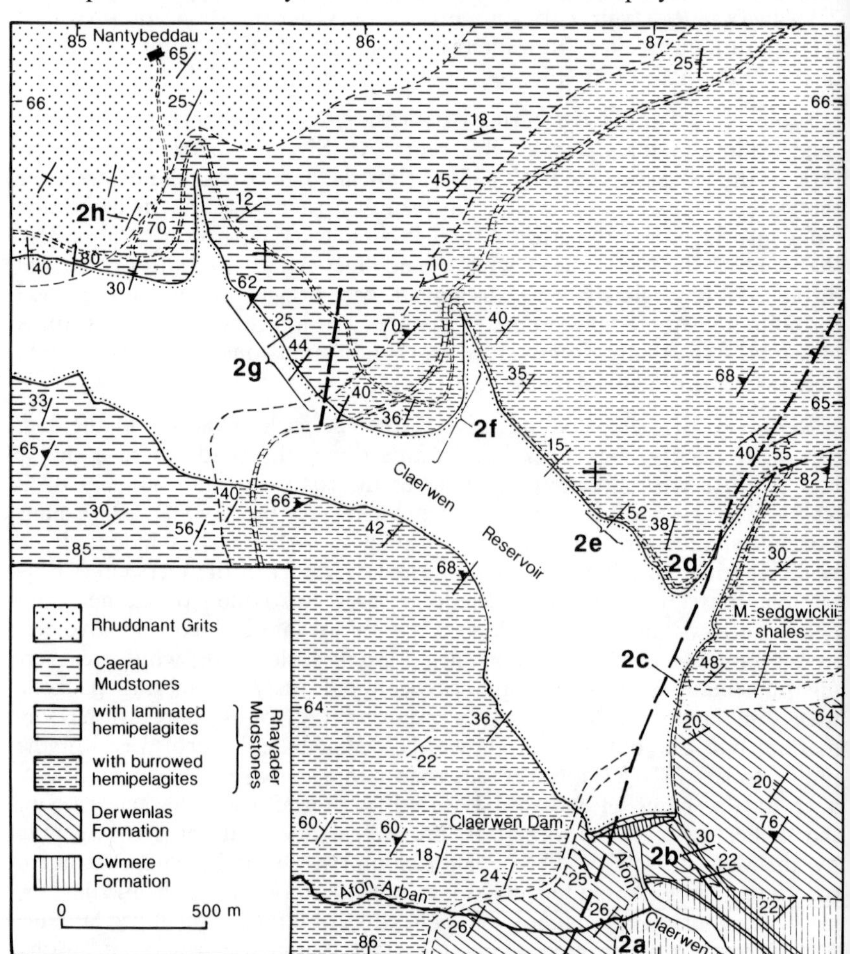

FIG. 6. Geological map of the Claerwen Reservoir area (Locality 2).

lithologically very similar to the mudstones of the upper Derwenlas Formation but are a vivid green colour and contain few siltstone laminae. Between SN 8687 6468 and SN 8675 6475, (Locality **2e**) a few very thin units containing laminated hemipelagites, exhibiting rusty and sulphurous staining, yield *turriculatus* Biozone graptolites. About 100 m north-west along the track, siltstone laminae become common and progressively more abundant.

The highest part of the Rhayader Mudstones is well exposed between the head of the next inlet (SN 8629 6535) and SN 8610 6503, (Locality **2f**). The lithology comprises thinly interbedded, medium grey mudstones and pale grey, burrow mottled hemipelagites, similar to the lower Derwenlas Formation. Thin siltstone laminae are abundant, and packets of thin turbidite sandstones are locally developed. A thin unit with laminated hemipelagites is exposed at the northern end of the locality.

The overlying Caerau Mudstones are best exposed along the water-washed shore of the reservoir. Walk to the shore at SN 8590 6501, (Locality **2g**), where the base of the formation is taken at the appearance of thick turbidite mudstone packets in which the laminated type of hemipelagite is dominant. The Caerau Mudstones are characterized by packets of turbidite mudstone interbedded with laminated hemipelagite alternating with packets of turbidite mudstone interbedded with burrowed hemipelagite. This alternation is well displayed between the base of the formation and the stream at SN 8556 6540. This sandstone beds and laminae, characteristic of the upper part of the Rhayader Mudstones, die out rapidly in the Caerau Mudstones. Graptolites of probable upper *turriculatus* Biozone age occur at the top of the formation at SN 8508 6545.

Return to the track and proceed around the next inlet to SN 8522 6549; here take the track to Nant-y-beddau, where **permission** to visit crags, 60 m north-west of the track at SN 8519 6567, should be sought from the tenant farmer. The crags (Locality **2h**) are formed by SE-dipping high matrix sandstones, the appearance of which defines the base of the Rhuddnant Grits, part of the Cwmystwyth Grits Group (Fig. 3). *Crispus* Biozone graptolites have been recovered from the lower part of the formation in the vicinity.

3. Caban-coch. From the Claerwen Reservoir return along the Rhayader road to the Caban-coch Dam and turn left into the **quarry car park** (SN 9240 6460). The quarry and crags above (Fig. 7), expose the western part of a channel complex, filled contemporaneously by conglomerates and sandstones of the Caban Conglomerate Formation. West of the Foel Fault, these coarse clastics are replaced by interbedded turbidite mudstones and sandstones ('Dyffryn Flags', '*M. sedgwickii* Grits' and

'Gafallt Shales' facies), which represent the contemporary channel margin deposits (Fig. 4).

Exposed in the quarry (Locality **3a**) are cobble and pebble conglomerates with subordinate coarse sandstones, part of the 'Lower Conglomerate' of Lapworth (1900). Individual conglomerate units exhibit scoured channelized bases. One such channel, at the base of a cobble conglomerate, in the lower part of the north-east side of the quarry, can be traced for *c.*20 m, locally exhibiting steep sides and undercutting. Both normal and reverse grading and a variety of loading phenomena can be observed. Large vertically orientated mudstone clasts up to 2 m across occur in one bed in the north-east of the quarry. These conglomeratic units are the product of high-concentration turbidity currents in the apical part of a small submarine fan. Although the conglomerate and underlying 'Dyffryn Flags' facies only yield long-ranging graptolites, the conglomerate is correlated with the upper unit of Caban Conglomerate

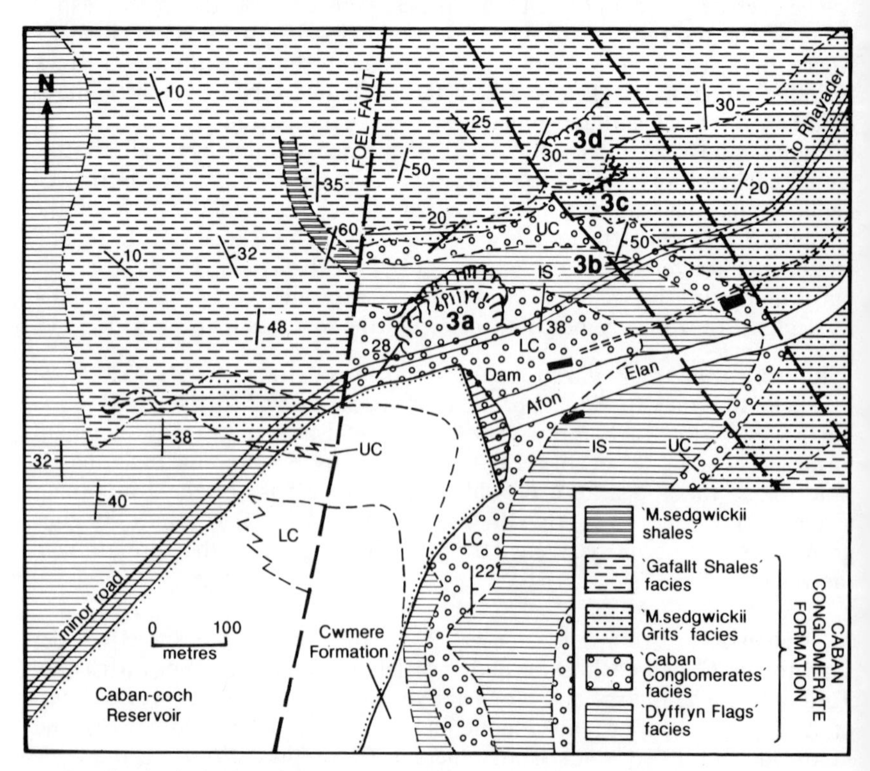

FIG. 7. Geological map of the Caban-coch area (Locality 3). (LC – Lower Conglomerate, IS – Intermediate Shales, UC – Upper Conglomerate of Lapworth 1900.)

observed in Nant Paradwys (Locality **1**) and therefore is probably of late Rhuddanian age.

Leave the quarry and walk 200 m east down the road to where the 'Upper Conglomerate' (Lapworth 1900) descends to the road as a prominent line of crags (SN 9267 6469), (Locality **3b**). Follow the conglomerates up the hill, observing their channelized base and the underlying 'Dyffryn Flags' facies (termed 'Intermediate Shales' by Lapworth 1900). The latter comprise interbedded turbidite sandstones, up to 0.3 m thick, and mudstones. Sparse, very thin laminated hemipelagites yield graptolites, which from localities on the south side of the valley prove the *triangulatus* Biozone.

Below a fault-guided gulley, scramble over the crags of the 'Upper Conglomerate' to the base of the large buttress (SN 9260 6477), (Locality **3c**). Here the '*M. sedgwickii* Grits' facies, gradationally overlying the conglomerate, can be examined by walking round the base of the buttress. The turbidite sandstones are up to 40 cm thick and commonly exhibit shelly bases. They are interbedded with thin turbidite mudstones, but hemipelagites are not common. Palaeocurrent data collected from the Caban-coch area by Kelling and Woollands (1969) suggest a south-easterly derivation for the two conglomerates and the '*M. sedgwickii* Grits' facies.

At the eastern end of the buttress, scramble up the hillside to Craig Gigfran (SN 9260 6488), (Locality **3d**), a prominent line of crags in the 'Gafallt Shales' facies. These pale green-grey turbidite mudstones, with burrowed hemipelagites and numerous siltstone laminae, contain scattered fine to coarse sandstones up to 4 cm thick. In this area the facies spans the *sedgwickii* to early *turriculatus* Biozones. The '*M. sedgwickii* Grits' and 'Gafallt Shales' facies together comprise an extended thinning and fining upwards sequence and record the gradual decline of coarse clastic supply and abandonment of the Caban channel complex.

4. Cwmystwyth. The upper reaches of the Afon Ystwyth (Fig. 8) provide almost unbroken exposure through the upper part of the Cwmystwyth Grits Group on the western limb of the Central Wales Syncline. Here the group reaches its greatest thickness in the Rhayader area and all the component formations can be examined. Approach via the Aberystwyth–Rhayader Mountain Road, leaving vehicles at the car park at Blaen-y-Cwm (SN 8540 7582).

Proceed westwards along the mountain road for *c*.1.5 km to the first fence line on the right (SN 8377 7529) and follow this down to the river. Those wishing to see only the lithological features of the Rhuddnant Grits may proceed up-river from this point, but about 450 m down-river at Locality **4a** (SN 8335 7545) graptolites of the *crispus* Biozone make their first appearance in the sequence (**permission** to visit this locality should first be

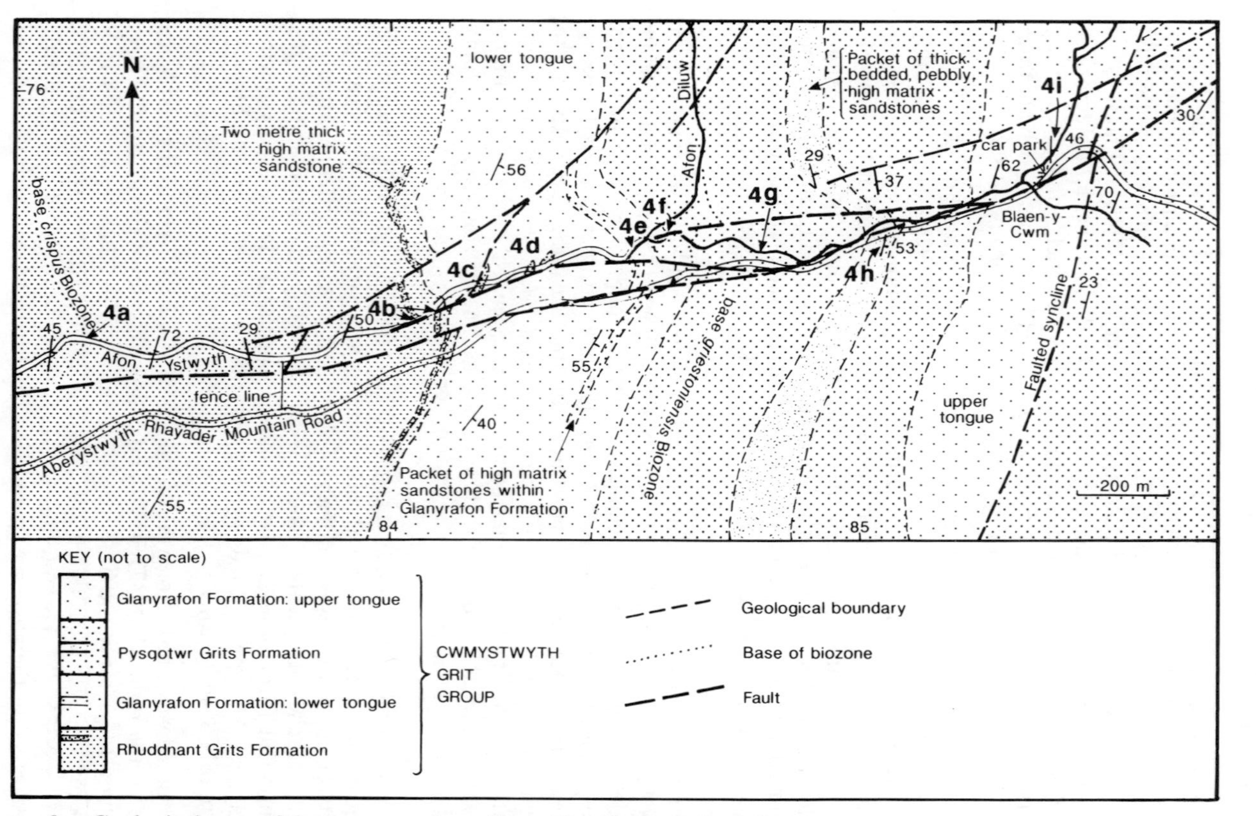

FIG. 8. Geological map of the upper reaches of the Afon Ystwyth (Locality 4).

sought at Ty-llwyd Farm, SN 8230 7530). It is worth considering that Rhuddnant Grits in excess of 1 km thick, all within the *turriculatus* Biozone, underlie this point.

Walk eastwards, following either the Afon Ystwyth or the abandoned Cwmystwyth Mine leat to the north, through virtually continuous exposure of the upper third of the Rhuddnant Grits. Here, and typically, the formation comprises interbedded sandstone and mudstone turbidites. The sandstone turbidites are of two distinct types. The first are coarse to medium-grained, argillaceous (high-matrix) sandstones, commonly graded, occurring in beds greater than 0.1 m and up to 2 m thick. The second are cleaner, more quartz-rich sandstones, less than 5 cm thick, which, interbedded with turbidite mudstones, form a background facies to the first. The thinner sandstones are readily recognized as classical turbidites with sharp, fluted bases and typical Bouma (1962) waning flow internal structures (Fig. 2). The high-matrix sandstones, which are commonly cleaved, display no internal structures and are less easily interpreted. They compare with facies C1.1 of Pickering *et al.* (1986) and probably represent high-concentration mud-rich turbidites or very muddy debris-flow deposits. Many display large rope-like grooves on their bases, but many of these are sub-parallel to cleavage and may represent tectonically modified load features. Sole structures, more confidently diagnosed as being of sedimentary origin, confirm the southerly derivation of these deposits. The high-matrix sandstones display no systematic variation in thickness or grain size throughout the section, but do occur in 5 to 15 m thick bundles separated by thick units of the thinly bedded background facies. Smith (1987a) interpreted comparable high-matrix sandstone packets within the overlying Pysgotwr Grits as delta-fed turbidite sand lobes. He suggested that sea floor topography generated by the aggradation of these features repeatedly led to their abandonment as the high concentration flows were deflected into adjacent hollows. The thinly-bedded turbidite background facies, possibly generated by storms and sourced from a much wider area of shelf, then blanketed the abandoned lobe features. Laminated hemipelagites in the background facies are best seen on water-worn surfaces and yield *crispus* Biozone graptolites throughout the section.

At Locality **4b** (SN 8410 7550) a 2 m thick high-matrix sandstone, exposed in both banks of the river, is displaced by a fault along the river. Rope-like sole structures are well displayed on the base of this bed. Multiple grading suggests it represents the amalgamation of two or more mass-flow units. A short distance upstream, thinly interbedded turbidite sandstones and mudstones, free of high-matrix sandstones, represent the succeeding Glanyrafon Formation. The high-matrix sandstones of the Rhuddnant Grits are brought to surface again, 100 m upstream (Locality **4c**), in a faulted anticline and again in two further anticlines developed near the old dam (SN 8432 7560), (Locality **4d**).

For *c*.300 m upstream beyond the dam, the Afon Ystwyth provides excellent exposures in the Glanyrafon Formation. These are rhythmically interbedded sandstone and mudstone turbidites with laminated hemipelagites. A discrete packet of high-matrix sandstones in the upper half of this sequence (Locality **4e**) records an initial reintroduction of high-concentration turbidites which culminated in the succeeding Pysgotwr Grits. The basal high-matrix sandstones of the main part of the latter formation are exposed just upstream from the confluence of the Afon Ystwyth and Afon Diluw (Locality **4f**). Graptolites recovered from this point are still of the *crispus* Biozone. However, 220 m to the east along the Afon Ystwyth (Locality **4g**), some 50 m above the base of the Pysgotwr Grits, thinly interbedded sandstones and mudstones exposed in an anticlinal core, have yielded *griestoniensis* Biozone taxa. These thinly bedded lithologies of Glanyrafon Formation type are well exposed upstream, both between and within packets of high-matrix sandstone. The section illustrates well the facies interrelationships of the Cwmystwyth Grits Group formations.

The acme of deposition of the Pysgotwr Grits is marked by the packet of thick, coarse, locally conglomeratic, feldspathic, high-matrix sandstones which form the prominent crags on the north side of the valley. These strata are conveniently examined along the track, 25 m south of the Mountain Road (Locality **4h**), where amalgamated beds over 3 m thick are exposed. The sedimentology of these sequences is discussed by Smith (1987a). Proceed eastwards along the road to the car park noting the feature forming packets of high-matrix sandstones on the valley side. Walk from the car park, via the wooden footbridge, to the other side of the stream (Locality **4i**) to observe thinly interbedded green turbidite sandstones and mudstones, part of a higher tongue of the Glanyrafon Formation (Fig. 3). Graptolites recovered from the basal beds of this unit further to the north indicate that these strata are also of *griestoniensis* Biozone age. Note the westerly dipping crags of high-matrix sandstones across the valley, that form part of the eastern limb of the Central Wales Syncline (Fig. 8).

5. Nant Caletwr. From Blaen-y-Cwm follow the Aberystwyth–Rhayader Mountain Road eastwards, crossing the axial trace of the Central Wales Syncline just beyond the head of Cwmystwyth, and continue east to Pont ar Elan. Taking the right turn off the mountain road, continue south for *c*.3 km to the bridge over Nant Caletwr (SN 8860 6885). **Vehicles may be parked just north of the bridge** in the picnic area on the west side of the road. This section (Fig. 9) demonstrates the dramatic thinning of the

Rhuddnant Grits that occurs across the Central Wales Syncline. Only *c*.280 m of Rhuddnant Grits are present beneath the Glanyrafon Formation and of that, in marked contrast to Cwmystwyth, only *c*.65 m are of *turriculatus* Biozone age, thus confirming that much of this thickness diminution records diachroneity at the base of the formation, a response to eastwards sidelap (Fig. 3).

Although the top of the Caerau Mudstones is not exposed, low beds of the Rhuddnant Grits can be examined on the northern bank of the inlet east of the bridge (SN 8874 6890), (Locality **5a**). Here the formation comprises thin turbidite sandstones 1–3 cm thick interbedded with much thicker turbidite mudstones capped by laminated hemipelagites, well seen on cleavage surfaces. Fine-grained high-matrix sandstones up to 0.3 m thick are common. These lithologies, dipping upstream, continue beyond the bridge and at (SN 8853 6886) the lowest graptolites of the *crispus* Biozone have been recovered from the hemipelagites. About 90 m upstream of the bridge (Locality **5b**), a packet of turbidite mudstones with burrowed hemipelagites, and lacking high-matrix sandstones, is developed. A higher packet of high-matrix sandstones, in beds up to 0.5 m thick, crosses the stream below the bend where the sandstone-free Caerau Mudstones are repeated by a thrust (Locality **5c**), (SN 8843 6882).

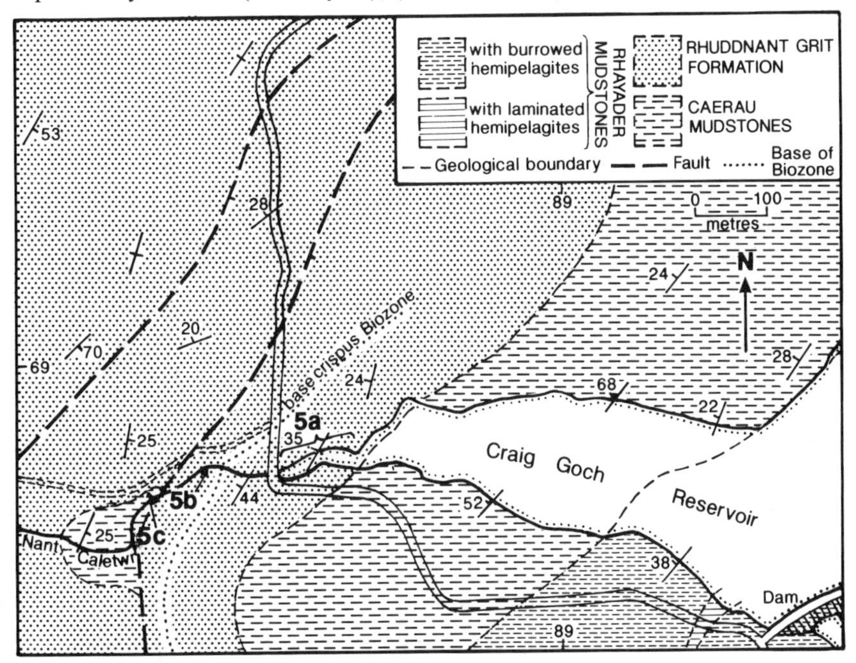

FIG. 9. Geological map of the Nant Caletwr area (Locality 5).

6. Nant-y-Ffald. From Nant Caletwr return to Pont-ar-Elan and follow the Aberystwyth–Rhayader mountain road westwards to the confluence of Nant-y-Ffald and Afon Elan (SN 899 720), where **vehicles can be parked at the roadside**. The sections at this locality demonstrate the higher part of the Cwmystwyth Grits Group, on the eastern limb of the Central Wales Syncline. Walk northwards along the track west of the stream, to a quarry (SN 8988 7240) in easterly dipping Glanyrafon Formation. Here, mudstone turbidites with laminated hemipelagites are overlain by rhythmically interbedded sandstones and mudstones more typical of the formation. These strata yield graptolites of the *crispus* Biozone.

Continue northwards along the track to where it crosses the southerly flowing Trawsnant (SN 9030 7320). Immediately to the north, the basal beds of the Pysgotwr Grits are observed in a N–S trending strike ridge. Several beds of vertical, easterly younging, high-matrix sandstone form a 20 m thick packet. The sandstones are coarse, feldspathic and up to 0.7 m thick. The underlying Glanyrafon Formation is well exposed along the Trawsnant. Laminated hemipelagites within the basal part of Pysgotwr Grits in this area yield *griestoniensis* Biozone graptolites and are therefore younger than in Cwmystwyth, demonstrating the easterly sidelap of the formation (Fig. 3).

7. Pant-y-dwr. From Nant-y-Ffald take the mountain road to Rhayader. At Rhayader take the B4518 to Pant-y-dwr and at the cross-roads turn right to the Cnych, where **permission to enter the roadside quarry** (SN 9854 7516), 300 m south of the farm, must be obtained. The quarry exposes the Caerau Mudstones of the eastern limb of the Tylwch Anticline. The lower part of the westerly dipping sequence comprises interbedded, medium grey turbidite mudstones and laminated hemipelagites. Burrowed hemipelagites predominate in the upper part. Sparse thin sandstones, up to 3 cm thick, occur throughout the sequence. In this area the base of the *crispus* Biozone occurs within the middle of the Caerau Mudstones; the quarry yields *crispus* Biozone graptolites. East of the quarry, the Glanyrafon Formation, of high *crispus* to *griestoniensis* Biozone age, succeeds and shows that the Rhuddnant Grits fail to cross the Tylwch Anticline (Fig. 3).

8. Fishpool quarries. From Pant-y-dwr turn left off the B4518 to Bwlch-y-sarnau. At the cross-roads, take the Abbeycwmhir road to Fishpool Farm (SO 0438 7245), (Fig. 10), where **permission must be obtained** to enter the quarries (SO 0429 7289), that lie 0.5 km north of the farm. Vehicles can be parked on the east side of the road by the gate leading to the quarries. The section exposes the lower tongue of Dolgau Mudstones (Fig. 3) and its contact with the overlying Glanyrafon Formation.

In the southern quarry, northward-dipping Dolgau Mudstones comprising pale green-grey turbidite mudstones, with abundant siltstone laminae and scattered thin sandstones up to 4 cm thick, are thinly interbedded with both burrowed and laminated hemipelagites. *Griestoniensis* Biozone graptolites are present.

Walk 60 m north to the northern quarry, where the base of the Glanyrafon Formation is exposed. Here, a lithology similar to that of the Dolgau Mudstones seen to the south, contains turbidite sandstones, up to 14 cm thick, increasing in abundance upwards. This quarry is also within the *griestoniensis* Biozone.

9. Bwlch-y-sarnau. From Fishpool return north to Bwlch-y-sarnau (Fig. 10). A small roadside quarry (SO 0296 7455), 30 m south of the cross-roads, is in the upper tongue of Dolgau Mudstones which overlies the Glanyrafon Formation (Fig. 3). Thinly interbedded grey-green turbidite mudstones and paler burrowed hemipelagites contain siltstone laminae and scattered turbidite sandstones up to 4 cm thick. Sparse thin laminated hemi-pelagites yield *crenulata* Biozone graptolites. Bedding dips steeply east.

10. Waun Marteg Quarry. Leave Bwlch-y-sarnau by the Pant-y-dwr road. Turn right at (SO 0087 7570) and follow the road to Waun Marteg. The quarry is situated on the east side of the road at (SO 0092 7707) and exposes the junction of the Dolgau Mudstones and early Wenlock Nant-ysgollen Mudstones. Pale grey-green turbidite mudstones and burrowed hemipela-gites of the Dolgau Mudstones pass up gradationally intc grey thinly inter-bedded turbidite mudstones and laminated hemipelagites of the Nant-ysgollen Mudstones. A *crenulata* Biozone graptolite fauna has been obtained from a thin unit of laminated hemipelagites 10 m below the top of the Dolgau Mudstones. The overlying Nant-ysgollen Mudstones have only yielded long-ranging taxa (*crenulata-riccartonensis* Biozones).

11. Cwm Barn Quarry. Take the A44 eastwards out of Rhayader and fork left on to the minor road, signposted Abbeycwmhir, on the outskirts of the town. Keep left at the first road junction, again signposted Abbeycwmhir, and 250 m beyond the junction turn left on to the track to Brynscolfa Farm (Fig. 10). Cwm Barn Quarry is on the right beyond the first set of farm buildings (SO 0082 7001), but **permission** should first be obtained from the farmhouse further up the track.

The quarry exposes the lower part of the north-easterly dipping Cwm Barn Formation. About 4 m of silty mudstones with floating discoidal and oblate cobbles comprise the lowest beds in the section. Most of the cobbles are carbonate concretions but their random orientation and the occurrence of rarer burrow-mottled mudstone clasts confirms the resedimented origin of these beds. They are overlain by thinner pebbly mudstones, each capped

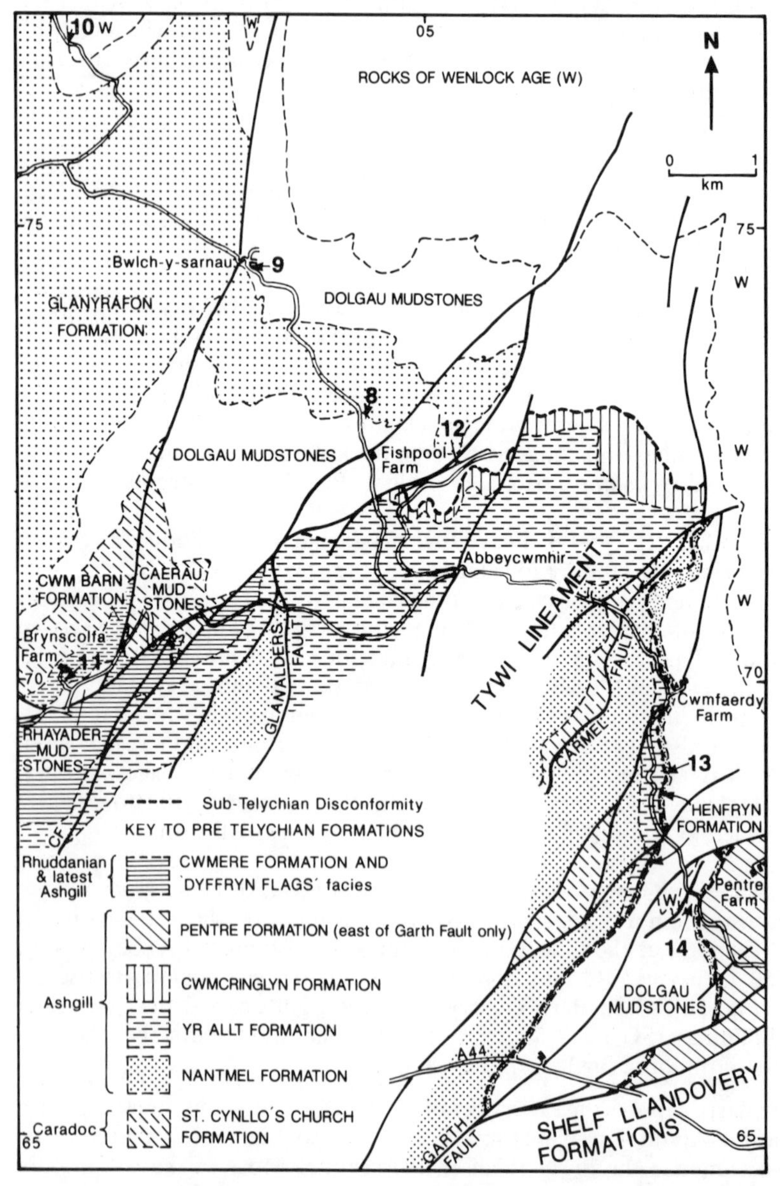

FIG. 10. Geological map of the Llandovery basinal and slope sequences around the hinge of the Tywi Lineament, showing the sub-Telychian overstep (see also Fig. 3) and the position of localities 8–14.

by planar laminated, rusty weathering sandstones. Convolutions seen in the rhythmically bedded mudstones and thin sandstones opposite the corner of the barn may represent slump or wet sediment deformation phenomena, but low-angle quartz veins suggest tectonic disturbance. Higher in the section pebbly mudstones give way to clast-supported conglomerates. Reworked concretions are still common but black granules and pebbles of phosphatized mudstone are also abundant. Quartz and sparse igneous pebbles have been recorded, and cobbles and boulders of green sandstone up to 70 cm in diameter are observed. More significantly, the conglomerates contain a well-preserved, diverse and abundant shelly fauna. The conglomeratic beds display sharp, locally fluted and gutter-casted bases. They fine upwards into the planar or more rarely cross-laminated, rusty weathering sandstones, rich in comminuted shell debris and in ostracode valves. There is little doubt, therefore, that these coarse-grained units represent mass-gravity transport deposits yet the abundance and preservation of the shelly material suggests only limited transportation and that contemporary benthic communities of shelf affinity flourished no great distance to the east.

Graptolite assemblages recovered from several levels in the section suggest a mid *turriculatus* Biozone age for these rocks. Yet strata of this age are absent from the region of Tywi Lineament to the east (Fig. 3). The Cwm Barn deposits appear, therefore, to include the detritus of shelf sequences previously deposited across the Tywi Lineament but shed into the adjacent basin during early Telychian mass-wasting of the basin margin.

12. Y Glog. From Cwmbarn drive to Abbeycwmhir where, at the Happy Union Public House, turn sharp left on to a forestry track (Fig. 10). After *c.*500 m take the right-hand fork and continue to the junction of five tracks (SO 0480 7163), from which point the main NW-trending track should be followed. Where the track turns east (SO 0470 7195), sandstones of the Hirnantian Cwmcringlyn Formation are exposed, dipping north-east (see Excursion 7). Beyond a 50 m wide exposure gap, unfossiliferous grey turbidite mudstones and burrowed hemipelagites form part of the overlying Dolgau Mudstones (SO 0479 7198), (Fig. 10). However, sections slightly higher in the sequence yield long-ranging Telychian graptolites. The plane of the sub-Telychian disconformity is nowhere exposed in the district.

Proceed north-east along the track and continue straight on at the junction of five tracks (SO 0502 7218), until (SO 0550 7243), where a forestry track exposes NW-dipping Dolgau Mudstones, at a level *c.*50 m above the disconformity. Medium grey, turbidite mudstones with abundant siltstone laminae and sparse turbidite sandstones, up to 10 cm thick, are thinly interbedded with both burrowed and laminated hemipelagites. Graptolites of the lower *griestoniensis* Biozone have been obtained from the section, whilst in a nearby section (SO 0568 7235) close to the disconformity, possible *crispus* Biozone graptolites are present.

13. Henfryn Quarry. Return to Abbeycwmhir and continue eastwards along the minor road from the village. Cross the bridge near Clywedog Brook and take the second track on the left to Cwmfaerdy Farm (Fig. 10) to gain **permission** to enter the quarry. Return to the minor road and continue south for about a kilometre. Vehicles can be parked in the roadside quarry on the left (SO 0776 6881) which is in the upper Ashgill Yr Allt Formation. Walk back along the road for about 400 m to the track on the right and follow this up through the wood to Henfryn Quarry (SO 0779 6903), (Fig. 10).

The southernmost face of the quarry provides a section through the whole of the Henfryn Formation, the basal Llandovery division on this eastern flank of the Tywi Lineament. The lower 10 m comprises a chaotic sequence of silty mudstones with cobbles and boulders of mudstone and coarse sandstone, reworked carbonate concretions and large, tabular rafts of silty mudstone. The latter resemble the underlying Yr Allt Formation, which may be seen *in situ* in the entrance into the older, northern bay of the quarry. Both clasts and matrix, within this heterogeneous Llandovery deposit, have yielded graptolites of the *griestoniensis* Biozone.

The unit appears to represent the deposits of a huge single debris flow but may in detail comprise a series of amalgamated mass-flow units. Over a metre of coarse, planar laminated, micaceous sandstone, rich in grains of mudstone, sharply succeed these lower deposits. The complex nature of the contact between the two lithologies records the effects of tectonic disturbance.

Upper beds in the section comprise thinly interbedded coarse sandstones and mudstones and form a transition into the overlying pale grey and green-grey Dolgau Mudstones observed in the stream beyond the quarry.

In the older northern bay of the quarry, the sandstone unit, let down by minor faulting, is seen again but is much thicker and locally conglomeratic, illustrating the rapid lateral variations exhibited by the Henfryn Formation along its outcrop.

14. Coedtrewernau. From Henfryn Quarry proceed southwards along the minor road, for just over a kilometre, crossing the northernmost splay of the Garth Fault system. Vehicles can be parked on the roadside just beyond the entrance to Coedtrewernau Farm (SO 0807 6765), (Figs. 10 and 11). Having gained **permission** at the farm, walk down to Clywedog Brook, where an almost complete section through the highly attenuated Llandovery sequence of the eastern flank of the Tywi Lineament may be examined. This section should only be visited when the river level is low but even then wellington boots are essential. It starts at the first bend downstream from the old farmhouse (Locality **14a**), where westerly dipping beds of dark sandstone, striking across the stream, form the upper part of the Henfryn Formation. Upstream for about 120 m, the entire local sequence of the succeeding Dolgau Mudstones is displayed. Pale to medium grey and greenish-grey colour-banded mudstones, commonly burrow mottled, are continuously exposed in the stream bed and banks. Easterly attenuation has reduced the strata to a thickness of only 55 m. Carbonate cone-in-cone nodules become abundant towards the top of the sequence and about 30 m upstream from the first fence line on the northern bank (Locality **14b**), a concretionary bed of carbonate mudstone, up to 15 cm thick, crosses the stream. Internally this unit displays complex veins and rosettes of calcite. A further *c*.25 m upstream (Locality **14c**) laminated hemipelagites of the early Wenlock Nant-ysgollen Mudstones appear. Interbedded burrow-mottled mudstones demonstrate the gradational lithological transition from the underlying Llandovery succession. Proceed a further 20 m upstream (Locality **14d**) to observe, in the northern bank, a thrust which repeats the Dolgau Mudstones. The Nant-ysgollen Mudstones, yielding graptolites

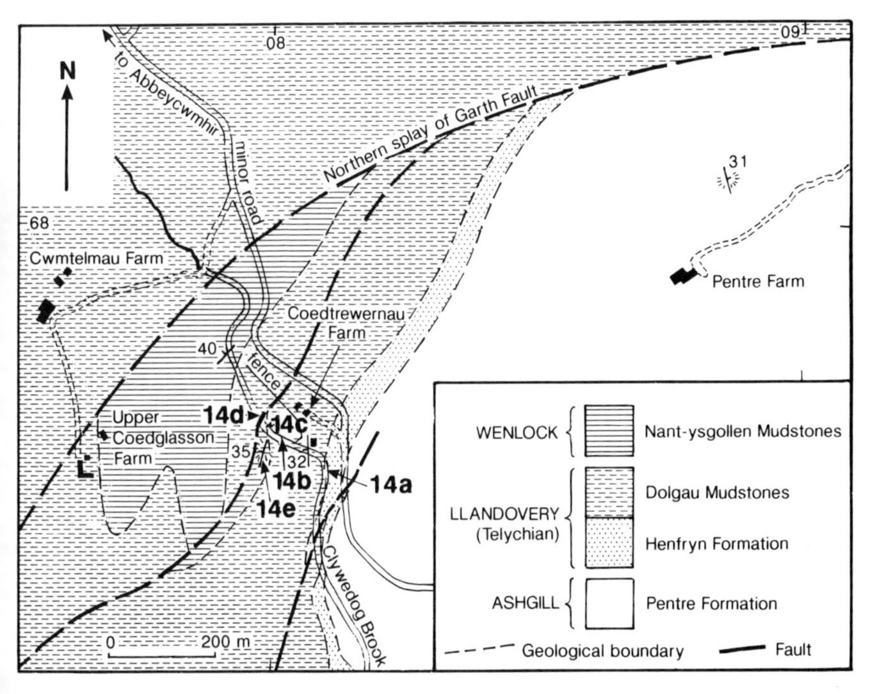

FIG. 11. Geological map of the Coedtrewernau Farm area (Locality 14); the quarry north-west of Pentre Farm is Locality 6 of Excursion 7.

referable to either the *crenulata* or *centrifugus* biozones, are better examined in the quarry 50 m to the south (Locality **14e**) where the same thrust contact with overlying buff weathered Dolgau Mudstones is observed. To visit this quarry **permission** should first be sought at Upper Coedglasson Farm (SO 0765 6755), (Fig. 11).

Acknowledgements. Mr S. P. Tunnicliff is duly acknowledged for his identifications of late Ordovician and early Silurian graptolite and shelly faunas. This excursion is published by permission of the Director of the British Geological Survey (NERC).

7. THE ORDOVICIAN OF THE RHAYADER DISTRICT

by D. WILSON, J. R. DAVIES, C. J. N. FLETCHER, *and* R. A. WATERS

Maps *Topographical*: 1:50 000 Sheet 147, Elan Valley and Builth Wells
1:25 000 Sheets SN 86, SN 96, SO 06, SO 07
Geological: 1:50 000 Sheet 179 (Rhayader)

THE Ordovician rocks of the Rhayader district are exposed mainly along the Tywi Lineament and in the core of the Rhiwnant Anticline to the west (Fig. 1). The Tywi Lineament is one of the major structural features of the region. It is a broadly anticlinal structure in Ordovician rocks, closing to the north near Abbeycwmhir, but in detail representing a complex zone of folding and faulting (Figs. 1 and 2). Although it is regarded conventionally as marking the transition in Silurian rocks between shallow-water shelf facies in the east and the basinal turbidite sequences to the west (Jones 1912; Smallwood 1986a; but see Excursion 6), its role during the Ordovician was, until recently, less well understood. However, it has been suggested that extensive fault movements occurred during the Ordovician along the lineament (Smallwood 1986) and on contiguous structures comprising the Welsh Borderland Fault System (Woodcock 1984a,b; Woodcock & Gibbons 1988; Lynas 1988).

Recent work by the British Geological Survey in the Rhayader District has established an Ordovician stratigraphy for the northern part of the Tywi Lineament (Fig. 3). In this area, the oldest exposed Ordovician strata are of Caradoc age (*multidens* Biozone). The top of the Ordovician, defined by the base of the Silurian *acuminatus* Biozone, falls within the lower part of the Cwmere Formation, and the contiguous Caban Conglomerate Formation (Excursion 6; Fig. 3 herein).

The Ordovician rocks are predominantly mudstones with subordinate sandstones of distal shelf and slope aspect. The anticlinal core exposes dark grey mudstones and sandstones, the St Cynllo's Church Formation, with intercalated volcaniclastic horizons including bentonites and coarser ash bands, spanning the *multidens* to *clingani* biozones of the Caradoc. The upper part of the formation comprises a coarsening-upward sandstone sequence, overlain by a thin unit of black mudstones locally yielding

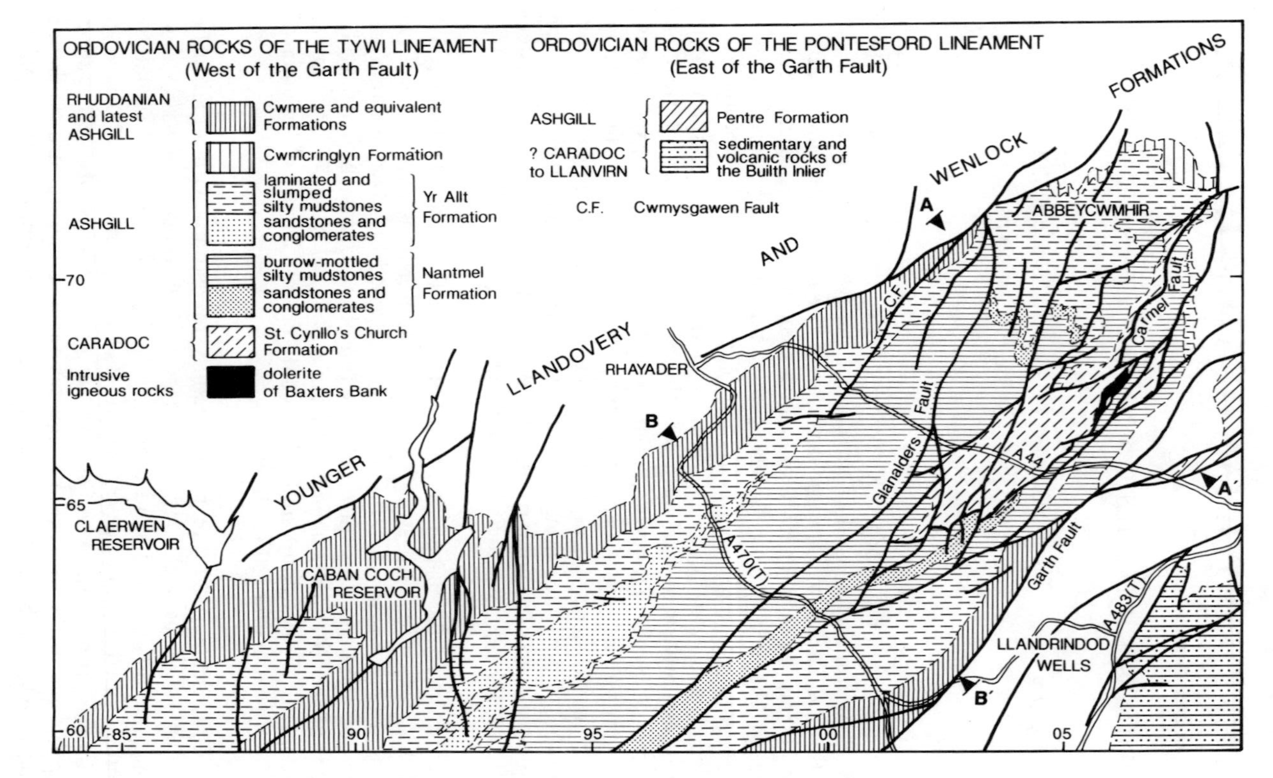

FIG. 1. Generalized geological map of the Ordovician rocks of the northern part of the Tywi Lineament; taken from BGS 1:50 000 Sheet 179 (Rhayader). A–A′ and B–B′ show the positions of cross-sections in Figure 2.

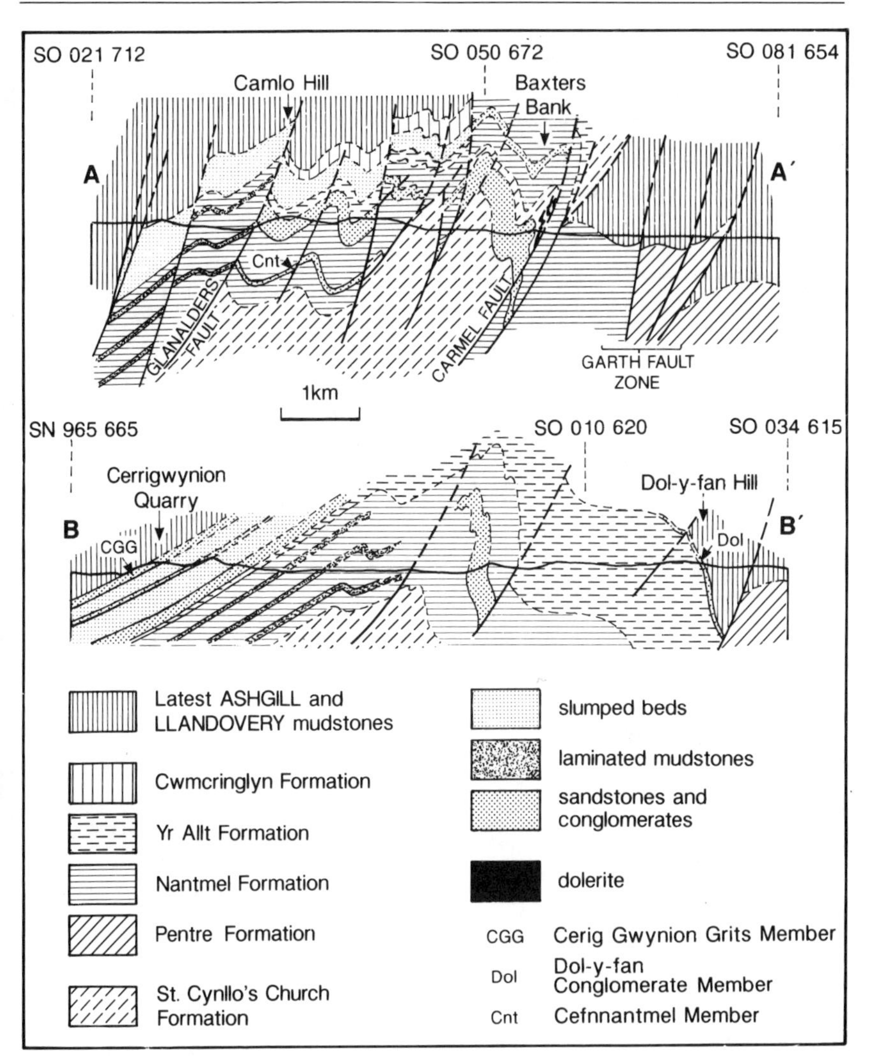

FIG. 2. Detailed cross-sections through the Ordovician rocks of the northern part of the Tywi Lineament along the lines shown in Figure 1; taken from BGS 1:50 000 Sheet 179 (Rhayader). Horizontal scale = Vertical scale.

graptolites of *clingani* Biozone age. Intruded into these mudstones are several thin dolerite sills forming the prominent ridge of Baxters Bank (SO 0580 6727).

The overlying Ashgill strata, occupying most of the Ordovician outcrop of

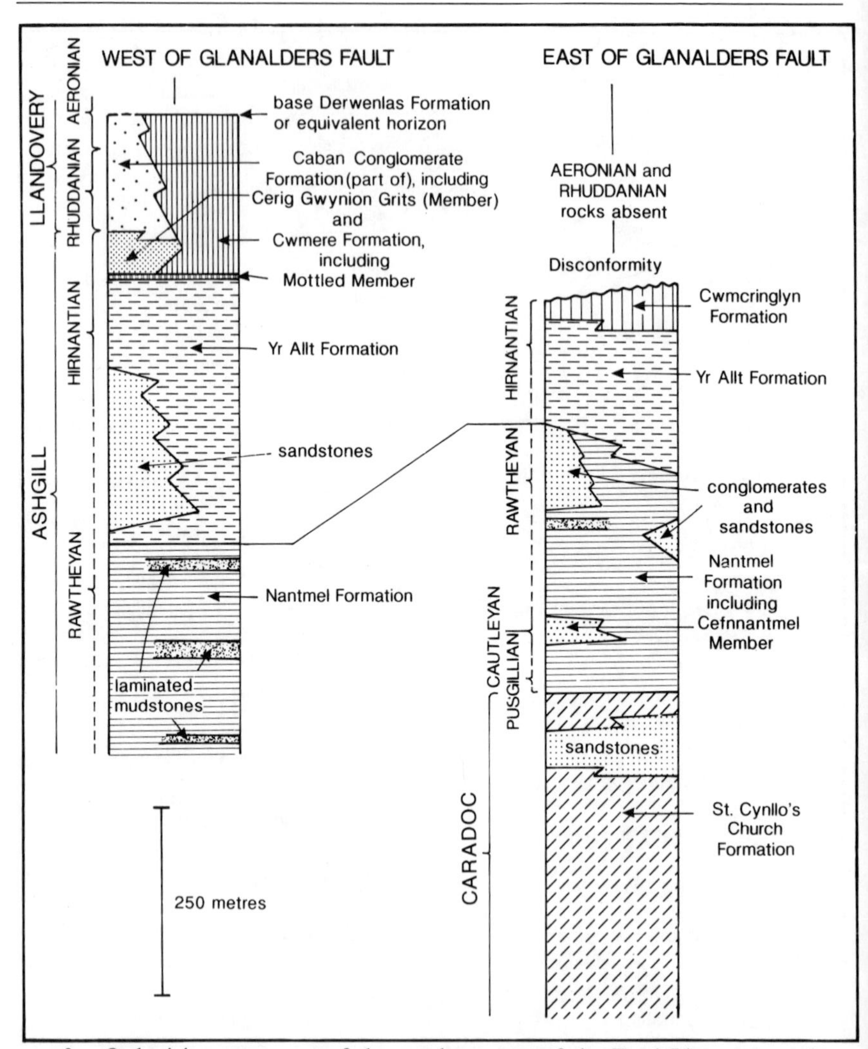

FIG. 3. Ordovician sequences of the northern part of the Tywi Lineament; taken from BGS 1:50 000 Sheet 179 (Rhayader). (NB Dol-y-fan Conglomerate Member and Pentre Formation not shown.)

the area, comprise sequences of silty mudstones, slumped extensively in their upper part, with beds of conglomerate, pebbly mudstone and sandstone representing the products of contemporaneous mass-flow deposits. Several formations are recognized (Fig. 3). The Nantmel Formation is the oldest and is represented by a monotonous sequence of massive, extensively bioturbated ('mottled') mudstones, recording deposition under predominantly oxic conditions. It is generally unfossiliferous but contains shelly sandstones in its lower part (the Cefnnantmel Member), yielding a trilobite fauna of mid-Cautleyan age (D. Price, personal communication). In the upper half of the formation three horizons of laminated mudstones record the introduction of anoxic bottom waters. These strata contain a sparse graptolite fauna of possible *anceps* Biozone age, and correlate with the Red Vein laminites of the Corris area (Pugh 1923). The overlying Yr Allt Formation comprises a thick sequence of unbioturbated and poorly fossiliferous coarse silty mudstones, commonly with abundant thin siltstone laminae. Sandstones up to 10 cm occur in packets, notably in the higher parts of the formation.

The top of the Yr Allt Formation in the south-east is represented by the Dol-y-fan Conglomerate Member, a locally thick but laterally impersistent suite of channelized conglomerates containing a variety of exotic clasts. However, in the west and in the hinge zone around Abbeycwmhir (SO 0544 7126), the bulk of the formation is slumped or has suffered *in situ* destratification, and locally contains thick sequences of coarse sandstone, representing mass-flow deposits. In the north-east around Abbeycwmhir these slumped beds are overlain by the Cwmcringlyn Formation, a coarsening-upward sequence of shallow-water sandstones, containing a distinctive cold-water Hirnantian fauna (Brenchley & Cullen 1984). The Cwmcringlyn Formation correlates with the shallow subtidal Scrach Formation of the Llandovery area (Woodcock & Smallwood 1987), recording the acme of the late Ordovician glacio-eustatic regression in the region (Brenchley & Newall 1984). The formation is not developed in the west, where it is assumed that deeper water conditions endured throughout the late Ashgill, and there is mainly conformable passage into the overlying Llandovery strata. In this area, slumped strata of the Yr Allt Formation are overlain by the late Ashgill Mottled Member, a thin but persistent unit of burrow-mottled mudstone at the base of the Cwmere Formation, containing a bed of laminated hemipelagite yielding *persculptus* Biozone graptolites.

The Mottled Member is overlain in turn by the Cerig Gwynion Grits Member, of proximal turbiditic aspect (Kelling & Woollands 1969). The latter lies entirely within the *persculptus* Biozone in the Wye Valley and is the lowermost member of the Caban Conglomerate Formation, of mainly Llandovery age (Excursion 6).

The sequence of Ashgill strata established on the western side of the Tywi Lineament is essentially similar to that exposed in the core of the Rhiwnant Anticline (James 1986) to the west. However, east of the main Ordovician crop, in the vicinity of Pentre Farm (SO 0882 6790), an Ashgill shallow-water assemblage of trilobites and bivalves has been obtained from isolated exposures of silty mudstones and thin sandstones (the Pentre Formation). Correlation of the latter with the Ashgill sequence of the Tywi Lineament is uncertain, as the two areas are separated by a major splay of the Garth Fault (Figs. 1 and 2).

The establishment of an Ordovician stratigraphy has facilitated the recognition of several hitherto unknown structures (Fig. 1). These include the Glanalders Fault, restricting essentially conformable Ashgill and Llandovery sequences to the western part of the Tywi Lineament, and the Garth Fault and its splays, east of the main Ordovician crop, bringing into proximity dissimilar Ashgill strata. In addition, there are other structures, no less important, which exercise a profound effect on the overall outcrop patterns. Amongst these is the Carmel Fault, recognized previously by Roberts (1929), with a large reverse component of throw, and a set of unnamed faults dislocating the western margin of the main Caradoc outcrop.

Many faults show evidence of late Caledonian (Acadian) reactivation (Woodcock 1984a; Woodcock *et al.* 1988), forcing folds above the propagating fault tips in Upper Llandovery and Wenlock strata. However, the total Acadian fault displacement over the hinge of the Tywi Lineament is small in comparison with the stratigraphical offsets and mismatches that occur across major faults within the Ordovician sequences; thus a pre-Upper Llandovery episode of faulting is indicated. It is likely that a significant amount of this faulting was intra-Llandovery, being influential in the development of Llandovery sidelap and overstep relationships and facies (Excursion 6). However, Woodcock (1984a,b, 1987a; Woodcock & Gibbons 1988) has suggested that faulting occurred during the late-Caradoc to early Ashgill, involving a significant amount of dextral strike-slip and leading to the widespread Pusgillian unconformity of the Welsh Basin (Price 1984). It has also been argued (Smallwood 1986a; Mackie & Smallwood 1987) that much of the faulting was syndepositional, being responsible for the generation of the mass-flow deposits which characterize parts of the Ashgill sequence. The evidence is mainly circumstantial, but both are additional possibilities in view of the disparities in formational thicknesses and juxtaposition of dissimilar stratigraphies that occur across major faults and preclude detailed Ashgill correlation.

ITINERARIES

The itineraries are intended to provide an introduction to the stratigraphy and structure of the Ordovician in the northern part of the Tywi Lineament and adjacent areas. They are arranged as three broad traverses each occupying at least a half day. The localities are numbered consecutively (Fig. 4) and generally arranged in stratigraphical order, although for

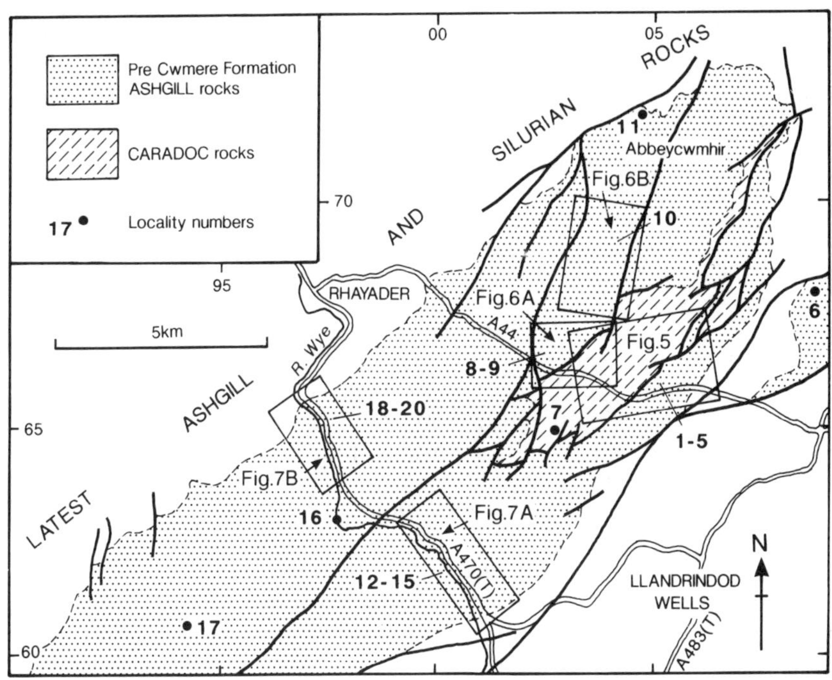

FIG. 4. Sketch map indicating the positions of localities described in the three itineraries.

reasons of time and travel some, particularly those of itinerary 3, are non-sequential. Access to most localities is by way of main roads and country lanes. However, many of the latter are unsuitable for vehicles larger than a minibus, and there are localities along tracks and in forest plantations where vehicular access is restricted; these are indicated where appropriate. Large parties or those with limited time can gain an introduction to the geology of the area by visiting easily accessible localities 1b, 3a, 8, 13, 14, 15, 18 and 20.

The stratigraphical nomenclature used in the itineraries is that shown on BGS 1:50 000 Geological Sheet 179 (Rhayader).

ITINERARY **1: Caradoc and Ashgill sequences between Nantmel** (SO 0341 6616) and **Pentre Farm** (SO 0882 6790)

The village of **Nantmel** lies on the A44, 7 km east-south-east of Rhayader, and may also be approached from Llandrindod Wells via the A483(T) and A44. **Larger vehicles can be parked by the War Memorial.**

1. Maesygelli Farm (SO 0374 6614). The farm lies 300 m east of the war memorial (Fig. 5) and provides sections in the lower part of the St Cynllo's Church Formation of Caradoc age. **Permission should be sought at the farmhouse** for access and parking of smaller vehicles. Good exposures in rusty weathering, dark grey mudstones with scattered dark, fine-grained sandstones, occur **in and around the farmyard** (Locality **1a**). Bundles of thin (1 – 2 cm) white bentonitic clay seams are particularly conspicuous and characterize the lower part of the Caradoc sequence in this area. Bentonites are also seen in exposures along the track leading to fields behind Nantmel School (SO 0347 6618), forming composite units up to 0.4 m thick. They are commonly quartz veined and slickensided, and have obviously been a focus for bedding-parallel slip during folding. The rocks are strongly folded, with a NW-inclined cleavage, variably developed in the vicinity of the farm. Steep fold limbs are locally inverted, as demonstrated by cleavage/bedding relationships in the farmyard, with small thrust faults disrupting sandstone beds. Contiguous exposures of the formation can be examined in the stream section running north-north-east from the school, and in the A44 cutting 200 m west of the war memorial (Locality **1b**).

2. Pen-y-lan Firs and **Rhos Esgairannos** (SO 0347 6582 to SO 0415 6535), (Fig. 5). **The long track through the forestry plantations** on the south side of the River Dulas affords almost continuous exposure through the middle and upper parts of the St Cynllo's Church Formation. Access to the western part is by a footpath leading from the minor road, 100 m south of its junction with the A44 at Nantmel; alternatively, enter the eastern end by a track immediately south of Caerfagu Quarry (SO 0444 6516). The plantation is owned by Fountain Forestry Ltd (Hay-on-Wye) from whom **permission should be sought**; vehicular access is difficult towards the west end of the section. Dark greenish-grey mudstones with subordinate thin sandstones and bentonites, similar to those at Maesygelli Farm, are intermittently **exposed at the western end** (Locality **2a**), succeeded eastwards by a sequence of fine-grained, thinly bedded sandstones, forming the high ground that spans the Dulas valley at this point. The contact is gradational and poorly exposed on the western side, but the sandstones are well displayed in a **quarry** (at Locality **2b**), occupying the core of a large syncline. They are generally 2–3 cm thick, occurring in bundles of 2–3 m, and interbedded with dark grey, greenish weathering silty mudstones. The sandstones are planar- and diffusely cross-laminated with micaceous partings, and biogenic structures are evident on the bases of a few units.

Approximately 75 m along the track to the east, the gradational top of the underlying mudstone division is exposed in the core of a large asymmetrical anticline, with the sandstones reappearing on the steeply inclined eastern limb. Examination of the sequence at this point reveals several pale grey

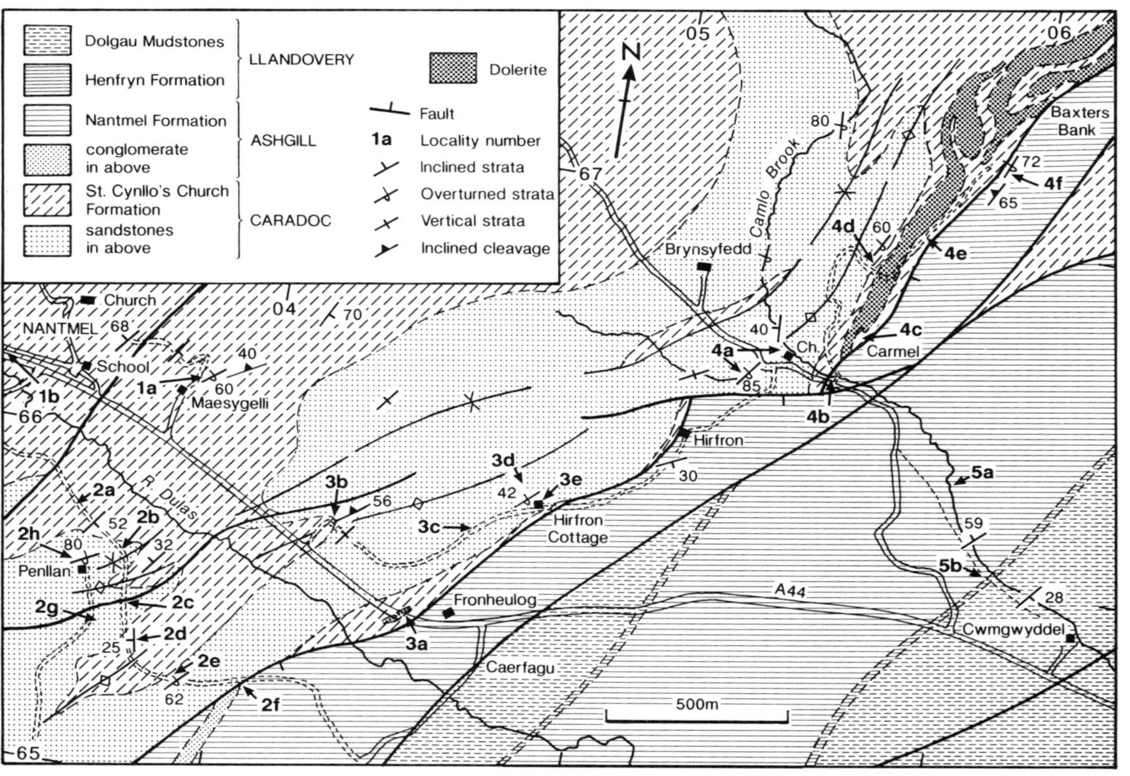

FIG. 5. Geological map of the Nantmel to Baxters Bank area (Localities 1 to 5).

ash horizons and hard splintery sandstones, probably of volcaniclastic origin. The sandstones occupy the core of a second large syncline (Fig. 5) which is disrupted by a fault with reverse displacement, exposed **in the trackside section** at Locality **2c**.

The section continues to the east in interbedded sandstones and mudstones, as far as Locality **2d**, where the underlying mudstone division of the St Cynllo's Church Formation reappears. The sequence is inclined west and is right way up as determined by cross-bedding and bioturbation structures on the bases of a few sandstone beds. Locality **2d** lies at the **eastern end of the Pen-y-lan plantation.** The mudstones exposed here contain scattered thin sands and sparse bentonitic seams up to 4 cm thick. They crop out in the core of a second large anticlinal structure, forming the conspicuous slack which runs up the hill between the two plantations. Continue eastwards through steeply inclined and overturned mudstones on the eastern limb of the anticline to Locality **2e**, where the overlying sandstones are again brought down. From here to the eastern end of the plantation the sequence is mostly sandstone-dominated and inverted, although there are gaps in the section.

About **80 m from the eastern entrance to the plantation**, at Locality 2f, dark grey, fine-grained sandstones of the St Cynllo's Church Formation, overturned to the east, are succeeded by brownish-grey silty mudstones of the Nantmel Formation (Ashgill), which are rather blocky in appearance and exhibit faint bioturbation mottling. The junction between the two formations is not exposed, but is probably tectonic, representing the southern continuation of the Carmel Fault (the Carmel Thrust of Roberts 1929).

If time permits, it is possible to include an additional two localities near the forestry plantations to examine Caradoc volcaniclastic rocks within the sandstone sequence. From Locality **2d** climb the hill to the track at the top which leads to **Penllan** (SO 0358 6560). A **small pit by the side of the track** (Locality **2g**) exposes a total of 12 m of dark grey, fine-grained sandstones, mostly 2 or 3 cm thick, but up to 0.3 m, interbedded with appreciable amounts of splintery tuffaceous sandstones and mudstones, ash bands and bentonites. **In the yard at Penllan** (Locality **2h**), from where **permission must be obtained**, a good section through 8 m of grey splintery siliceous mudstones and subordinate sandstones is exposed, overturned to the south-south-east, with an irregular spaced cleavage inclined at a low angle to the north-north-west.

3. Fronheulog (SO 0461 6564) to **Carmel** (SO 0556 6652), (Fig. 5). Return to the A44 road at Caerfagu (SO 0471 6563). Fronheulog Farm (SO 0461 6564) lies 100 m west of the road junction. **Request permission here for access** to the first part of the section, which is via the bridle path to Hirfron Cottage (SO 0482 6600); **the path is unsuitable for most vehicles.** Parking on the A44 is possible for smaller vehicles and for a limited period on the wide verges near Fronheulog; **larger coaches may have difficulty.** Whilst in the vicinity of Fronheulog, visit the **roadside section** 150 m west of the farm (Locality **3a**) which exposes the highest Caradoc strata. These are black graptolitic mudstones, intruded by a small dolerite body at the east end of the cutting.

Take the track where it leaves the A44 immediately east of the Grange (SO 0419 6577), (Fig. 5). About 100 m north-east of the road **a small quarry** (Locality **3b**) exposes rusty weathering, dark grey Caradoc mudstones with packets, up to 0.4 m thick of interbedded fine-grained planar- and cross-laminated sandstones. The mudstones are sheared and thrust in the core of a large anticline, but contain graptolites indicative of a *multidens* Biozone age (Zalasiewicz, personal communication). They probably represent the lowermost division of the St Cynllo's Church Formation, close to the junction with the overlying sandstones, which outcrop to the east. Follow the track through intermittent exposures in these sandstones to Locality **3c**. From here to Hirfron Cottage an almost continuous section through the upper part of the sequence is **exposed in the trackside**, comprising sandstones, in beds up to 0.2 m thick, with subordinate mudstones. The strata are overturned, younging to the south-east, and are lithologically similar to, but stratigraphically higher than the sequence in Pen-y-lan plantation. Follow the **farm track** up the hill to Locality **3d**, where dark grey mudstones with thin sands, tuffaceous horizons, bentonites and ash bands up to 0.9 m are exposed. Ripple-marked surfaces show that the beds are overturned south-eastwards and stratigraphically lower than the previous locality. They are inferred to lie at a comparable stratigraphical level to the volcaniclastic horizons on the hillside above Pen-y-lan plantation (Localities **2g** and **2h**).

Rejoin the main bridle path to Hirfron Cottage, and **ask permission** to examine the sequence in **the old quarry** immediately behind the house (Locality **3e**), which exposes about 15 m of overturned fine to medium-grained sandstones with subordinate dark grey, coarsely micaceous mudstones. The sandstones are generally thinly bedded but are locally up to 0.5 m thick, with several beds displaying a planar laminated lower part and a well-defined, ripple-marked top; grooves and trails occur on the bases of a few beds. These represent the highest strata of the sandstone sequence within the St Cynllo's Church Formation, as they are succeeded by black mudstones in the trackside about 40 m east of the cottage. The latter are poorly exposed but apparently overlie grey, silty mudstones with thin sandstones, representing the Nantmel Formation. It is assumed that the contact between the two formations is faulted at this locality, from the thickness variations along the mapped crop of the black mudstones.

Follow the **bridle path to Hirfron Farm** (SO 0516 6629) examining small trackside exposures of Nantmel Formation on the way. These are buff weathering, silty mudstones, locally micaceous and laminated in places with scattered thin 5 cm sandstones; small exposures also occur in the floor of the farmyard. The track then leads from the farm down to the road at Carmel (SO 0539 6653).

4. Carmel (SO 0539 6653) and **Baxters Bank** (SO 0580 6727), (Fig. 5). Carmel is approached by a narrow lane leading north from the A44, 650 m west of the Gwystre Inn (SO 0665 6567). **The road is unsuitable for large coaches** and parking in the vicinity of Carmel is difficult for more than one or two vehicles. However, it is a convenient collection point for those visiting localities 3a–e. **Permission for access to** all the following localities must be obtained at Brynsyfedd Farm (SO 0514 6679).

Caradoc sandstones, equivalent to those seen at Hirfron Cottage, are exposed in **Camlo Brook** adjacent to Carmel Chapel (SO 0542 6658) and in a tributary stream 150 m to the south-west (Locality **4a**) where they form beds up to 0.75 m thick, overturned to the south-east. To examine these sandstones, enter the tributary at the sharp road bend 50 m west of the track from Hirfron and walk 30 m upstream to the first exposures; alternatively, enter Camlo Brook at Bailey-Walter (SO 0554 6653) and walk 150 m upstream to the chapel. A continuous sequence through the highest sandstones interbedded with black mudstones is exposed in Camlo Brook, downstream from the chapel. Thin horizons of dark grey, burrow-mottled mudstone occur locally within the sandstone sequence. About 100 m downstream from the chapel the sandstones are overlain by black mudstones, representing the uppermost division of the St Cynllo's Church Formation, and equivalent to the mudstones exposed in the road cutting at Fronheulog. The contact is sharp and appears conformable although the sequence at this point may be repeated by faulting. The overlying mudstones are black and dark grey, faintly colour laminated with a strong, primary bedding-parallel fabric which has probably been enhanced during deformation. The absence of bioturbation and presence of graptolitic material in the mudstones suggests that they largely represent deposition under anoxic bottom conditions. Irregular shear zones cut the mudstones in places.

Continue downstream in Camlo Brook to where the junction of the black mudstones with the Nantmel Formation is exposed **immediately below Bailey-Walter** (Locality **4b**). The latter are grey, silty mudstones, with diffuse dark grey burrow-mottling and scattered thin sandstone beds. The contact, a reverse fault inclined steeply westward, is the Carmel Fault (the Carmel Thrust of Roberts 1929). Although there is little evidence of disturbance, the throw on the fault may be substantial as stratigraphical evidence along the outcrop suggests that a significant thickness of Ashgill strata is missing in Camlo Brook.

From the bridge at Bailey-Walter, take the farm track to **Pen-y-Banc Quarry** (SO 0560 6665), (Locality **4c**). The east face of the quarry exposes the Caradoc black graptolitic mudstones, overlain by a dolerite sill. The contact is locally irregular, sheared and veined with quartz, and in places truncates the strong bedding-parallel fabric of the mudstones. The latter show little evidence of alteration adjacent to the contact, and have yielded a low diversity fauna indicative of the *multidens* Biozone (J. Zalasiewicz, personal communication). The dolerite is massive, unlayered and broadly concordant within the Caradoc stratigraphy. Three or four similar intrusive dolerite bodies, with intervening black mudstones, outcrop to the north on Baxters Bank, where they are folded into a tight, overturned anticline/syncline couplet truncated to the east by the Carmel Fault (Fig. 5).

Follow the track northwards past the uninhabited buildings of Pen-y-Banc Farm (SO 0549 6686). About 100 m to the east, the dark, fine-grained sandstones of the St Cynllo's Church Formation crop out in **the trackside** at Locality **4d**, and in contiguous small exposures in the coppice to the north. Ripple-marked and bioturbated surfaces establish that the sandstones are overturned and young eastward. The conspicuous slack immediately to the east hides the overlying black mudstones, with the dolerite occupying the ridge beyond. **Continue along the track**, past small exposures of dolerite to Locality **4e**, where the contact between the sill and mudstones is again exposed. Local thin sandstone ribs within the mudstones display dewatering structures, possibly resulting from intrusion of the dolerite and demonstrating that the sequence, including the sill, is overturned to the east.

Continue east along the track for 250 m to (SO 0584 6721), where baked mudstones adjacent to the contact can be examined. The zone of alteration is up to 1 m wide and cut by several strong shear zones which locally transpose the bedding-parallel fabric of the mudstones; a number of small folds with sheared limbs are also exposed. The contact itself is complex and interleaved representing a zone of high strain with shearing and localized *en-echelon* quartz veining.

At this point leave the track to visit a **small farm quarry in the field** to the east, which exposes the faulted contact between Caradoc and Ashgill strata (Locality **4f**). Most of the pit is occupied by grey, silty, bioturbated mudstones of the Nantmel Formation, with the black Caradoc mudstones exposed in the degraded west face. Burrowing is recognizable in the Nantmel Formation as scattered diffuse mottles of darker mudstone, locally up to 1 cm diameter. Thin, fine-grained, planar-laminated sandstones occur at intervals and, in places, Fe-carbonate concretions with uncompacted burrow systems. Bedding in the Nantmel Formation is steeply inclined west-north-west but cleavage/bedding relationships reveal that the sequence is overturned eastward. The junction with the underlying black mudstones is sharp and inclined north-west at 58°, discordant by about 10° to bedding in the Nantmel Formation; otherwise there is little direct evidence that it is faulted.

5. Stream section north-west of Cwmgwyddel Farm (SO 0635 6590), (Fig. 5). This locality is readily accessible from Carmel, but for those wishing to confine their excursion to Caradoc sequences, it is best omitted, together with the next locality (6), and Locality 7 visited instead. The farm lies immediately north of the A44, 350 m north-west of Gwystre Inn. It is possible to park in the farmyard, but first **obtain permission at the house.**

Take the footpath that runs north-west from the farm, on the south-west side of Camlo Brook; the section begins about 600 m north-west of the farm. Grey silty mudstones of the Nantmel Formation are well exposed on water-worn surfaces on the **stream bed** at Locality **5a**. A streaky, impersistent colour banding and lamination pervades the mudstones and burrowing mottling is seen locally. In a few places the lamination is persistent, but discontinuities and variations in dip and strike suggest that the beds may locally be slumped. Concretionary Fe-carbonate bands and nodules are present in places. The Ashgill sequence is a continuation of that exposed below the Carmel Fault at Bailey-Walter (Locality **4b**). It generally dips north-west, but the direction of younging is uncertain.

Follow the section downstream for 300 m to Locality **5b**. Here, pebbly mudstones are well exposed in a **trackside cutting close to the stream.** Clasts range locally up to 50 cm in diameter and include sandstones, mudstones, volcanic rocks, some vein quartz and conglomerate; the mudstone matrix also contains abundant lithic grains. Rafts of dark grey laminated mudstone can be observed in places, one of which has yielded acritarchs indicative of an Upper Llandovery (Telychian) age (H. Barron, *personal communication*). These strata represent the Upper Llandovery Henfryn Formation (Excursion 6) They appear to dip east and are thus markedly unconformable on the Nantmel Formation. Overlying the Henfryn Formation are smooth, pale grey mudstones with thin laminated hemipelagites and rare burrows, representing the Dolgau Mudstones also of Upper Llandovery age (Excursion 6). Slump folding in these rocks is well displayed in the cutting and intermittent exposures in the formation occur in Camlo Brook as far as the farm.

6. Quarry at Pentre Farm (SO 0882 6790), Fig. 4; Excursion 6, Fig. 11). From Cwmgwyddel Farm take the A44 east to Crossgates, turning left onto the A483 at the roundabout. Follow the A483 for 3 km passing the junction to Abbeycwmhir and turn left onto a minor road at (SO 0960 6753). Follow the minor road for 750 m to a turning on the left for Pentre Farm. **Seek permission at the farm** for parking and access to the quarry.

The quarry (SO 0886 6808) is situated in the field on the north-east side of the farm and is the type section for the Pentre Formation. The section exposes a 6 m thick sequence of buff-weathered fine-grained argillaceous sandstones, overlain by silty mudstones. The sandstones yield a shallow water fauna dominated by bivalves and trilobites but also including crinoid columnals and orthocones. The trilobites *Brongniartella* cf. *sedgwickii*, *Gravicalymene* cf. *pontilis* and *Mucronaspis*? sp. are suggestive of an Ashgill age (A. W. A. Rushton, personal communication). However, there is no obvious lithological correlative within the Ashgill sequence to the west of the Garth Fault.

7. **Gwerncynydd Farm** (SO 0270 6490). This locality can be visited as an alternative to localities 5 and 6, but could be included in the main itinerary if time permits. It exposes the youngest Caradoc strata in the area, containing an abundance of well-preserved graptolites. Take the minor road on the south side of the A44 opposite Nantmel village, following it for about 1.75 km to Gwerncynydd Farm (Fig. 4). **Obtain permission at the farm** to examine black, rusty weathering mudstones exposed in a **cutting behind the old farmhouse** (SO 0270 6490). These are equivalent to, but at a slightly higher stratigraphic level than the black mudstones exposed in Camlo Brook and on Baxters Bank. There is no need to hammer as graptolites are readily gathered from the spoil. Several specimens of *Dicranograptus clingani*, confirming the biozonal age, have been obtained from these rocks, together with numerous examples of *Climacograptus spiniferus* (J. Zalasiewicz, personal communication).

The black mudstones crop out in Nant Treflyn and on the track running eastward along the south side of the stream, 75 m north of the farm. Graptolites indicative of a level within the *clingani* Biozone have been recovered from the stream at (SO 0280 6500). Beyond this point exposure is impersistent, but the overlying bioturbated mudstones of the Nantmel Formation can be observed in the stream 130 m to the east at (SO 0293 6499).

ITINERARY **2: Ashgill stratigraphy and structure between Nantmel** (SO 0342 6606) **and the Abbeycwmhir area** (SO 054 713)

8. **Road cutting** (SO 0245 6630), on the **A44 near Dolau** (Fig. 6A). Take the A44 to Nantmel. The cutting is approximately 1 km west of the village. **Parking is possible next to Dolau Chapel** (SO 0230 6646) and at the junction of the disused road on the opposite side of the A44; alternatively, park smaller vehicles on the wide verges of the cutting.

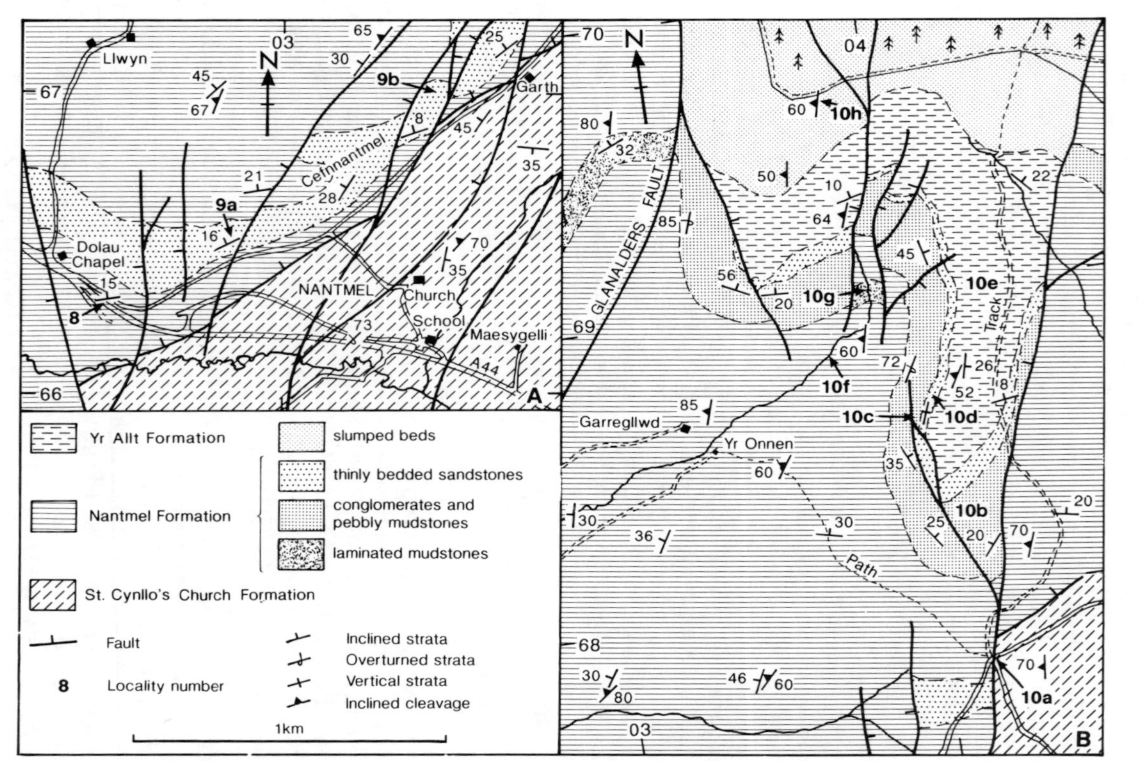

FIG. 6. *A*, Geological map of Cefnnantmel (Localities 8 and 9); *B*, Geological map of Camlo Hill (Locality 10).

Good exposures in the Nantmel Formation can be examined on **both sides of the cutting.** These comprise medium to dark grey, burrow-mottled silty mudstones, colour banded in places and with scattered fine sandstone laminae and thin bands. The effects of bioturbation are conspicuous throughout the sequence and include both large aligned burrows and clusters of the smaller *Chondrites* burrow systems.

9. Cefnnantmel (SO 0270 6636) to (0375 6719), (Fig. 6A). **Permission for access** to the section should first be sought from the farms at Llwyn (SO 0250 6718) and Garth (SO 0374 6701). Take the minor road that joins the A44 between Dolau Chapel and the road cutting (Locality 8), and runs north-east along the lower slopes of Cefnnantmel; **note that the road is unsuitable for coaches and has no parking places.** Follow the road for 500 m, noting exposures of Nantmel Formation in the roadside, to the first **crags on the western end of Cefnnantmel** (Locality **9a**).

Ascend the crags which expose the Cefnnantmel Member, a sandstone sequence within the Nantmel Formation. The sandstones are 1–2 cm thick but, locally, up to 10 cm, generally fine grained, rusty weathering (?decalcified), and thinly laminated. A number of beds display coarse shelly basal lags which have yielded poorly preserved trilobite fragments tentatively assigned to the mid Cautleyan. The intercalated mudstones are less conspicuously mottled than the underlying parts of the Nantmel Formation (it is worth noting that at this point at least 75 m of the Nantmel Formation underlie the Cefnnantmel Member). As the sequence is ascended, burrow mottling increases and the sandstones become more sporadic until, about 100 m north of the crags, they have almost disappeared, being represented by a few thin disrupted beds.

Follow the crags north-east, noting more of the sandstones which form the prominent ridge of Cefnnantmel. Locality **9b** is a point at which to examine the broad topographical features of the area. Many of the localities visited in Itinerary 1 can be observed from here. The lower undulating ground immediately to the south-east is formed by the mudstones of the St Cynllo's Church Formation, with the overlying sandstone division occupying the wide ridge beyond. It can be seen from here that the Caradoc strata are not contained within a simple anticlinal structure (cf. Roberts 1929), for their overall younging direction is eastwards and there is no repetition of the sandstone division on the west limb; it must therefore be faulted out (Fig. 1). One such fault, running along the foot of Cefnnantmel, separates the Caradoc and Ashgill strata. It can be demonstrated by the apparent difference in stratigraphic interval between the Cefnnantmel Member and the Caradoc mudstones. The latter are **exposed in the road and farmyard** below this vantage point, with no more than 30 m of the intervening Nantmel Formation at crop; compare this to the thickness exposed on the south-west end of the ridge. From the viewpoint (Locality **9b**) note also the marked features running obliquely north-north-east across the ridge, denoting the position of faults which disrupt the crop of the Cefnnantmel Member.

10. Camlo Hill (SO 0414 6775 to SO 0403 6972), (Fig. 6B). **Permission for access** to Camlo Hill can be gained from the following farms: Glanalders (SO 0289 6933), (NW slopes); Garreglwyd (SO 0327 6865) and Yr Onnen (SO 0336 6856), (NW and W slopes); Bwlch-Mawr (SO 0431 6714), (S slopes); Brynsyfedd (SO 0514 6679), (S, E and W slopes).

From Garth follow the minor road north-east to the **road junction** (Locality **10a**). A large fault crosses the junction and its trace is marked as a conspicuous slack on the hillslope to the north. It represents one of the suite of NNE-trending faults previously described on Cefnnantmel, and can be located at the junction by subjacent small exposures of Ashgill and Caradoc mudstones.

Take the steep bridle path north up the hill, through crags of bioturbated mudstones of the Nantmel Formation, crossing the fault slack to Locality **10b**. Here, examine crags of pebbly mudstone, argillaceous and pebbly quartzose sandstones, representing part of a thick sequence of mass-flow deposits forming most of the western part of Camlo Hill. Clasts within the conglomeratic deposits include intraformational sandstones and mudstones, as well as exotic vein quartz and igneous rock types.

Climb to the **vantage point** at Locality **10c**, looking north to view crags of conglomerate and sandstone spectacularly displayed in a large anticline, flanked by corresponding synclines, which wrap around the west slopes of the hill. The anticline is asymmetrical, with its steep limb overturned to the south-east; walk north along this limb, through steeply inclined beds of conglomerate to Locality **10d**. Here, examine **crags** of thinly interbedded sandstones and silty mudstones with abundant sandstone laminae, capping the conglomerate sequence. Note the strong pressure-solution cleavage disrupting the sandstone beds and preferentially developed in these lithologies on the steep limb of the anticline.

Cross the hillside to Locality **10e** where the overlying Yr Allt Formation is **exposed in small crags** in the core of the easternmost syncline. These strata comprise dark grey, silty mudstones with abundant siltstone and fine sandstone laminae.

Traverse west to the stream section in the anticlinal core to Locality **10f**, crossing the steep fold limb and noting the vertical beds of sandstone and conglomerate well displayed on the hillside. The **stream contains small**

exposures in medium to dark grey, silty bioturbated mudstones of the Nantmel Formation, containing scattered 5 cm sandstones. Follow the stream north to Locality **10g** where, within the upper part of the Nantmel Formation, dark grey, rusty weathering laminated mudstones are poorly exposed. Immediately to the north, silty mudstones with diffuse sandy streaks and local faint burrow-mottling supervene. The mudstones are slumped in places and represent the highest beds of the Nantmel Formation in the area.

Walk north upstream through the anticlinal core, noting how faulting has disrupted the crop, juxtaposing the Yr Allt Formation on the west bank of the stream with conglomerate to the east. Follow the boundary fence north to Locality **10h**, viewing, on the way, the strong features on the hillside immediately to the west, which represent a series of slump sheets overlying undisturbed parts of the Yr Allt Formation. At Locality **10h** examine one of these slumped units, comprising grey silty mudstones with abundant diffuse sandstone knots and wisps, and scattered small sandstone blocks. A thin lamination, comparable with that observed in the undisturbed parts of the Yr Allt Formation is preserved locally, and a few slump fold hinges can be recognized in places. The main tectonic fabric is a spaced cleavage, irregularly developed on and around the slump fabrics. The variability of this superimposed cleavage probably reflects the degree of differential induration and transposition of the sedimentary fabrics during the slumping process.

To avoid returning by the same route, arrange for a collection point for vehicles on the minor road at (SO 0258 7062); a forestry track, immediately north of Locality **10h** leads north-west directly to this point.

11. Y Glog (SO 0480 7185), (Fig. 4; Excursion 6, Fig. 10). Follow the minor road from Rhayader north-east to Abbeycwmhir, turning sharply left onto a forestry track at the Happy Union public house. Follow the track for 500 m to the point at which it forks, taking the right hand fork and continuing to the junction of five tracks (SO 0480 7163). Take the main track leading north-west to where **the section begins in a large trackside quarry** at (SO 0474 7175). The quarry exposes coarse silty mudstones with abundant siltstone laminae and scattered thin beds of sandstone, up to 3 cm thick, within the upper part of the Yr Allt Formation; some slumped units are evident. Walk up through the section, noting how the percentage of thin sandstones gradually increases until, 100 m north-north-west of the quarry, they account for approximately 25 per cent of the sequence.

The track bends sharply east at (SO 0470 7195) and at this point there is a rapid transition into the overlying Cwmcringlyn Formation. The lower part of the formation comprises thin sandstones with mudstone partings. The sandstones are cross-laminated and their upper surfaces exhibit a range of undulating bed forms including probable wave ripples. At higher levels, parallel-sided, fine to medium-grained sandstones enter the sequence. The highest exposed beds are massive sandstones, but the top of the formation is not seen. About 50 m beyond the end of the exposure, mudstones of Telychian age are encountered. These are the Dolgau Mudstones and they disconformably overlie the Ashgill strata (see Excursion 6, Locality 11). The Cwmcringlyn Formation at this locality is poorly fossiliferous, but nearby, at (SO 0585 7219), it has yielded abundant Hirnantian brachiopods. This distinctive, cold-water fauna, and the coarsening/shallowing-upwards motif of the formation is thought to record the worldwide late-Ashgill glacio-eustatic regression (Brenchley 1988).

ITINERARY 3: The Ashgill sequences of the Wye Valley

The Wye Valley south-east of Rhayader exposes Ashgill sequences on the north-west and south-east sides of the Tywi Lineament. This itinerary begins in the Ashgill on the south-east side where, for convenience, localities 12–16 are described in descending stratigraphical order.

12. Dolyfan Hill (SO 0514 6117), (Fig. 7A). Take the A470(T) road south-east from Rhayader, turning onto the A4081 and **park at the entrance to a forestry track** (SO 0225 6076) on the north side of the road. **Permission to climb Dolyfan Hill** should first be obtained from Pistyll Gwyn Farm (SO 0245 6132). Continue up the track through the forestry plantation and then across the open ground to the summit. From here walk south-west along the ridge and then downhill, keeping to the west side of a prominent **line of crags** which expose vertical and overturned strata in the Dol-y-fan Conglomerate Member, the highest division of the Yr Allt Formation in this area. Thick, highly lenticular channelized units, containing both clast and matrix-supported conglomerates can be examined in a faulted syncline on the south-west face of the hill. Further down the hill, at (SO 0115 6088), the strata underlying the conglomerates is exposed, comprising silty micaceous mudstones with abundant siltstone laminae and packets of thin cross-laminated sandstones. Individual sandstones are up to 6 cm thick and commonly rich in finely comminuted bioclastic debris.

13. Road cuttings on the A470 between (SO 0083 6136) and (SO 0022 6218). Return towards Rhayader, turning onto the A470 and **park in the lay-by** 350 m north-west of the junction with the A4081. This well-exposed section begins about **200 m north-west of the lay-by** (Fig. 7A) in dark grey, thinly laminated silty mudstones of the Yr Allt Formation. The abundant, closely spaced laminae, which characterize this otherwise monotonous formation, are generally of siltstone or fine sandstone and commonly micaceous. Rusty weathering coarse-grained sandstones, a few centimetres

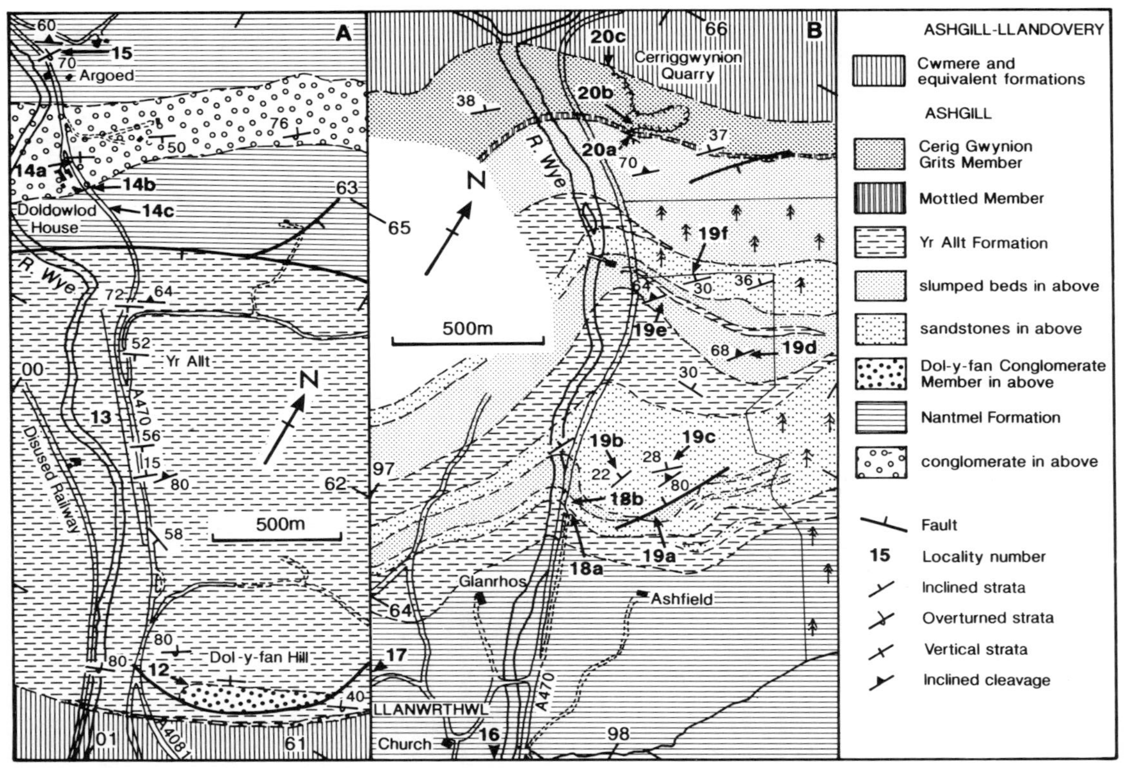

FIG. 7. *A*, Geological map of the Yr Allt area of the Wye Valley (Localities 12 to 15); *B*, Geological map of the Cerriggwynion area of the Wye Valley (Localities 18 to 20).

thick, occur sporadically along the section. In places they appear decalcified and at intervals contain indeterminate bioclastic debris. The sequence is mainly inclined southward at moderate to high dips and cut by a NW-dipping cleavage which, locally, is partitioned into anastomosing domains, leaving intervening lozenges of relatively uncleaved mudstone.

14. Road cutting on the A470 near Doldowlod House (SN 9974 6258). Drive north-west along the A470(T) to Doldowlod House. The road cutting is on the north-east side of the road opposite the house (Fig. 7A), but it may be more convenient to **park vehicles in the lay-by 300 m to the north-west at Argoed** (SN 9952 6284) and walk back. **The cutting** (Locality **14a**) exposes a vertical to steeply overturned, SE-younging sequence of coarse-grained and pebbly sandstones, conglomerates and pebbly mudstones with thin silty mudstone interbeds. Clasts include intraformational mudstone and sandstone, as well as exotic vein quartz and scattered igneous lithologies; fragments of crinoid and coral may also be seen. The sandstones form beds up to 0.75 m thick, commonly parallel-laminated and locally graded, with pebbly bases and fine-grained, faintly cross-laminated tops. The bases of a few sandstone beds exhibit load structures. The sandstones and conglomerates thin and fine upwards at Locality **14b** into a sequence of interbedded grey silty mudstones and fine to medium-grained quartzose sands which are slumped in places. Immediately to the east the cutting exposes till deposits, but a further 100 m beyond this, higher strata can be seen (Locality **14c**), representing an interval in the upper part of the Nantmel Formation. These comprise dark grey mudstones with sporadic, diffuse bioturbation mottling and scattered, disrupted sandstones up to 5 cm thick.

The sequence represents another example of mass-flow deposits within the Nantmel Formation, with obvious similarities to those exposed on Camlo Hill. However, it is difficult to make direct correlation between the two areas because of dissimilarities in the contiguous Ashgill stratigraphy.

15. Road cutting north-west of Argoed (SN 9940 6292). Walk 100 m west of the lay-by at Argoed to the cutting (Fig. 7A), which exposes bioturbated mudstones of the Nantmel Formation underlying the sandstones and conglomerates of the Doldowlod road cutting. The characteristic dark burrow mottling is readily seen against the paler background of these friable mudstones. Bedding is defined by differential weathering of the mudstones and, at the east end of the cutting, it youngs and dips eastward.

16. Section along River Wye (SN 9785 6279) (Fig. 4). This section is only exposed when the water level in the River Wye is low; it can be omitted and Locality 17 visited as an alternative. From Argoed, drive north-west along the A470 to the Vulcan Arms and **park in the adjacent lay-by. Request**

permission at the Caravan Club site (**note**: this is the furthest site south of the lay-by) for access to the east bank of the river; a path leads from the site to the river bank. A wide **rock platform in the river** bed exposes a sequence of dark grey laminated mudstones with rare silty mudstone beds. Contiguous exposures in the opposite bank, which can be approached via a minor road leading south from Llanwrthwl village (SN 9759 6374), have yielded a sparse graptolite fauna of possible *anceps* Biozone age (J. Zalasiewicz, personal communication).

17. Cwm Pistyll stream section (SN 9432 6050 to 9420 6082), (Fig. 4). This section is remote from the main traverse, but can be included if time permits, and as a substitute for Locality 16; it is **only accessible in smaller vehicles and on foot**. Turn off the A470 at the signpost for Llanwrthwl and follow the minor road through the village. At the west end of the village, about 200 m beyond the church, a minor road leads south-south-west. Follow this road as far as the National Trust monument, where an unmetalled track continues; it is advisable to **leave vehicles at this point** (*c*. SN 9580 6146). Continue south-south-west on foot for about 1.8 km along the track and path which gradually climbs the scarp slope to the north-west, until the steep gully of Cwm Pistyll is reached. The gully below the track provides a continuous section through the upper part of the Nantmel Formation on the west side of the Tywi Lineament.

Descend the hillside to the foot of the gully where the lowermost beds are exposed (SN 9432 6050). These are interbedded, grey, laminated and bioturbated mudstones with bundles of fine-grained sandstones locally up to 30 cm thick. They form a transition into a laminated mudstone, the highest of three such units within the Nantmel Formation, which is exposed 250 m upstream in the **steep sides of the gully** at (SN 9423 6071). Here, it comprises 16 m of dark grey, rusty weathering mudstones, laminated at less than 1 mm scale, with scattered, pale grey, mudstone bands, rarely 1 cm thick. Climb 30 m up the gully to (SN 9421 6074) where the background burrow-mottled mudstones are again exposed. Follow the section up through these mudstones to the top of the gully where, at (SN 9418 6083), the first appearance of horizons of thinly laminated, silty mudstone define the base of the overlying Yr Allt Formation. Small isolated exposures of the latter occur **beyond the top of the gully** and on the escarpment to the north-east.

18. Road cutting at (SN 9750 6450), (Fig. 7B). Return through Llanwrthwl, turning left onto the A470 and **park on the verge** about 700 m along the road; parking may also be possible in a disused road entrance on the right of the A470, about 500 m from the turn. **At the disused road junction**, dark grey, thinly laminated, silty mudstones of the Yr Allt Formation are well displayed (Locality **18a**). The boundary with the underlying Nantmel Formation occurs about 75 m to the south-east but is not exposed. Walk up through the sequence to Locality **18b**, where several thick sandstones are interbedded with the laminated silty mudstones. The sandstones are medium to coarse grained, locally graded, in beds up to 0.4 m thick, with planar and cross-lamination in places. The bases of a number of beds exhibit load structures. Well displayed slump folds in the interbedded mudstones are exposed in a small quarry at the north-west end of the road cutting. Contiguous exposures in the sandstones occur **in the River Wye immediately below the road**. They represent one of several thick sandstone sequences comprising a series of mass-flow deposits within the Yr Allt Formation on the west side of the Tywi Lineament, which have no obvious correlative in the east.

19. Hillside south-west of Caerhyddwen Forestry Plantation (SN 9748 6462 to SN 9747 6531), (Fig. 7B). This section is a continuation of the previous one, but for access to the hillside **permission must be sought at Ashfield** (SN 9784 6440). Climb the hill to the **east of the small quarry**, noting the dip slopes of massive sandstones, to Locality **19a**. Here, at a horizon similar to that of the quarry, cuspate dewatering structures, convolute laminae and numerous slumps are well exposed in a sequence of thinly laminated silty micaceous mudstones with thin interbedded fine-grained sandstones.

Traverse north-west through the sequence, noting in this upper part of the crop, the massive beds of medium and coarse-grained pebbly sandstone, up to 2.5 m thick, and the faint lamination on weathered surfaces defining convolutions and dewatering structures. At Locality **19b** examine a 1.5 m thick sandstone, possibly an amalgamated unit, with cuspate dewatering structures along which cleavage has been enhanced. Note the spaced rough cleavage which pervades the sandstones and, at Locality **19c** examine pressure solution effects on pebbles within a 2.3 m thick bed of coarse sandstone.

Cross the col marking the upper boundary of the sandstone sequence, to the steep hillside beyond. The hillside is formed of well-featured thick slump sheets within the Yr Allt Formation, alternating locally with horizons of thinly laminated, micaceous, silty mudstones. Climb to Locality **19d** to examine pervasively slumped and locally disaggregated silty mudstones with thin sandstone beds. Traverse the hillslope to the crags at Locality **19e** where well exposed slump sheets display transposed bedding fabrics, slump folds and small, randomly distributed blocks, knots and wisps of sandstone. Climb up through the sequence of slump sheets to Locality **19f**, where several thick sandstones are exposed. These form part of a higher sandstone sequence within the slumped Yr Allt Formation on the western side of the Tywi Lineament. The sandstones are massive and coarse grained, displaying coarse feldspathic bases with intraformational mudstone clasts; parallel lamination and load structures are evident in places. Interbedded with these sandstones are units of thinner bedded, finer-grained, planar and cross-laminated sandstones and interbedded silty mudstones with localized slump structures.

20. Cerrigwynion Quarry (SN 9710 6570). The quarry is on the north-east side of the A470, 3.5 km south-east of Rhayader (Fig. 7B). It is presently disused but **permission to visit must be obtained in advance** by writing to the current owners who are Tarmac Roadstone Ltd (Western), Whitehall House, Whitehall Road, Halesowen, West Midlands B63 3LE.

The sequence in the quarry comprises:

Caban Conglomerate Formation	{'Dyffryn Flags' facies	8 m +
	{Cerig Gwynion Grits Member	70 m
Cwmere Formation	Mottled Member	6 m
Yr Allt Formation		4 m +

The whole succession is **exposed in the main (lowest) face.** At the south-east end (Locality **20a**), the junction between the Yr Allt Formation and Cwmere Formation can be examined in detail. Unbedded (?slumped) and irregularly cleaved, dark grey, silty micaceous mudstones of the Yr Allt Formation are gradationally overlain (over a few cm) by grey turbidite mudstones and pale grey burrow-mottled hemipelagites of the Mottled Member. Standing at a distance from the face note how, immediately above the contact, bedding becomes apparent and the cleavage becomes uniformly spaced. An 8 cm thick, rusty weathering bed, 1.5 m above the base of the Mottled Member, contains laminated hemipelagites which yield abundant pyritized graptolites of the *persculptus* Biozone.

The base of the Cerig Gwynion Grits rests on the major bedding plane, which defines the **eastern side of the main quarry face** (Locality **20b**). The member comprises thick, commonly lenticular, turbidite sandstones with mudstone partings and beds. In the upper half, some sandstones are pebbly or conglomeratic at their base. Most of the thick sandstones are weakly laminated and poorly graded. Dish structures, the result of dewatering, can be examined in loose blocks below the main face at the eastern end of the quarry. Sole structures are sparsely developed but indicate transport directions to the north-north-east (Kelling and Woollands 1969); by contrast, local cross-stratification suggests reworking by currents flowing towards the north-north-west.

At the **western end of the main face** (Locality **20c**), there is a rapid transition up into the interbedded, thin turbidite sandstones and mudstones, with laminated hemipelagites of the 'Dyffryn Flags' facies. This part of the sequence is not accessible. The base of the *acuminatus* Biozone (base Silurian System) has been proved about 20 m above the top of the Cerig Gwynion Grits.

Acknowledgements. We thank Dr A. W. A. Rushton, Dr J. A. Zalasiewicz, and Mr S. P. Tunnicliff for their identification of Ordovician graptolite and shelly faunas. This paper is published by permission of the Director of the British Geological Survey (NERC).

8. THE LUDLOW AND PŘÍDOLÍ OF THE RADNOR FOREST TO KNIGHTON AREA

by N. H. WOODCOCK *and* J. E. TYLER

Maps *Topographical:* 1:50,000 Sheet 148 Presteigne & Hay-on-Wye

1:25,000 Sheets 950 (SO 27/37) Knighton and Brampton Bryan, 970 (SO 06/16) Llandrindod Wells, 971 (SO 26/36) Presteigne and area, 992 (SO 05/15) Builth Wells.

Geological: 1:250,000 Mid Wales and Marches

THE Ludlow Series marine rocks comprising most of this itinerary are transitional in facies between those of the main Welsh Basin to the west and those of the shallower Midland Platform to the south-east. The first part of the itinerary illustrates this spatial transition on a basinward traverse through Gorstian (lower Ludlow) rocks, although Ludfordian (upper Ludlow) sequences record a similar palaeogeographical transition. The second part explores another important transition, through time rather than space: the replacement of marine by non-marine conditions. Ludfordian (upper Ludlow) and Přídolí sediments record this shallowing-upward sequence that occurs in basin and platform sequences alike. This is interpreted as the result of sediment supply exceeding subsidence, prior to the contractional inversion of the Welsh Basin and its margins during the late Caledonian (Acadian) deformation. The regional context is outlined in the introduction to this volume (Woodcock & Bassett 1993).

GEOLOGICAL SETTING

The area (Figs. 1, 2) lies within the Welsh Borderland Fault System that defines the old basin margin (Woodcock & Gibbons 1988). Beyond the Church Stretton Fault on its south-east flank are developed the classic Silurian shelf facies of the Midland Platform. Beyond the Tywi Lineament to the north-west of the area are the thick clastic sequences deposited by basinal Silurian turbidite systems. The intervening Ludlow sediments have been interpreted variously as products either of a turbidite system or a more complex depositional system dependent on storm-generated processes. In all models the area represents a platform to basin transition zone during Gorstian time, though its abruptness and depth are in doubt.

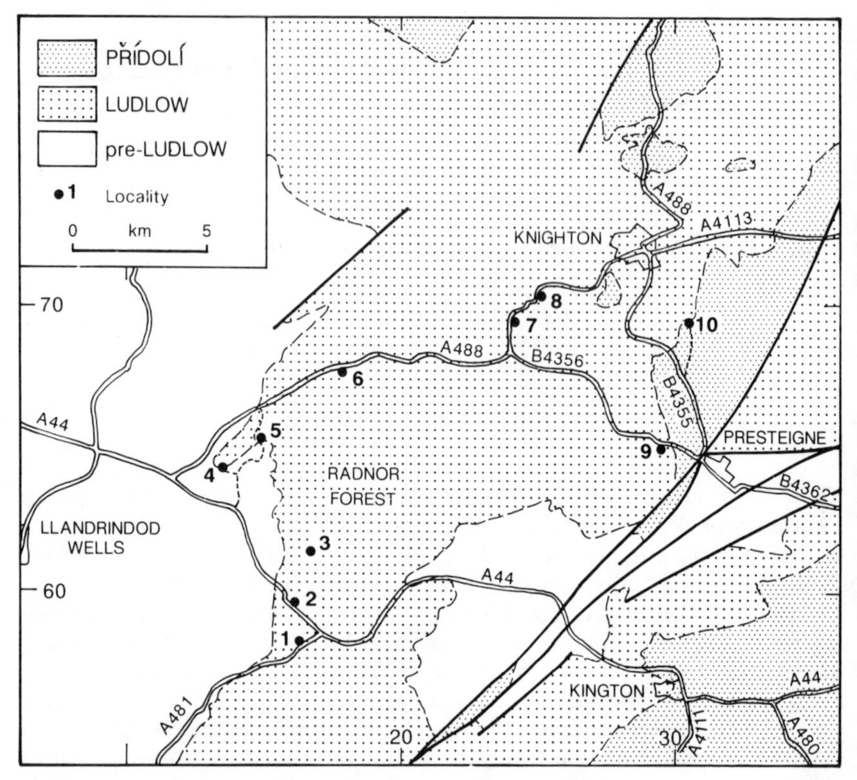

FIG. 1. The area west of Knighton and Presteigne showing the outline geology and localities covered in the itinerary.

Extensive submarine slumping indicates that the fault zones underlying the basin margin were active during Ludlow sedimentation. The fault zones were re-activated during the late Caledonian (Acadian) deformation, culminating in early Devonian time. This produced NE–SW-trending upright folds and steep faults (Fig. 2). This deformation is seen in the area as steep dips in the north-west (Localities 5 and 6), along the Clun Forest Disturbance, a component of the Pontesford Lineament, and in the south-east (Localities 9 and 10) close to the Church Stretton Lineament. The intervening area preserves gently dipping rocks lacking marked folds.

STRATIGRAPHY

The lithostratigraphy of the area is illustrated on transects (Fig. 3) showing restored thicknesses for the Gorstian and Ludfordian intervals. Units are characterized lithologically on Fig. 4.

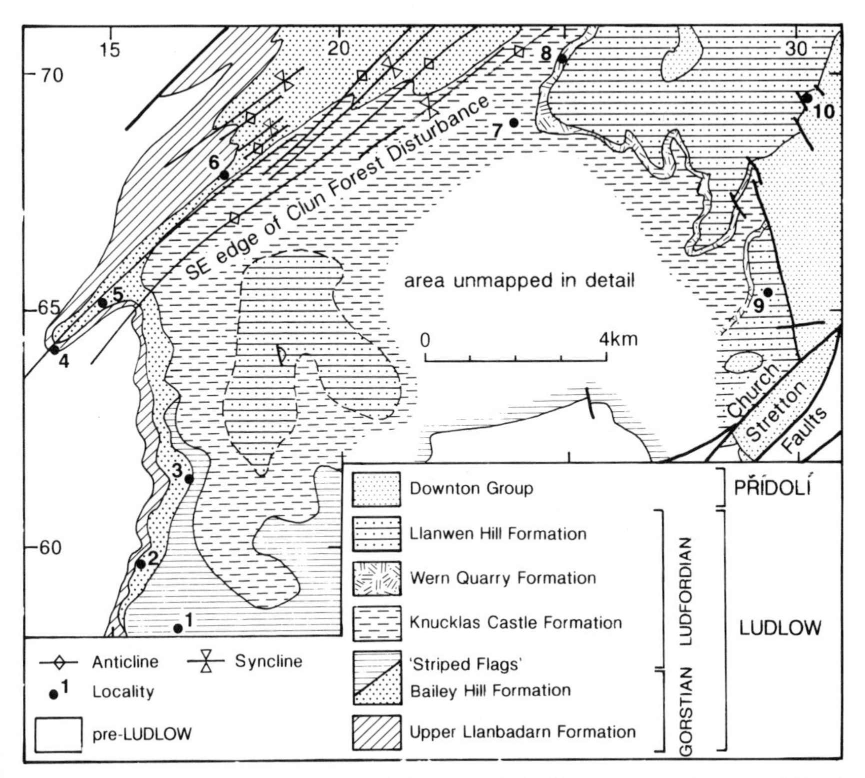

FIG. 2. Provisional geological map of the area of the itinerary, partly after Holland (1959) and Kirk (1947). See Fig. 1 for geographical setting.

At the western end of the transect the lithostratigraphy is essentially that of Holland (1959), with his 'Beds' accorded Formation status. This is an appropriate amendment to current practice, all these 'Beds' in the Knighton area being lithologically distinctive and mappable units. Holland's subdivisions of some of these units rely more heavily on essentially biostratigraphical criteria and are therefore not incorporated in the lithostratigraphical scheme. The Bailey Hill Formation has already been redefined and used by Tyler & Woodcock (1987). The underlying Llanbadarn Formation is that defined by Dimberline & Woodcock (1987), and characterized by the abundance of finely laminated carbonaceous hemipelagic mudstones. It spans the Wenlock–Ludlow boundary, and correlates with the Nantglyn Flags further north. Its upper part corresponds to the Lower Ludlow Graptolitic Shales of Holland (1959).

In the central part of the transect the informal term 'Striped Flags' (Kirk

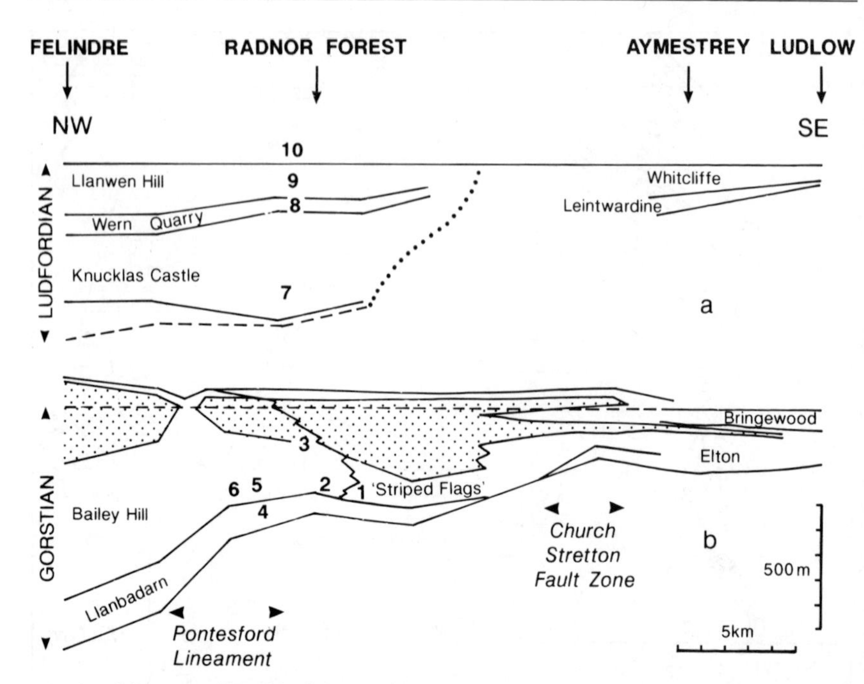

FIG. 3. Lithostratigraphical formations shown on basin-platform transects for a) Ludfordian and b) Gorstian rocks: vertical scale is restored stratigraphical thickness.

1947, 1951b) is retained for a large volume of mainly Gorstian rocks characterized by smooth, multi-compositional bedding-parallel lamination, but lacking discrete units of finely laminated hemipelagite of the Llanbadarn Formation. The 'Striped Flags' are the *Cyrtoceras* Mudstones and *Wilsonia* Shales of Straw (1937, 1953), and approximate to the 'laminated siltstone facies' of Holland & Lawson (1963) and the 'finely flaggy siltstone facies' of Bailey (1969). Although all sequences included within the 'Striped Flags' are superficially similar, they actually contain subtle facies variations. Linking these variations with the facies found in adjacent formations provides an important key to the interpretation of Ludlow palaeogeography. Tyler (1987) has made a case for a coherent 'Cwm Graig Ddu Formation' to include the 'Striped Flags', but this awaits formal definition.

The eastern part of the transect (Fig. 3b) is on the platform beyond the itinerary area, but shows the approximate correlation with the type Ludlow Series area. A debate about the lithostratigraphical viability of this scheme (Holland *et al.* 1963, Lawson 1971, 1982) remains to be resolved.

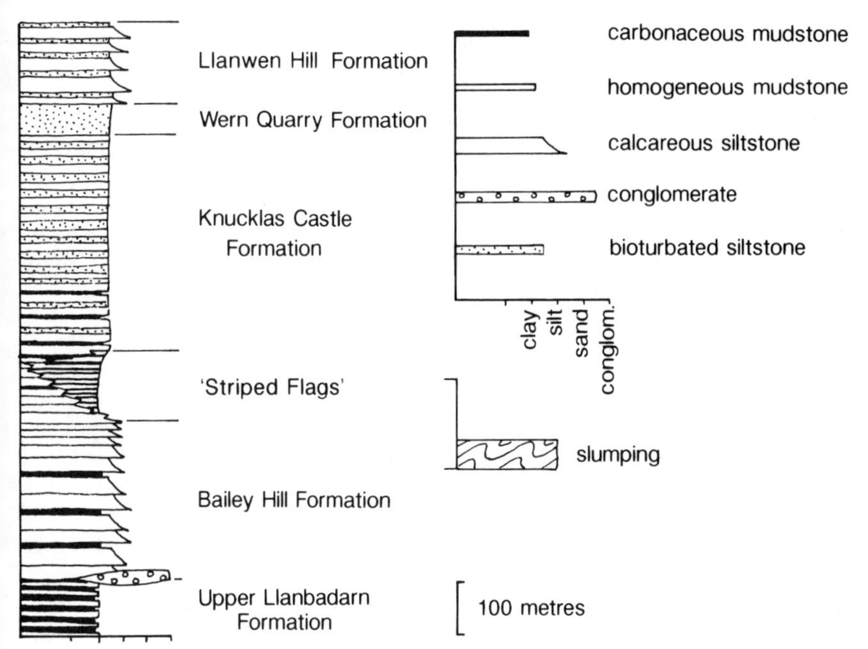

FIG. 4. Schematic log showing the lithological components of the Ludlow stratigraphy in the area.

Submarine slumping affects significant volumes of both the Bailey Hill Formation and the 'Striped Flags'. These slump sequences are shown as stippled areas on the stratigraphical transects (Fig. 3b), transgressing lithostratigraphic boundaries. Tyler (1987) has accorded them member status, but these remain to be defined formally.

FACIES DESCRIPTION AND INTERPRETATION

The various Ludlow facies can be specified (Fig. 4) in terms of five outcrop-scale primary lithological components.

a) *Carbonaceous mudstones* are dark grey to black units with the carbon either as mud-grade flakes or as silt-grade aggregates. The most conspicuous occurrence of this lithology is in the Llanbadarn Formation. Here it forms laminated units averaging 25 mm in thickness, containing alternating carbon-rich and quartz, silt-rich laminae with an average of three to four carbonaceous laminae per millimetre. The lamination was preserved due to lack of a burrowing infauna. The laminated facies is interpreted as the product of vertical fallout into bottom waters with low oxygen content (Tyler & Woodcock 1987; Dimberline & Woodcock 1987; Dimberline *et al.* 1990). The lamination reflects periodical, possibly approximately annual, variations in silt versus organic fallout.

Units of this laminated carbonaceous mudstone continue through the Bailey Hill Formation and up into the Knucklas Castle Formation. However, the carbon-rich laminae tend to be separated increasingly by thicker quartz-silt laminae, which eventually assume the status of discrete homogeneous mudstones (below). In the 'Striped Flags' the carbonaceous mudstones occur wholly as isolated laminae separated by these homogeneous mudstones.

b) *'Homogeneous' mudstones ± sandy siltstone bases* occur as 1–25mm thick units throughout the Gorstian and into the Ludfordian sequence. Although apparently homogeneous at outcrop, thin and polished sections reveal fine-scale sedimentary structures. Thicker units are conspicuously graded, a feature accentuated when based by a sandy siltstone. These bases may be parallel or ripple cross-laminated. The units are interpreted as the products of discrete depositional events, either turbidity flows or the residual components of storm-induced fluid-gravity flows (Fig. 6a; Tyler & Woodcock 1987; Dimberline & Woodcock 1987, Tyler 1987). The absence of infaunal bioturbation during the Gorstian preserves even the finest laminae of this facies. However, with the onset of bioturbation during the Ludfordian this facies is progressively obscured.

c) *Calcareous siltstones ± shell-rich bases* typically occur as 20–300mm thick units grading from laminated, calcareous, sandy siltstone up to poorly laminated or homogeneous muddy siltstone (Fig. 6a, b). Shell-rich bases to units enhance the grading where present. The sandy siltstones display a range of sedimentary structures from parallel lamination to migrant ripple-form sets. Convolute lamination occurs in many thicker units (Fig. 6b). Palaeoflow estimated from cross-lamination and ripple-crest orientations was between north and west. The bases to units show rare NE-directed flutes.

The calcareous siltstones have been interpreted widely as the deposits of discrete depositional events. The grading and sedimentary structure sequences are characteristic of deposition during a waning in flow intensity, indicating the progressive inability of the flow to carry and transport sediment. The facies is interbedded with units of hemipelagic sediment (Gorstian) or intensely bioturbated intervals (Ludfordian). This indicates that the facies was deposited from individual flow events, and that these events were widely separated in time.

The calcareous siltstones that comprise most of the Bailey Hill Formation were interpreted as turbidites by Cummins (1959a), Holland & Lawson (1963), Bailey (1964, 1969) and Woodcock (1976*b*). However, more recent work, summarised by Tyler & Woodcock (1987), has favoured a storm-influenced origin. The model envisages entrainment of sediment above storm wave base on the inner shelf, offshore transport in the storm-forced flow, and redeposition below storm wave base under the residual influence of this same flow. Sporadic wave ripple forms are present in the south-east of the area, but the storm interpretation rests more on the fine details of internal bed structure and on the lateral and vertical facies relationships (Tyler & Woodcock 1987; Tyler 1987).

Calcareous siltstones occur sporadically in the 'Striped Flags' and Knucklas Castle Formation. They again form a major component of the Llan-wen Hill Formation, where they have been interpreted as shallow marine in origin (Holland & Lawson 1963; Bailey 1969). Calcareous siltstones also occur in the main shelf sequence east of the Church Stretton Fault, although their presence is often obscured by intense bioturbation. The facies is most conspicuous in the Upper Elton Formation and the Whitcliffe Formation (Holland & Lawson 1963; Bailey 1969; Tyler 1987).

d) *Conglomerates* occur sporadically at the base of the Bailey Hill Formation. The distribution of this facies is extremely limited both stratigraphically and geographically, suggesting that a unique set of conditions led to their development at this horizon. The conglomerates comprise mostly pebble- and cobble-grade clasts of calcareous siltstones and mudstones in a matrix of shelly or sandy siltstone containing some black phosphatic pebbles. They occur in both matrix- and clast-supported lenticular beds up to 1.0 m thick (Fig. 7b). Between one and six beds are interbedded with shelly gravels and calcareous siltstones to form the conglomeratic sequence.

Individual cobble clasts have a variety of shapes, but two main suites can be identified: flat, rectangular, discoidal forms and ellipsoidal forms. Cross-sections through these clasts reveal them to be exhumed carbonate nodules nucleated on lithologies identical to those of the Llanbadarn Formation.

The conglomerates have been interpreted as slide conglomerates (Bailey 1964, 1969) or debris flows (Bailey & Woodcock 1976). An alternative favoured here is that the conglomerates are a condensed facies (Tyler 1987). A low sedimentation rate promoted early lithification and calcareous nodule formation. Erosion exhumed the nodules and concentrated them with newly formed phosphatic nodules and shell debris.

e) *Bioturbated siltstones* are almost entirely absent from the Gorstian sequence, but are increasingly common in the Ludfordian. They dominate the Wern Quarry Formation, where little trace of original lamination remains. In the Llan-wen Hill Formation depositional remnants of graded calcareous siltstones remain, but their tops and any intervening sediment are strongly bioturbated. This clear evidence of a burrowing infauna indicates oxic bottom waters and contrasts strongly with the prevailing anoxia through Gorstian time.

A sixth lithological feature is clearly of post-depositional origin.

f) *Slump sequences* comprise multiple units of the primary lithologies affected by deformation due to downslope submarine sliding (Fig. 7a). The geometry of typical structures is described below at Localities 1 and 3. There is no evidence that the slump sheets are strongly allochthonous. Tighter stratigraphical control and a greater resolution of facies types indicates that, contrary to previous interpretations (Bailey 1964, 1969; Woodcock 1976*b*), the slumps are not synchronous with the onset of the 'turbidite' deposition of the Bailey Hill Formation. Instead, they represent a discrete episode of mid Ludlow reactivation of elements of the Welsh Borderland Fault System (Tyler 1987), possibly on slopes of local rather than regional significance.

ITINERARY

The itinerary described here could be completed in a long day, but requires one and a half to two days for a proper study of all localities. Some localities **are not convenient for parties with a large coach.** A suggested one-day programme for them comprises Localities 2, 3, 6, 7 and 9.

Some good sections in this area occur in gullies and stream sections some distance from the road. Some of these are identified for the benefit of small parties with more time, although space does not permit detailed description.

The first locality can be approached from the east via Kington and the A44 road, from the north-west via Rhayader and the A44, or from the south-west via Builth Wells and the A481.

1. Pool House Quarry (SO 1656 5825). **Vehicles can be parked by the small lake** of Llynheilyn on the south side of the A481, about 600m south-west of its junction with the A44, or in the quarry itself on the north side of the A481. The quarry exposes the 'Striped Flags' of late Gorstian age, dipping gently south-eastward. This unit equates with the laminated siltstone facies of Holland & Lawson (1963) and the finely flaggy siltstone facies of Bailey (1969).

The main lithology here is pale grey, homogeneous mudstone (1–30 mm thick). The thicker units are often graded, with a thin (2–5 mm) base of laminated or microrippled clean sandy siltstone. The mudstones are interspersed with isolated laminae ($<<$1mm thick) of carbonaceous mudstone. As elsewhere in the 'Striped Flags' the facies sequences preserved in this quarry record the regular interbedding and amalgamation of thin, homogenous mudstone units.

The 'Striped Flags' crop out over a large area, between the shelf sequences to the south-east of the Church Stretton Fault Zone and the deeper water sequences to the north-west. However, it has been suggested that the sequence progressions within the 'Striped Flags' do not simply record the transition from inner shelf to further offshore, but also record an increase in sedimentation rate in each palaeogeographical setting during mid Ludlow times (Tyler 1987).

The quarry also displays a number of slump units (0.5–2.0 m thick) in which lamination is folded or destroyed due to deformation of sediment soon after its deposition. Slumping affects the upper part of the 'Striped Flags' extensively, and because of its lithological homogeneity tends to destroy the primary sedimentary structures. Transitional features at the base and the top of each slump unit are well displayed in this quarry, but better examples of internal structures can be examined in the Bailey Hill Formation at Locality 3.

The top of the 'Striped Flags' in this area is **exposed in a gully section in Cwm Crawnlwyn** just south of Beilibedw Mawn Pool (SO 1600 5640). This locality shows a transition to thicker laminated and more bioturbated mudstones equivalent to the Knucklas Castle Formation further north.

From Pool House Quarry take the A481 north-eastward to its junction with the A44. Turn left onto the A44 and drive north-west for about 1.2 miles.

2. Gwernargllwyd road cutting (SO 1558 5978 to SO 1583 5955, Fig. 5) is an extensive exposure on the north and east of the road. **Parking is available on the adjacent verge** at either end of the exposure. The Gorstian rocks dip gently eastward, and are all assigned to the Bailey Hill Formation. This has an upper *nilssoni* Biozone fauna at its base and an *incipiens* fauna at its highest horizon in this section (Tyler 1987).

The base of the Bailey Hill Formation is here marked by a 2 metre thick interval containing three lenticular units of cobble conglomerate, exposed near the northern end of the section (Locality **2a**, SO 1557 5975). The conglomerates are clast supported, with cobbles of calcareous siltstone up to 300 mm in diameter. The conglomerates interdigitate with silty shelly limestone and calcareous siltstones in a way that suggests amalgamation of thinner beds to form each conglomerate unit (Fig. 7b). Discrete layers of pyritized components and phosphatic clasts also occur in the sequence.

The basal units of the Bailey Hill Formation are marked by a line of old quarries running up the opposite hillside to the north (SO 1565 6010). Here conglomerates appear to be absent, but conglomeratic units occur sporadically in sections through this stratigraphical horizon over a wide area.

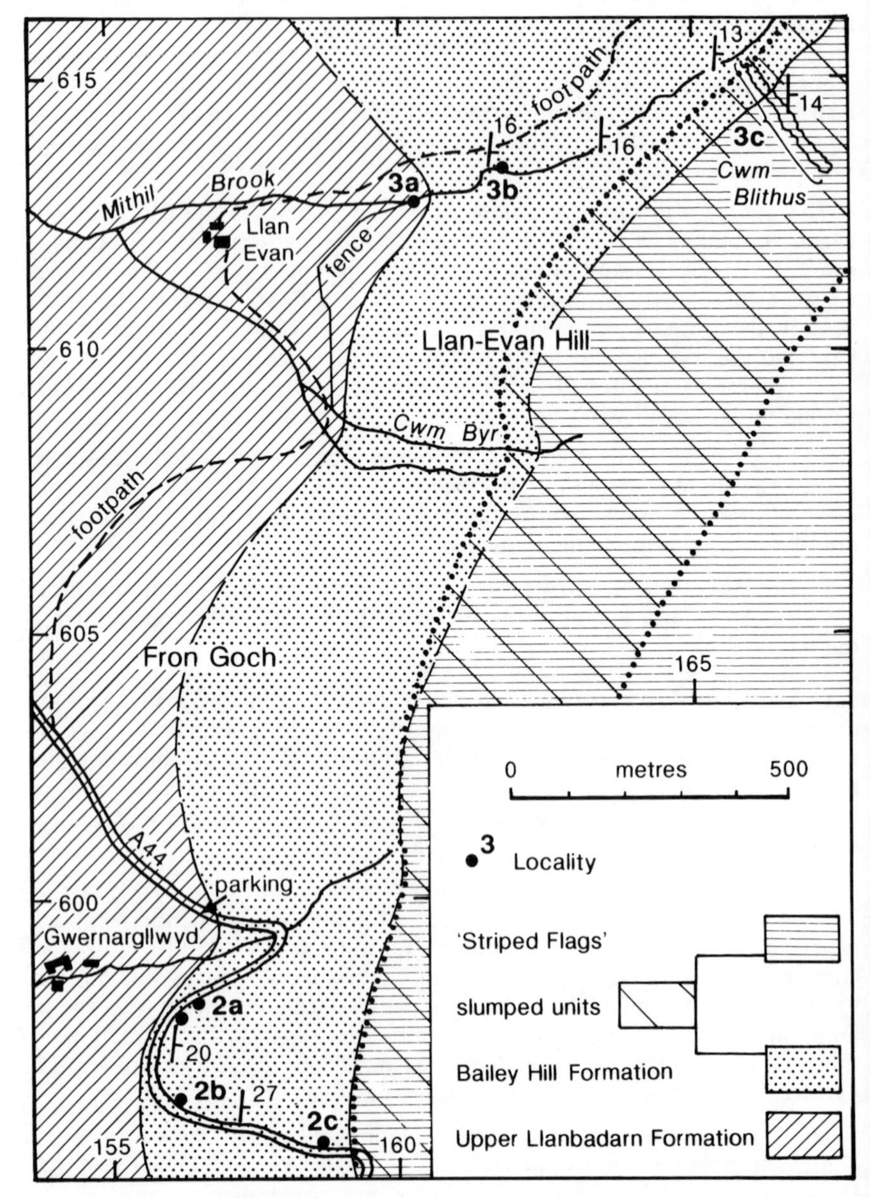

FIG. 5. Map of the geology around localities 2 and 3, after Bailey & Woodcock (1976) and Siveter *et al.* (1989).

FIG. 6. a) Calcareous siltstone bed (below lens cap) interbedded with laminated siltstones of the 'Striped Flags' facies; Locality 2B (SO 1560 5963). b) Calcareous siltstones showing convolute laminated lower parts and weakly parallel laminated upper parts; Bailey Hill Formation; Locality 5 (SO 1473 6516).

The conglomerates have been interpreted previously as slide conglomerates or debris flows (Bailey 1964, 1969; Bailey & Woodcock 1976). However examination of the contained clasts reveal them to be exhumed carbonate nodules enclosing the lithologies characteristic of the underlying Llanbadarn Formation (Tyler 1987). This and associated observations led Tyler (1987) to propose an origin by traction-current winnowing. On this interpretation the conglomerates are condensed deposits related to low- or negative-sedimentation rates.

Above the basal conglomerates the Bailey Hill Formation comprises mainly units of calcareous sandy siltstones grading up into homogeneous muddy siltstone (e.g. Locality **2b**, SO 1560 5962). These graded units average about 50mm thick and contain a common sequence of sedimentary structures. A typical unit has a lower, parallel-laminated division, a middle division of migrant ripple form sets, and an upper division of finely laminated to homogeneous mud (Fig. 6a). In rare cases ripples appear to have features characteristic of wave-generation: form symmetry, form-discordant lamination, and thin form-drape laminae. However, unidirectional traction currents dominated, with a palaeoflow average to the north-north-west (Tyler 1987).

In the lower part of the section there are occasional thin intervals of the finely laminated carbonaceous mudstones, interpreted as the products of hemipelagic fallout. In the upper half of the section the graded units become thinner bedded and more amalgamated to produce a facies transitional to the laminated siltstones of the 'Striped Flags' (e.g. Locality **2c**, SO 1583 5955).

The grading and sequences of sedimentary structures within individual units can be interpreted as the result of deposition from a single flow event. These features are characteristic of a progressive waning of flow strength and record the progressive decline in the flow's ability to transport sediment. The sediment-laden flows responsible for these units may have been turbidity currents or more complex sediment-gravity flows related to storm processes (Tyler 1987).

From Locality 2, head north-west on the A44 for about 500 m.

3. Mithil Brook and Cwm Blithus (SO 1600 6127 to SO 1677 6131, Fig. 5) expose an excellent section through the whole Bailey Hill Formation and the lower part of the overlying 'Striped Flags', both strongly affected here by submarine slump structures. The section is approached on a track leading north from the A44 at SO 1538 6030. **Vehicles can be left at SO 1555 6000.** Walk north and then east along the track, noting **small quarries** in Llanbadarn Formation mudstones. At the first valley, Cwm Byr, leave the track and follow the fence northwards

FIG. 7. a) Slump folds affecting calcareous siltstones of the Bailey Hill Formation; Locality 3C (SO 1662 6153). b) Conglomerate overlain by bedded siltstones; base of Bailey Hill Formation; Locality 2A (SO 1557 5975).

over the spur of Llan-Evan Hill and into the large valley of the Mithil Brook.

Mudstones of the Llanbadarn Formation, dipping gently east, are **exposed where the fence crosses the brook** (Locality **3a**, SO 1600 6127). Their age has not yet been verified here, but to the south at Locality 2, and 2 km to the north at Fron Las Dingle, the top of the Llanbadarn Formation yields upper *nilssoni* Biozone graptolites. The lithology is better examined at Locality 4.

Just upstream there is a rapid transition to the Bailey Hill Formation. The conglomerate exposed at this horizon at Locality **2a** is absent, and the formation base is marked by 100–200 mm thick beds of calcareous sandy siltstone, often with bases rich in transported shells. Further up-section (e.g. Locality **3b**, SO 1618 6124), the average bedding thickness is reduced to about 50 mm. In addition to ripple cross-lamination, Locality **3b** displays convolute lamination and sole marks. Palaeoflow was between north and west.

Further Bailey Hill Formation **exposures occur up the valley** as far as the bottom of Cwm Blithus (Locality 3C, SO 1662 6153), a dry gully on the south-east side of the valley. A few metres up the gully the calcareous siltstones are affected by spectacular slump structures (Fig. 7a) marking the base of a 200 m thick sequence dominated by soft sediment deformation. The whole sequence can be examined by ascending the gully, but a good impression of the geology can be had from the lower 50 m alone.

The slump sequence is described in detail by Woodcock (1976 *a,b*), and is summarised in Fig. 8. Slumped units are intercalated with units of unslumped sediment, typically on a scale of 2–10 m. The tops of some units show erosional truncation of folds below overlying sediment, demonstrating the synsedimentary origin of the deformation. Slump structures in the lowest 30 m of the sequence are particularly clear, because they are marked out by the thick graded units of the Bailey Hill Formation (Fig. 7a). The upper part of the sequence consists of more finely laminated lithologies included within the 'Striped Flags'.

The most conspicuous slump structures are folds, typically recumbent, with limb lengths from 10–80 cm and interlimb angles from 0–45°. Most folds have hinges trending NE–SW and are overturned towards the north-west, indicating formation on a north-west dipping slope. Many of the slump units show a lineation due to microfolds or microshears of the sedimentary lamination. This lineation is re-folded around some of the slump folds. Some recumbent folds have an apparent axial planar cleavage, but this is likely to be an effect of later compaction.

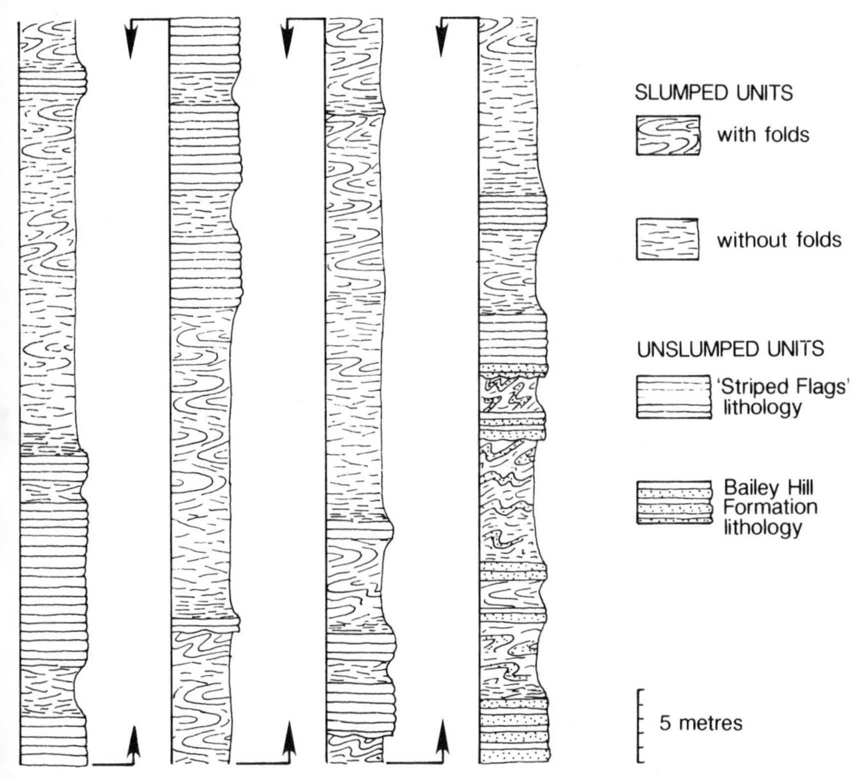

FIG. 8. Log of the slump sequence in Cwm Blithus (Locality 3C).

Additional exposures of the slump sequence crop out in the Mithil Brook **upstream from Cwm Blithus** (up to SO 1740 6190).

Return to the A44, noting the good views of the Ordovician Builth Inlier to the west. Drive 4 km north-north-west to Llandegley. Localities 4 and 5 are **not accessible to coaches**, which should be left in Llandegley. Smaller vehicles should turn right 200 m beyond the church. After 1.2 km turn left to The Pales (Friends Meeting House) and continue through the gateway beyond.

4. Meeting House Quarry (SO 1372 6407) is on the north side of the lane, 80 m west of The Pales. **Vehicles may be parked in the disused quarry.** Llanbadarn Formation mudstones dip very gently north-east, with the boundary to the overlying Bailey Hill Formation occurring a short distance above the top of the quarry section. Whereas graptolite faunas from the uppermost

Llanbadarn Formation usually indicate the upper *nilssoni* Biozone, here they indicate the *scanicus* Biozone (Tyler 1987).

The Llanbadarn Formation lithology here comprises homogeneous silty mudstones alternating with finely laminated carbonaceous siltstones. The mudstones are from 1–20 mm thick, sometimes with a basal unit of clean sandy siltstone. The sequence is interpreted as an interbedding of graded units deposited from turbidity currents or storm generated flows, and laminated siltstones accumulated as hemipelagic fallout. The preservation of fine lamination indicates deposition below an oxygen-deficient water column (Dimberline *et al.* 1990). By analogy with Recent examples, the laminae may reflect approximately annual variations in plankton productivity and clastic influx.

A low-diversity fauna occurs, mainly in the laminated siltstones. Graptolites dominate, with orthoconic nautiloids, bivalves, ostracodes, rare planktic crinoids, phyllocarid crustaceans and small brachiopods also present (Siveter *et al.* 1989 p.74).

Return down the lane to the junction at SO 1405 6423. Keep straight on through Rhonllwyn Farm. In 1km turn left, and continue for a further 0.5 km.

5. Cwm Ffrwd Quarry (SO 1473 6516) is on the north-east side of the lane, 300 m west-north-west of Kilmanawydd Farm. There is parking here for several small vehicles only; **coaches cannot reach this locality**. The quarry exposes the Bailey Hill Formation (Gorstian) dipping at 45° to the north-west. Graptolites from here suggest a low *incipiens* Biozone age. The boundary with the underlying Llanbadarn Formation (e.g. Locality 4) probably lies less than 50 m down section to the south-east.

The lithology here comprises mainly units grading from calcareous sandy siltstone, often with a shell-rich base, up to a muddy siltstone top (Fig. 6b). Bed thickness, averaging 100 mm and ranging up to 300 mm, is greater than in much of the formation in this area. In particular, bed thickness and the thickness of the whole formation decreases steadily to the south towards Localities 2 and 3. This facies change is taken to mark progessively more inshore shelf conditions.

The calcareous sandy siltstones display a wealth of cross-lamination and convolute-lamination. Short trains of climbing ripple form sets and isolated current ripple form sets record components of vertical and lateral bedform growth. Syndepositional erosion surfaces are common within event beds, demonstrating a greater complexity to the depositing flows than can be explained by a simple turbidity current mechanism. A component of storm-induced flow is suspected (Tyler & Woodcock 1987). Palaeoflow was between north and west.

The high dips at this locality mark the south-eastern edge of a 5 km-wide belt of folding and faulting, the Pontesford Lineament, contrasting strongly with the gently dipping rocks of Radnor Forest (Fig. 2). The lineament probably overlies a major basement fault belt which had a strong influence on sedimentation across the margin of the Welsh Basin.

Proceed northwards along the lane from Cwm Ffrwd Quarry to the junction with the A488. Turn left and head north-east on the A488 for 3.5 km.

6. Graig road cut (SO 1760 6800) lies on the south side of the A488, **opposite a lay-by where vehicles may be parked**. The exposure shows the lowest part of Bailey Hill Formation. It therefore provides an alternative to Locality 5, inferior though more accessible for parties with a large vehicle. The beds dip steeply south-eastward due to involvement in the upright angular folds of the Pontesford Lineament. Erosion into these NE–SW-trending folds is responsible for the prominent ridged topography that characterises this flank of Radnor Forest.

The road cut displays beds of calcareous siltstone, 50–200 mm thick, interpreted as the deposits of storm-influenced waning-flow events. Their sedimentary structures are detailed at Locality 5, although convolute lamination is uncommon here. There is one thin slump bed near the centre of the exposure, with folds overturned to the north-west. The base of the Bailey Hill Formation is drawn in the poorly exposed bank at the western end of the cutting, at the horizon where calcareous siltstones become much thinner and more sporadic (Bailey 1964, Bailey & Woodcock 1976).

Drive eastwards on the A488 for 7.5 km.

7. Monaughty road cut (SO 2393 6915) is on the east side of the A488, several hundred metres north of its junction with the B4356. Vehicles can be **parked in the lay-by on the west side of the road** here. The road cut exposes a 10 m thick section of the Knucklas Castle Beds (Ludfordian), the unit overlying the Bailey Hill Formation. The dip is gently to the east, the locality lying some 2 km south-east of the Pontesford Lineament. Graptolites have not been recorded in the Knucklas Castle Formation here, though *B. bohemicus* occurs in correlative rocks 9 km to the north-west (Holland & Palmer 1974). In the area of Locality 7, *leintwardinensis* Biozone forms are already present in the uppermost Bailey Hill Formation (Holland 1959).

The upward transition to the Knucklas Castle Formation is marked by a progressive thinning of the event units of the Bailey Hill Formation, a loss of shell-based units, an increase in mica content, and more conspicuous bioturbation. The Knucklas Castle Formation in the road cut has 10–50 mm thick event units of homogeneous mudstone, often with a basal division of parallel-to cross-laminated siltstone. Interbedded with these are finely laminated (1–3 mm) units comprising thin, continuous carbonaceous laminae separated by laminae of homogeneous mudstone. The facies is interpreted as hemipelagic sediment analogous to that in the Llanbadarn Formation (Locality 4), but with a higher periodic input of silt. Its preservation implies persistence of the low bottom water oxicity that prevailed during Bailey Hill Formation deposition, although common preservation of discrete-bedding parallel burrow systems suggests that oxicity was slowly improving.

A comparable section through the Knucklas Castle Formation crops out on the **forestry track above the A488** to the east (SO 2399 6910 to SO 2416 6950). The track leaves the road at the junction of the B4356 with the A488 (SO 2390 6860).

Drive to the north-north-east on the A488 for about 2 km.

8. Wern Quarry (SO 2495 7040) is on the east side of the A488. There is **limited parking space in the quarry**, and this is **not suitable for a large coach. Permission to view the quarry must be sought** from Wern Cottage at the north end of the quarry. This is the type locality for the Wern Quarry Formation (Holland 1959), here sub-horizontal. The unit is of Ludfordian age.

The quarry shows massive siltstones with bedding on a 0.2–1.0 m scale. This does not reflect depositional events, as the original lamination has been destroyed mostly by intense bioturbation. Only wispy remnants of lamination remain, together with the occasional thin (< 10 mm) continuous calcareous siltstone beds displaying parallel- or weak cross-lamination. The rapid upward transition from the underlying laminated Knucklas Castle Formation represents a marked improvement in bottom water oxicity and a resulting increase in benthic activity (Holland 1959; Tyler 1987).

Return southward on the A488 for about 2.5 km and then turn left on the B4356. Continue in the direction of Presteigne for 8 km.

9. Rock Cottage Quarry (SO 2935 6535) is on the south side of the B4356, just south of Rock Bridge over the River Lugg. **Vehicles can be parked on the verge by the quarry**. The western end of the exposure is on Rock Cottage land, but the main features of the geology can be seen adequately at the eastern end. A 40 m thick sequence correlating with the Llan-wen Hill Formation (Ludfordian) dips moderately south-westward.

The sequence is dominated by event units, averaging 150 mm thick, not unlike those in the Bailey Hill Formation. They comprise calcareous siltstones, weakly graded from shelly or sandy at the base to muddy at the top. They display both parallel- or ripple cross-lamination and hummocky cross-stratification in their lower parts, but their upper parts are massive or bioturbated. Convolute lamination is absent. A major difference from the Bailey Hill Formation is the absence of any laminated hemipelagic intervals, and the abundance of a benthic fauna, particularly brachiopods. Bottom water oxicity was obviously high, continuing the trend seen in the underlying Wern Quarry Formation. The depositional events are interpreted as storm-induced waning flows. A shallowing-up trend in the Llan-wen Hill Beds may be reflected by an increase in micaceous sandstones.

Drive eastwards on the B4356 for 1.8 km to its junction with the B4355 on the outskirts of Presteigne. Turn left and take the B4355 northward for 3 km to Norton. **Small vehicles only** should take the right turn 250 m beyond the church, bear left at the fork after 1 km, and continue up the lane for a further 1 km.

10. Meeting House Lane (SO 3009 6993 to 3030 6930) is the unmade track heading north-north-westwards from Meeting House Farm onto Llan-wen Hill. **Small vehicles can be parked on the verge** about 200m south of the farm. **This locality is inaccessible to coaches**. Parties must disembark either in Norton village or at the road junction on the B4355 at SO 2817 6974.

The lane exposes a moderately south-east-dipping sequence from the Llan-wen Hill Formation into the lower Downtonian units recognized by Holland (1959). Exposure is not good enough to display sedimentary structures very clearly, but the lithological transition can be studied. The Llan-wen Hill Beds at the northern end of the lane section are more micaceous than those at Locality 9, possibly reflecting a shallowing trend. Parallel- and cross-lamination can be seen along with hummocky bedforms. A colour change upward from greenish to yellowish grey marks the transition to the Yellow Downtonian. These flaggy, sandy siltstones form the raised banks of the lane just south of a barn. The abundant marine fauna of the Llan-wen Hill Beds is now reduced to rare lingulide brachiopods.

After about 10 m of section, a further colour change to greyish or yellowish green marks the base of the Green Downtonian. These are irregularly bedded calcareous siltstones, containing occasional calcareous nodules. The Green Downtonian passes up into the red and green siltstones and mudstones of the Red Downtonian, which underlies the extensive area of red-soiled ground from Norton north-eastward to Bucknell. However, exposure is very poor and no convenient outcrops can be suggested. Reeves Hill (SO 3180 6940) and Stonewall Hill (SO 3170 6860) have overgrown pits in the occasional sandstone units in the Red Downtonian, and display dry stone walls built from this material.

The Meeting House Lane section exposes the marine to marginal or non-marine transition brought about by oversupply of sediment to a Welsh Basin perhaps already subsiding less rapidly than before, and destined to shorten and be uplifted by the end of Early Devonian time.

Supplementary localities can be recommended through Ludlow units further to the north. They are geographically more dispersed and require a greater investment of time than for the basic itinerary described above. Details of some of these localities have been published elsewhere and are listed below.

11. Water-break-its-neck (SO 1818 7353 to 1834 7367): Llanbadarn Formation to Bailey Hill Formation (Earp 1940; Bailey 1964).

12. Beacon Hill (SO 1823 7655 to 1808 7672): Knucklas Castle Formation (Siveter *et al.* 1989, loc 4.5).

13. Ring Hole (SO 1200 8395 to 1240 8370): slumped facies (Bailey & Woodcock 1976).

9. THE PRECAMBRIAN AND SILURIAN OF THE OLD RADNOR TO PRESTEIGNE AREA

by N. H. WOODCOCK

Maps	*Topographical:*	1:50 000	Sheet 148 Presteigne & Hay-on-Wye
		1:25 000	Sheets 971 (SO 26/36) Presteigne and 993 (SO 25/35) Kington
	Geological:	1:250 000	Mid Wales and Marches

THIS itinerary covers the area in Powys that best exposes the Church Stretton Fault System. Fault-bounded slivers of Precambrian igneous and sedimentary rocks have been uplifted through Silurian cover in the Old Radnor area. The Precambrian sediments are overlain by lower Wenlock limestones, contrasting with the upper Wenlock mudstones outside the fault zone. The stratigraphical relationship of these Silurian facies is seen in the Nash area near Presteigne. Here the mudstones overlie the limestones, which are underlain by upper Llandovery sandstones in the core of an anticline in the fault zone. Throughout the area there are excellent exposures of faults in the Church Stretton system, mainly steep strike-slip faults of post-Wenlock and probably post-Přídolí age.

GEOLOGICAL SETTING AND STRATIGRAPHY

The area (Fig. 1) straddles the Church Stretton Fault Zone, the south-eastern element of the Welsh Borderland Fault System (Woodcock & Gibbons 1988). This system controlled the Early Palaeozoic transition from the shallow facies of the Midland Platform in the south-east to the deeper facies of the Welsh Basin to the north-west. Repeated later movements on the Church Stretton zone, in particular of Acadian (late Caledonian) age, have uplifted the shallow-buried Precambrian basement. Two composite fault-bounded inliers have resulted.

The Old Radnor Inlier exposes almost wholly sedimentary rocks. These can be subdivided into the dominantly sandy Strinds Formation and the finer Yat Wood Formation (Fig. 2; Woodcock & Pauley 1989). Although direct evidence of their age is lacking, the facies of both formations have analogues in parts of the type Longmyndian sequence further north and are assumed to be of Precambrian age. The two formations are in fault

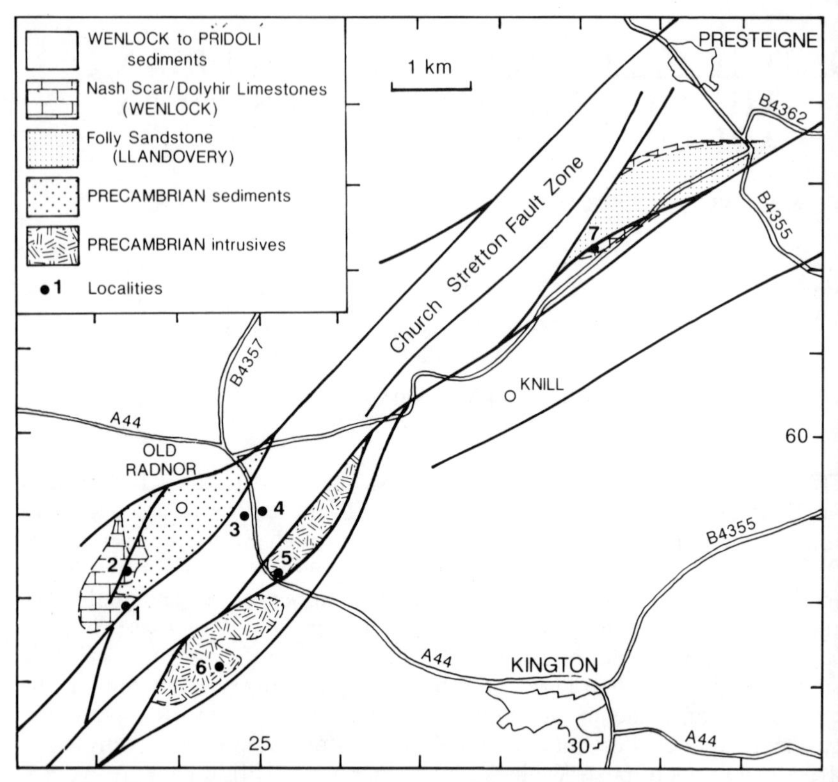

FIG. 1. Location map for the itinerary with outline geology.

contact in the Old Radnor Inlier and only the match with the type sequence suggests that the Yat Wood Formation may be the older. The same analogy implies a braided alluvial environment for the Strinds Formation and an alluvial flood plain or subaqueous delta environment for the Yat Wood Formation, though brecciation of most of the Old Radnor sequences precludes direct diagnosis.

The composite Stanner-Hanter Inlier comprises wholly intrusive igneous rocks, mostly dolerites and gabbros, but with minor acid components. Field relationships are not well-exposed, but Holgate & Hallowes (1941) deduced the following intrusive sequence: 1. dolerites intruded into an unknown country rock, probably as thin sills; 2. gabbros intruded into the dolerite complex, possibly as a laccolith; 3. fine granitic or granophyric textured acidic rocks; 4. quartz-porphyritic acidic rocks and dolerite dykes of uncertain relative age. All the intrusive rocks have suffered strong later alteration. An Rb-Sr whole-rock age of 702±8Ma (Patchett *et al.* 1980)

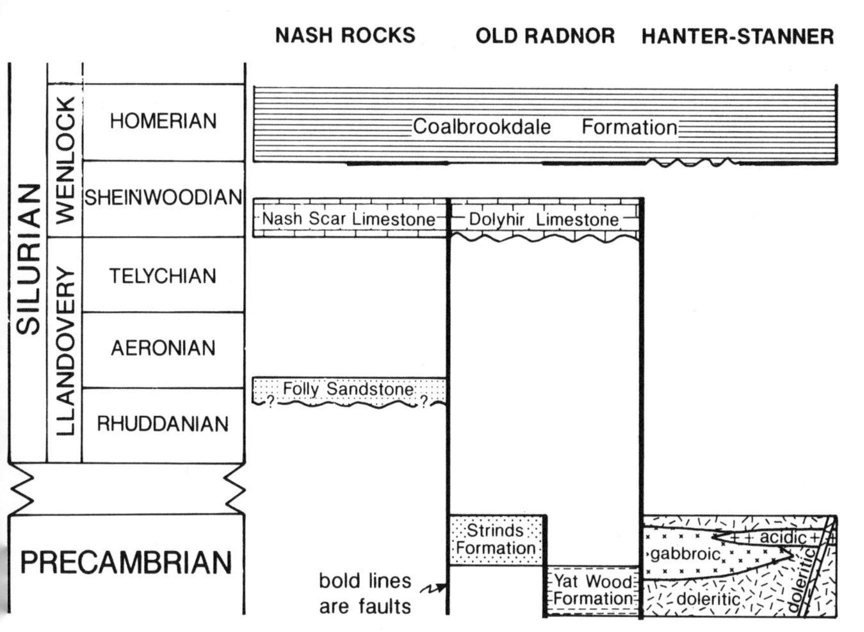

FIG. 2. Stratigraphy of the Old Radnor–Presteigne area. Wavy lines are angular unconformities.

confirms the Precambrian origin of the igneous sequence, and is the oldest date yet obtained from any part of the Precambrian basement in England and Wales (Thorpe *et al.* 1984).

Silurian sediments rest with angular unconformity on the Precambrian. This relationship is exposed clearly in the Old Radnor inlier (Garwood & Goodyear 1918) and strongly implied in the Stanner-Hanter inlier (Holgate & Hallowes 1941). In these localities the basal Silurian is probably of Wenlock age. Older, Llandovery age sediments are exposed in the Presteigne area to the north-east, although here the base is not seen. Disconformities may separate the three lithological components of the Silurian sequence (Fig. 2).

The Folly Sandstone contains a fauna of latest Aeronian to earliest Telychian (Llandovery) age (Ziegler *et al.* 1968). It is a thickly-bedded, moderately well-sorted sandstone with pebbly or shelly bases to beds in places. A brachiopod fauna of the *Pentamerus* Community, vertical burrows and sporadic hummocky bedforms suggest a shallow marine environment.

The overlying Nash Scar Limestone has yielded only early Sheinwoodian (Wenlock) faunas, so that most of the Telychian could be missing below it. It is seen in sedimentary contact above the Folly Sandstone in Nash Scar Quarry. No angular discordance is seen here, though a weak discordance was formerly visible near Presteigne (Ziegler *et al.* 1968). The lateral equivalent of the Nash Scar Limestone in the Old Radnor area, the Dolyhir Limestone, contains a fauna of a similar early Wenlock age. Both lithologies are massive, crystalline limestones, rich in algae and bryozoa, and containing brachiopods, trilobites and a conodont microfauna. They represent a shallow turbulent environment, possibly on a local submarine topographic high due to already partly uplifted basement slivers in the Church Stretton Fault Zone, or possibly part of a more extensive carbonate shelf continuous with the Woolhope Limestone in the inliers of the southern Borderland.

Wenlock mudstones assigned to the Coalbrookdale Formation (Siveter *et al.* 1989) overlie the Nash Scar Limestone at the eastern end of Nash Scar Quarry. Here they are of earliest Homerian (late Wenlock) age (Hurst *et al.* 1978). The time gap implied above the youngest dated limestone, spanning the later Sheinwoodian, may be represented by a hard ground at the top of the limestone. Elsewhere the limestone/mudstone contact is unexposed or faulted. Earlier mudstones, of Sheinwoodian age, may occur in places (Kirk 1951; M. G. Bassett 1974). The mudstones contain a deeper-marine brachiopod fauna than the limestones or sandstones. This muddy marine shelf characterized the later part of Wenlock time and persisted into the Ludlow.

Later Ludlow and Přídolí time saw the progressive elimination of marine conditions as the Welsh Basin to the north-west was oversupplied with sediment prior to basin shortening, deformation and uplift. The sedimentation history of the area to the north-west of the Church Stretton zone at this time is covered in Excursion 8.

STRUCTURAL GEOLOGY

The fragmentary stratigraphical scheme in the area (Fig. 2) is due largely to repeated displacements on its major structure, the Church Stretton Fault Zone. Outside this zone to both the north-west and south-east the rocks are only gently warped and cut only by faults with small displacements. The area lies just south-east of the front of strong Acadian deformation, and well to the north of the Variscan Front. But the fault zone was active in response to stresses of both these episodes, presumably due to the localizing effect of a pre-existing fracture in the underlying basement (e.g. Owen & Weaver 1983, Woodcock 1988).

The fault zone is now marked by the presence of basement slivers, localized folding, and a high density of faults including some with kilometre-scale displacement. The majority of exposed faults in the Old Radnor Inlier (Fig. 3) cut at least the Dolyhir Limestone (lower Wenlock) and have a dominantly strike-slip displacement (Woodcock 1988). The fault pattern

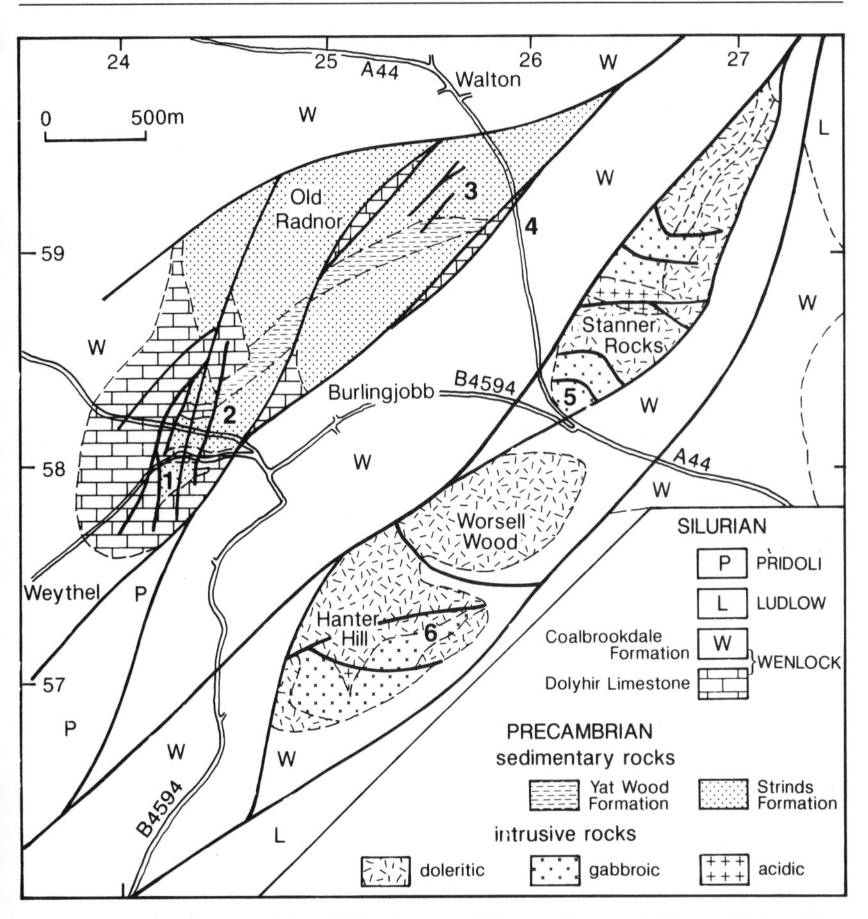

FIG. 3. Geological map of the Old Radnor and Hanter-Stanner inliers, with localities described in the text. After Holgate & Hallowes (1941), Kirk (1947), Woodcock (1988).

suggests sinistral displacement on the main NE–SW faults, probably during the Acadian (late Caledonian) climax in early to mid Devonian time. However an earlier episode of deformation is indicated by the strong angular unconformity of the Dolyhir Limestone on the underlying Precambrian. Later deformation, of Variscan age, is proved on the continuation of the Church Stretton line to the south-west (Owen & Weaver 1983) and cannot be ruled out here.

The Nash Scar quarry also exposes an abundance of post-Wenlock strike-slip faults, and suggests that the history deduced from the Old Radnor area may be applicable further along the fault belt.

ITINERARY

This itinerary begins with the Precambrian sedimentary rocks, where their relationship to the overlying Wenlock limestones and surrounding Wenlock mudstones can best be seen. Localities in the more poorly-exposed Precambrian igneous rocks follow. The excursion is completed by localities in a more continuous Silurian sequence from sandstones through limestones up into mudstones. The full itinerary will take about a day and a half. A convenient day trip would comprise localities **1, 2, 4, 5** and **7**. All localities are **accessible to a party with a large coach**, but such parties **should not be taken** into the quarry at Locality **5**.

The first locality is approached from the A44. Turn west on to the B4594 at SO 2620 5820. After 1.5 km turn right and then first left. **Smaller vehicles can be left in the car park** of Nash Rocks Ltd. by arrangement (SO 2433 5807). **Permission** to visit localities **1, 2** and **7** must be sought from the quarry office to the west of the car park. **Prior permission** should be obtained for parties from Nash Rocks Ltd., P.O. Box 1, Kington, Herefordshire, HR5 3LQ.

1. Strinds Quarry (entrance at SO 2443 5802) is at the southern end of the Old Radnor Inlier (Fig. 3). The lower faces expose the sandstones and pebbly sandstones of the Precambrian Strinds Formation, seen to be dipping steeply north-westward where bedding is not obscured by brecciation. These are overlain unconformably by the gently dipping Dolyhir Limestone of Wenlock age, exposed on the upper faces, with a basal rudite facies present in places.

The whole sequence is strongly faulted. Predominant are major NNE-striking faults dipping west-north-west with normal offsets (west-north-west downthrow). Lineations on the fault planes show that these offsets have arisen mainly from strike-slip rather than dip-slip displacement, though the sense is indeterminate or ambiguous. Less continuous NW-striking faults also show strike-slip slickenfibres, here with a more consistent dextral sense. They are interpreted as conjugate to the main NNE-striking faults, implying that these had a sinistral sense (Woodcock 1988). Minor dip-slip faults strike between north-north-east and north-east, some following bedding in the Precambrian.

The lithological and structural features can be seen at numerous places throughout the quarry and the most profitable route will depend on the state of working at the time. Sandstones of the Strinds Formation are exposed near the entrance to the quarry (Locality **1a**, Fig. 4). Traced along the lowest level of the south-eastern face, it becomes generally more pebbly (Locality **1b**). The western face of the lower level is formed by one of the

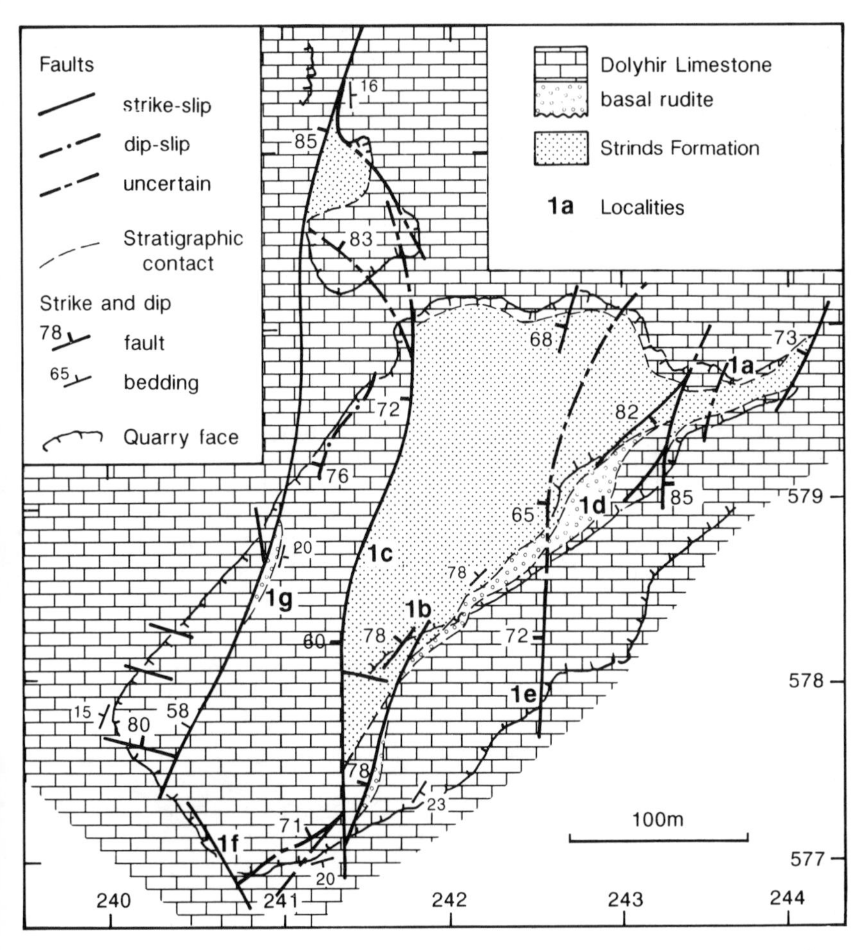

FIG. 4. Geological map of Strinds Quarry (Locality 1) with faces shown in their 1985 position. Key also applies to Figs. 5 and 6. After Woodcock (1988).

major strike-slip faults, and brecciation and fault plane structures can be examined (Locality **1c**).

The basal rudite to the limestone is exposed on the ramp to the upper level of the quarry on its south-east side (Locality **1d**). The rudite has angular sandstone and mudstone clasts of Precambrian derivation, together with more rounded quartz pebbles. The matrix contains pelmatozoan debris and favositid corals in growth position, and there is abundant calcite cement. The limestone itself is easily examined anywhere in the upper faces of the quarry. It contains abundant calcareous algae in white weathering, irregular nodular forms upwards of a centimetre in diameter. The algae have a porcellanous texture which contrasts with the coarsely crystalline texture of the remaining limestone. Also present are bryozoans and sporadic favositid corals and brachiopods.

Faults of all sets are well displayed. For example a N-striking strike-slip fault is seen on the south-east face (Locality **1e**) and a set of NW-striking faults parallels the south-west face (Locality **1f**). Another sliver of basal rudite is brought up against a fault in the floor of the upper bench at Locality **1g**.

From Strinds Quarry return towards the car park and walk northwards through the works complex to the inclined entrance road to Dolyhir Quarry.

2. Dolyhir Quarry (entrance at SO 2429 5820) duplicates many features of Strinds Quarry, but additionally displays the finer Precambrian sediments

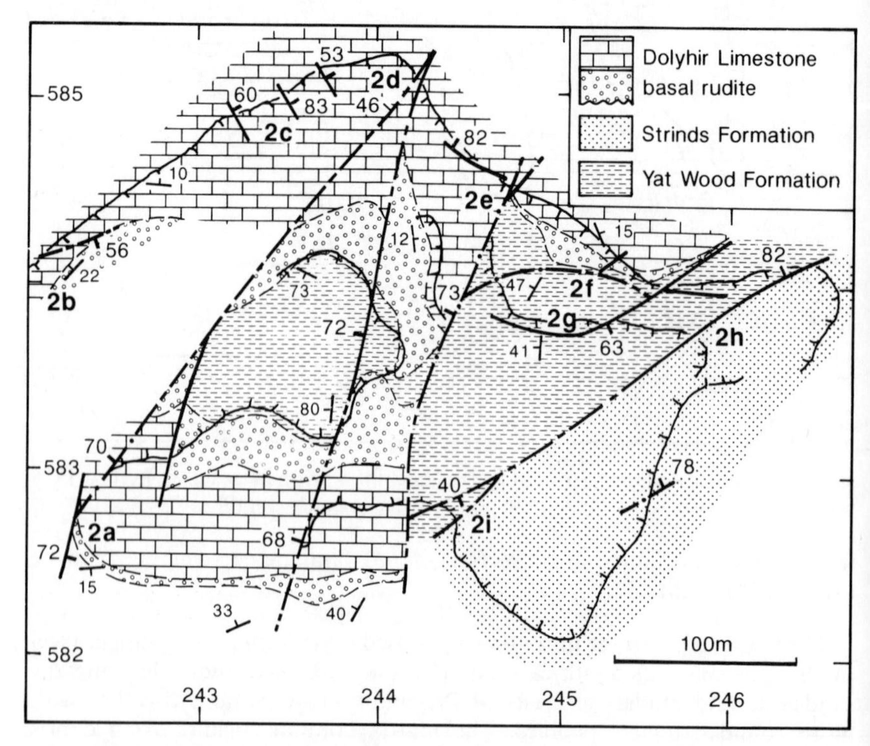

FIG. 5. Geological map of Dolyhir Quarry (Locality 2) with faces shown in their 1985 position. Key as for Fig. 4. After Woodcock (1988).

of the Yat Wood Formation. The quarry is worked intensively, producing fresh outcrops but a continually changing detailed outcrop pattern. Figure 5 shows faces in their 1985 positions. The suggested route may have to be modified to suit the prevailing configuration of faces.

The basal rudite to the Dolyhir Limestone is exposed in abandoned faces near the quarry entrance (Locality **2a**) and on its north-west side (Locality **2b**). Followed north-eastward, this face provides good exposure of the limestone, coarsely crystalline with abundant calcareous algae and bryozoans. The face is traversed by a number of steep NW-striking faults (e.g. Locality **2c**) marked by local brecciation and surfaces with fibrous calcite growths. These indicate dextral strike-slip displacements. At the northern corner of the quarry (Locality **2d**) a nodular shale horizon in the limestone is brecciated against a NW-dipping fault. Correlatives of this shale have formerly yielded abundant trilobites, gastropods, brachiopods, corals and crinoids (Garwood & Goodyear 1918).

The perimeter face can be followed south-eastward to the point (Locality **2e**) where a steep NNE-striking fault upthrows the unconformity surface again by about 30 m. The fault has indications both of normal dip-slip and of sinistral strike-slip. The angular nature of the unconformity can be seen in the upper faces on the east side of the fault (Locality **2f**), with the gently north-dipping rudite overlying Precambrian sediments dipping west-north-west. These sediments belong to the Yat Wood Formation, and can best be viewed in the northern face to the main bench in this eastern half of the quarry (Locality **2g**). They comprise fine-grained sandstones, siltstones and laminated mudstones and are conspicuously well-bedded by comparison with the Strinds Formation.

A NE-striking fault zone to the east (Locality **2h**) faults the Yat Wood Formation against the sandstones and pebbly sandstones of the Strinds Formation, which outcrop throughout the south-eastern corner of the quarry. The same fault contact can be seen at Locality **2i**.

A return should be made using the ramps on the north side of the quarry. The lowest worked levels usually show worthwhile exposures of Yat Wood Formation overlain by the basal Wenlock rudite.

Vehicles should now be collected and taken eastward on the B4594. Turn north at the junction with the A44 and in less than a kilometre turn left into Gore Quarry. **Small vehicles may be parked in the quarry car park by arrangement**, but a coach may need to be left at the northern end of the quarry slip road.

3. Gore Quarry (entrance at SO 2589 5915) is worked by Tilcon Ltd. from whom **permission for access should be sought**, in advance for large

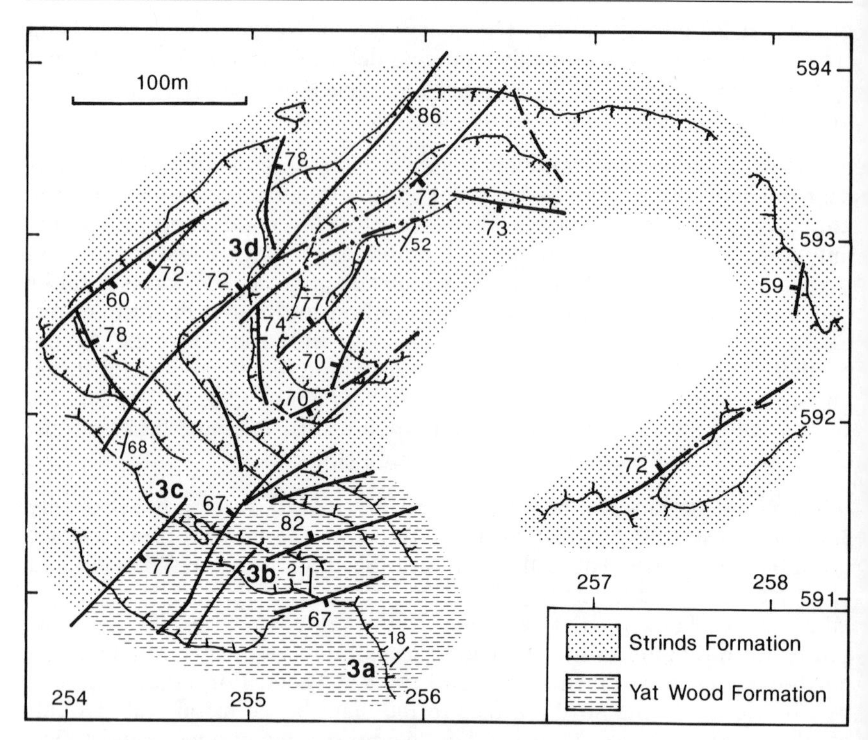

FIG. 6. Geological map of Gore Quarry (Locality 3) with faces shown in their 1985 position. Key as for Fig. 4. After Woodcock (1988).

parties. The quarry exposes only Precambrian rocks and is therefore of less general interest than localities **1** and **2**. Those interested in the fault pattern in the Old Radnor Inlier will find useful examples here, assumed to be dominantly of post-Wenlock age as at the two southern quarries. Strike-slip faults strike north-north-west and north-north-east (Fig. 6). The north-north-west faults show dextral slip indicators and are taken to be conjugate to the more continuous north-north-east set, assumed to be sinistral. Dip-slip faults strike NE–SW.

The character of the faults can be seen in any part of the quarry. They are perhaps best displayed in the Yat Wood Formation which occupies the southern part of the quarry. The faces in the extreme southern corner (Locality **3A**) show the fine sandstones and mudstones of the Yat Wood dipping gently westward. Followed north-westward the dip steepens and they are cut by a series of steep NE-striking faults, some with strike-slip indicators (e.g. Locality **3B**). The coarser Strinds Formation is faulted in further west (Locality **3C**) and occupies all the central and northern parts of the quarry. Complex relationships between the conjugate strike-slip faults can be seen here, and the interaction with a dip-slip strand is displayed around Locality **3D**.

Locality **4** is adjacent to the entrance to Gore Quarry. If Locality **3** has been omitted, **vehicles should be parked at the north end of the quarry slip-road or on the nearby verge to the A44** (SO 2592 5921).

4. Gore road cut (SO 2595 5920 to 2596 5899) displays the Wenlock mudstones, assigned to the Coalbrookdale Formation, which surround the Precambrian inliers and occupy much of the low unexposed ground in this area. The north end of the cut is within 100 m of the bounding fault on the south-east of the Old Radnor Inlier (Fig. 3). The mudstones are best seen on the eastern side of the A44 road. They are grey, weakly laminated mudstones with sporadic, 2–5 cm thick, fine sandstones and calcareous nodules. They dip moderately north-eastward, but are sheared and veined in places, presumably due to the same post-Wenlock fault episode seen in the inlier quarries. The mudstones in this area range upwards from the *rigidus* Biozone of the lower Wenlock (M. G. Bassett 1974) and are thought to be offshore shelf deposits (Hurst *et al.* 1978).

From Gore return south on the A44 to the junction with the B4594. **Cars can be parked on the verge** of the B4594 to the west of the junction.

5. Stanner Rocks Quarry (SO 2615 5828), on the north-east side of the A44 here, is the most convenient exposure of the basic rocks that make up much of the Stanner-Hanter Inlier. The quarry is, however, a botanical Site of Special Scientific Interest. Individuals and small parties should only visit the quarry floor and **should not hammer the outcrops** or ascend the slopes or faces. Large parties should not be taken into the quarry at all. A small roadside exposure is available immediately to the north-west (SO 2610 5835).

The locality shows various dolerites and gabbros, mostly highly sheared and altered. The primary mineralogy was plagioclase + augite but amphibolitization is extensive (Holgate & Hallowes 1941).

The Stanner ridge can be approached by the footpath along its south-east flank, but the afforestation makes access to most outcrops difficult. Small parties wanting to see more of the Precambrian igneous rocks are advised to return to the road junction and take the footpath south-westward skirting the north-west side of Worsell Wood. At Lower Hanter Farm, which can also be approached from Burlingjobb (SO 2505 5821), take the track south-eastwards to the east spur of Hanter Hill.

6. Hanter Hill (summit at SO 2520 5709) is composed mostly of dolerites and gabbros, extensively sheared and altered. Gabbros can be seen in a ENE-trending tongue along its eastern spur (Fig. 3). In places along its northern contact with the dolerites (e.g. SO 2530 5719) the gabbro has xenoliths of the dolerite and is clearly later (Holgate & Hallowes 1941). Later acidic intrusives can be seen just west of Hanter Hill summit (SO 2514 5702).

Vehicles should now be collected. Drive 1.4 km northwards on the A44 to Walton and turn right on the B4362 towards Presteigne. After 6 km park on the left of the road, beside the old crushing plant for Nash Scar Quarry (SO 3035 6227).

7. Nash Scar Quarry (entrance at S0 3024 6220) is operated by Nash Rocks Ltd. (address as for Localities **1** and **2**) from whom **permission for access must be obtained**. The quarry previously exploited the Nash Scar Limestone on the south-east limb of a gently SW-plunging anticline (Fig. 7). A recent extension has cut into the underlying Folly Sandstone occupying the core of the structure. The sandstone/limestone contact is faulted on the south-east limb but is seen intact on the north-west limb. The overlying Coalbrookdale Formation mudstones are seen on the south-east limb.

The Nash Scar Limestone can be examined at the south-west end of the old face (Locality **7A**), dipping gently south-eastward. The lithology is similar to the Dolyhir Limestone at Strinds and Dolyhir Quarries (Localities **1** and **2**), with calcareous algae and bryozoans dominant in a coarsely crystalline groundmass. Minor strike-slip faults here strike NW–SE. Followed

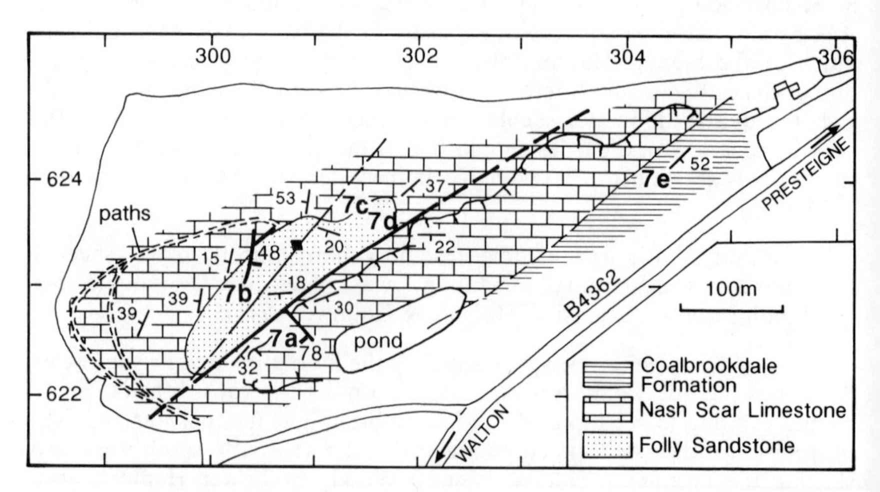

FIG. 7. Geological sketch map of Nash Rocks Quarry (Locality 7). Faces shown in their 1990 position.

downsequence to the north-west, the limestone is cut out by a prominent NE–SW fault. This dips north-westward and faults up the thick-bedded pebbly sandstones of the Folly Sandstone. These have yielded the brachiopod *Eocoelia hemispherica* in this area (Ziegler *et al.* 1968) indicating a latest Aeronian or earliest Telychian age (Llandovery). The sandstones dip south-eastward near the fault but followed across strike the dip shallows and reverses over the hinge of the plunging anticline. On the north-west face of the quarry here (Locality **7B**) an intact transition from the Folly Sandstone to the Nash Scar Limestone is preserved, though it is still cut by minor steep strike-slip faults.

The sandstone–limestone contact can be traced along the face to the south-west. To the north-east it gains height up-plunge on the anticline, and can be examined from the uppermost bench of the quarry. If this bench is not accessible from the quarry floor, it must be approached by the track around the west end of the quarry. The upper bench is approximately at the level of the base of the limestone, and remnants of the limestone closing over the anticline hinge can usually be seen at its north-east end (Locality **7C**). The south-east limb of the fold is again faulted here (Locality **7D**).

Descend on the path around the west end of the quarry, and regain the floor of the old limestone quarry (near Locality **7A**). The limestone face can be followed north-eastward; notice a number of fault planes with strike-slip slickensides. At the north-east end of the quarry (Locality **7E**) a sequence of the Coalbrookdale Formation nodular mudstones overlies the limestone. The upper surface of the limestone is a hard-ground (Hurst *et al.* 1978), suggesting a significant time gap in the later Sheinwoodian between carbonate deposition and the onset of muddy sedimentation in the Homerian (M. G. Bassett 1974).

A return can now be made south-west to the A44 (6 km) or north-east to Presteigne (3km).

10. ORDOVICIAN IGNEOUS ROCKS OF THE BUILTH INLIER

by R. E. BEVINS *and* R. METCALFE

Maps *Topographical*: 1:50 000 Sheet 147 Elan Valley and Builth Wells
1:25 000 Sheet SO 05/15 Builth Wells
Geological: 1:25 000 Llandrindod Wells Ordovician Inlier

ORDOVICIAN lavas and related pyroclastic and epiclastic rocks are exposed in a SW–NE-trending outcrop across the Builth Inlier, from the extensive quarries at Llanelwedd in the south, to Llandegley Rocks in the north (Fig. 1). These lithologies are responsible for much of the rather rugged, hilly ground comprising the Carneddau Range. Faunal evidence indicates a Llanvirn age for the volcanic activity, restricted principally to the *Didymograptus murchisoni* Biozone, although stratigraphical relationships indicate more than one phase of volcanism. It is likely that some of the basic and intermediate intrusions exposed across the inlier also belong to these Llanvirn volcanic episodes, although the fact that certain intrusions invade Llandeilo age strata suggests that igneous activity was not confined to Llanvirn times. This view is supported by palaeomagnetic data, which show that intrusions cutting Llandeilo strata have reversed remanence relative to those intruding Llanvirn strata (Piper & Briden 1973).

In the 1940s this volcanic succession was investigated and described extensively by O. T. Jones and W. J. Pugh (Jones & Pugh 1941, 1946, 1948a, b, 1949), studies that provided the basis for what must be considered as one of the earliest detailed palaeogeographical reconstructions (Jones & Pugh 1949). In their now classic model Jones and Pugh identified a fossil shoreline, with related beach deposits, sea stacks, and fault-defined cliff lines.

Recent studies of the volcanic succession of the Builth Inlier are few. Furnes (1978) provided a reinterpretation of the volcanic geology of the area as well as revising the stratigraphy, although unfortunately this work remains unpublished. Bevins *et al.* (1984) and Kokelaar *et al.* (1984) presented brief accounts of the petrography and geochemistry of the various lavas and intrusions, synthesising data from Furnes (1978) as well as from their own unpublished results. These studies concluded that the various magmas erupted were of calc-alkaline affinity with basic,

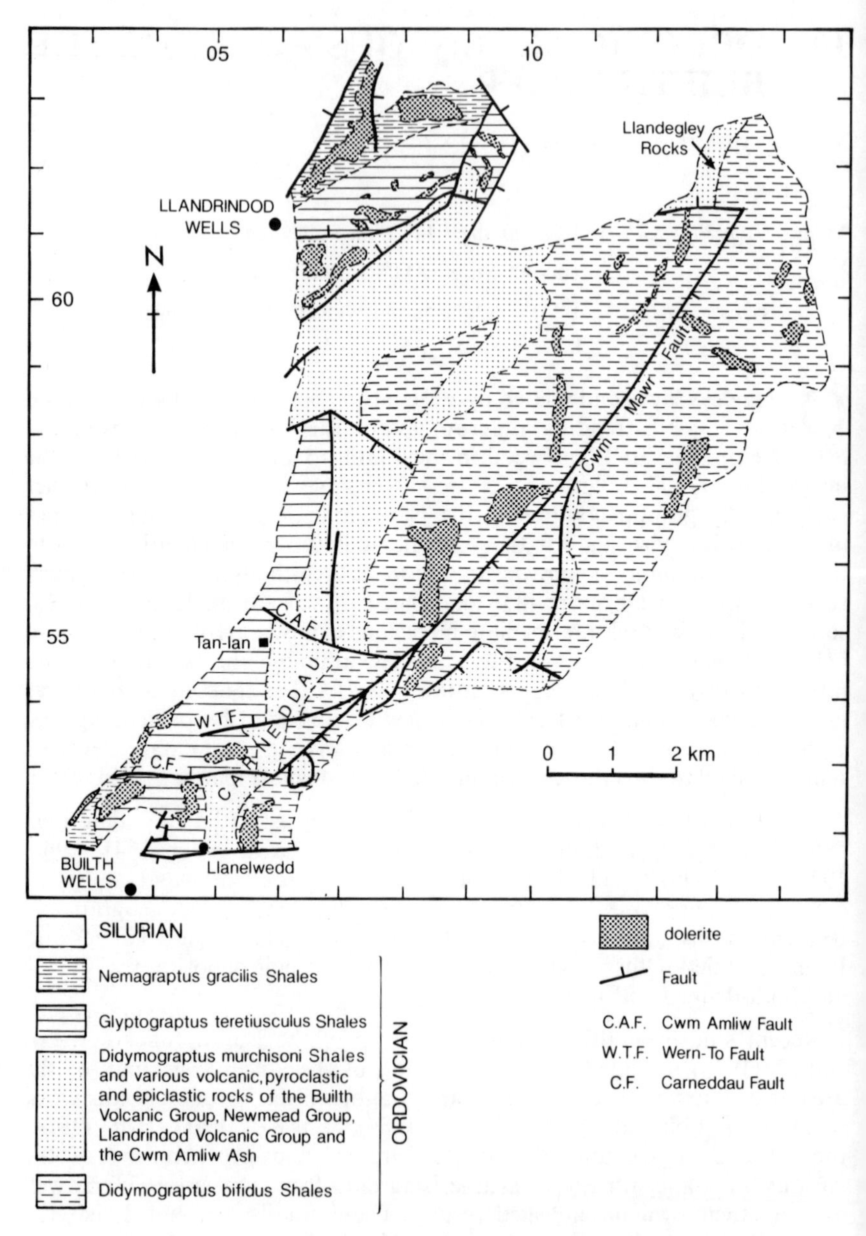

FIG. 1. Simplified geological map of the Ordovician rocks of the Builth Inlier.

intermediate and acidic chemistries, contrasting with the predominantly tholeiitic, bimodal magmatism present in the Ordovician successions elsewhere in Wales, in areas such as Pembrokeshire, the Aran Mountains and Llŷn (see Bevins *et al.* 1984). The basic lavas are characterized by being typically strongly feldspar-porphyritic, although intermediate to acid lavas or intrusions are usually sparsely phenocrystic to aphyric. The most recent study is that by Metcalfe (1990), who found that both calc-alkaline *and* tholeiitic lavas in fact were erupted at Builth, as has been confirmed also in the Aran Mountains region (R. E. Bevins and B. P. Kokelaar, unpublished data). In the Builth area, however, lavas of calc-alkaline affinity predominate, whilst in the Aran Mountains the lavas are chiefly tholeiitic.

The rocks of the Builth Inlier have suffered low-grade metamorphism, the basic igneous rocks showing development of chlorite, albite, prehnite, pumpellyite, titanite and secondary calcite, an assemblage considered by Bevins & Rowbotham (1983) to indicate prehnite-pumpellyite facies conditions of metamorphism. More recently, however, laumontite (Bevins & Horák 1985) and analcime (Metcalfe 1990), minerals more characteristic of the zeolite facies, have been identified in veins cutting basalts in Llanelwedd Quarry. The clay mineral assemblage in Ordovician mudstones of the Builth Inlier is characterized by interstratified illite/smectite (with < 10% smectite layers) and kaolinite, diagnostic of the diagenetic zone of metamorphism (Robinson & Bevins 1986). These data, combined with results of stable isotope geothermometry on $< 0.5 \mu m$ quartz and illite/smectite separates from Ordovician shales (Metcalfe 1990), indicate maximum temperatures in the region of 200°C; i.e. about the boundary

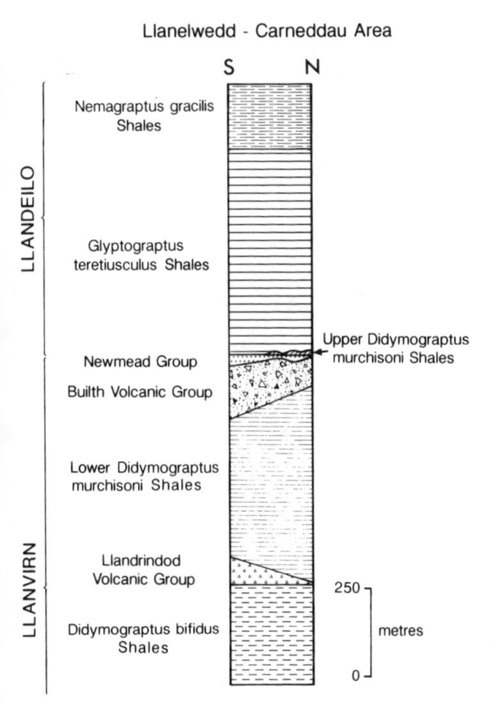

FIG. 2. Stratigraphy of the Llanelwedd–Carneddau area. Terminology is based on that of Williams *et al.* (1972) and Metcalfe (1990).

between the zeolite and the prehnite-pumpellyite facies. In addition to the regional low-grade metamorphism, alteration effects relating to a localized hydrothermal event are seen in lavas of the Llanelwedd area; this hydrothermal activity is thought to have occurred during Llanvirn times, being related to the original volcanic episode (Bevins 1985; Metcalfe 1989, 1990).

The inlier suffered significant faulting and gentle folding before the deposition of the unconformable Upper Llandovery to Wenlock cover. Steep strike-slip faults dominated (Woodcock 1987b). Some of these faults were reactivated during the Acadian (late Caledonian) deformation in Devonian time, and further gentle folding occurred.

The stratigraphy used in this chapter is shown in Fig. 2; it is based on that of Williams *et al.* (1972) and Metcalfe (1990).

Two itineraries are described below: the first concentrating on the various

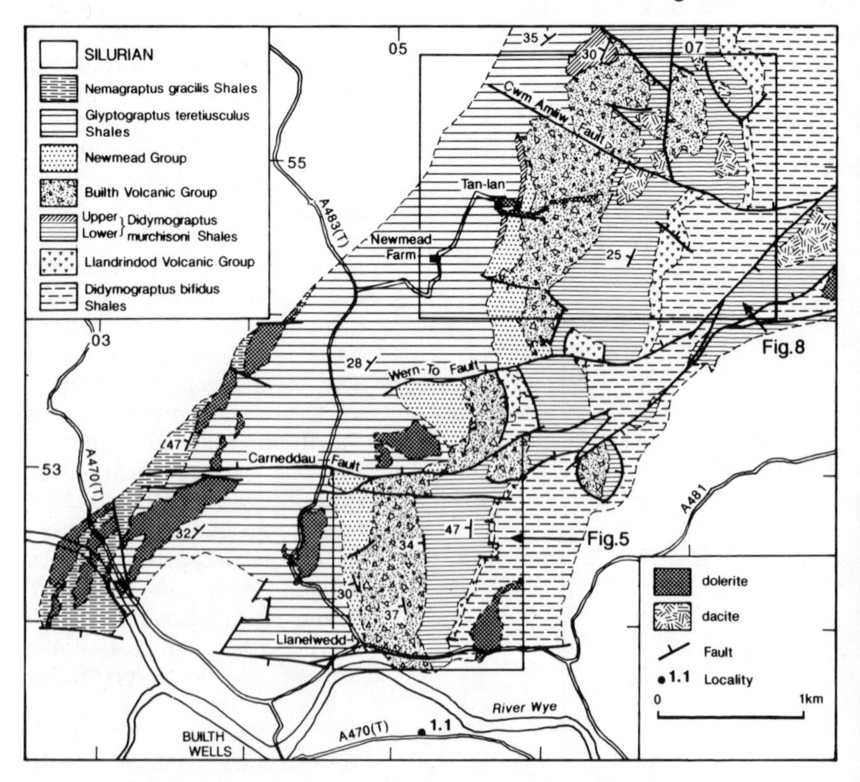

FIG. 3. Geological sketch map of the southern part of the Builth Inlier, based on the British Geological Survey 1:25 000 sheet Llandrindod Wells Ordovician Inlier. The areas for Fig. 5 (Itinerary 1) and Fig. 8 (Itinerary 2) are indicated.

basic to intermediate volcanic components of the Builth Volcanic Group exposed at the southernmost extremity of the Builth Inlier, in and around the Llanelwedd Quarries; the second in the Carneddau Range, traversing through the Builth Volcanic Group and its related intrusions further north, as well as examining rhyolitic tuffs and related rocks of the Llandrindod Volcanic Group. The location of these itineraries and access routes are shown in Fig. 3. Each itinerary will take one-half to three-quarters of a day to complete.

ITINERARY 1. LLANELWEDD QUARRIES AND ENVIRONS

This itinerary details an east to west traverse through part of the Builth Volcanic Group and overlying Newmead Group, of Llanvirn age, exposed at the southern extremity of the inlier (Fig. 3). Total walking distance is approximately 2.5 km and includes a visit to the working Llanelwedd Quarry.

Permission to enter this quarry **must** be obtained at least **2–3 days in advance**, from the *Quarry Manager, ARC Powell Duffryn Ltd, Llanelwedd Quarry, Builth Wells, Powys*

FIG. 4. View looking northwards from Locality 1.1, a lay-by, at SO 0510 5128 on the A470(T), approximately 800 m east of Builth Wells. Clearly seen are the extensive quarries in the basalts of the Builth Volcanic Group, whilst the small quarry (Gelli Cadwgan) at the right of the photograph is in silicified dolerite. The ridge between these two quarries is underlain by siliceous pyroclastic rocks of the Llandrindod Volcanic Group.

(tel. 0982 553608). Under normal circumstances permission to enter the quarry will be granted readily provided that indemnity forms are completed beforehand. **Extreme caution must be exercised** when working in the quarry.

Locality 1.1. An overview of the area to be visited can be gained from a lay-by on the A470(T), approximately 800 m east of Builth Wells (at SO 0510 5128). From this vantage point the extensive quarries of Llanelwedd (chiefly in basalts) and Gelli Cadwgan (chiefly in silicified dolerite) can be observed, as can the more rugged, craggy nature of the ground underlain by the various resistant igneous, pyroclastic and epiclastic rocks of the succession (Fig. 4). In contrast, the areas underlain by the weaker, variably tuffaceous, mudstones show little exposure, and have been eroded into by small streams orientated N–S, following the strike of the beds. The succession dips and youngs to the west.

Locality 1.2. Proceed to **Llanelwedd Quarry** by taking the A470(T) through Builth Wells, and following the A483(T) towards Llandrindod Wells. **Vehicles can be parked at the entrance to the quarry**, at SO 0472 5199. Fig. 5 illustrates the route to be followed for this itinerary. Enter the quarry via the main access road, turning to the right after 100 m, and passing the weighbridge on the left; be certain to register at the site office (second hut on the right). On leaving the site office take the track that leads upslope to the north-east. After 240 m, after passing gravel dumps on both the right and left, bear left at the T-junction, and then turn immediately to the right. Follow the track for 100 m, passing around a right-hand bend, to where the track levels out and is crossed by another track which runs NW–SE; enter **the small quarry** immediately opposite (sometimes containing dumped gravel).

Locality 1.2a (SO 0520 5205). This quarry exposes epiclastic sandstones and siltstones overlain by altered basaltic lavas of the Builth Volcanic Group. Massive, coarse to fine grained, thickly-bedded (10 cm to > 1 m) feldspathic epiclastic sandstones and siltstones, which dip to the west at approximately 40°, are well exposed in the **eastern face of the quarry.** In places the sandstones contain lithic clasts up to 20 mm across, as well as shale rip-up clasts up to 5 cm across. In thin section, the sandstones are seen to be clast-supported, and contain an admixture of both lithic clasts and crystals from 0.1 to 6 mm across. The lithics are equant and sub-rounded, and are principally of quenched basalt, now composed of tabular plagioclase (albite) crystals, showing dendritic terminations, set in a chloritic matrix. The crystals are predominantly albites, quartz being notably absent, whilst the matrix is chloritic. Impersistent shale horizons occur interbedded with the sandstones and siltstones. These various epiclastic rocks are thought to

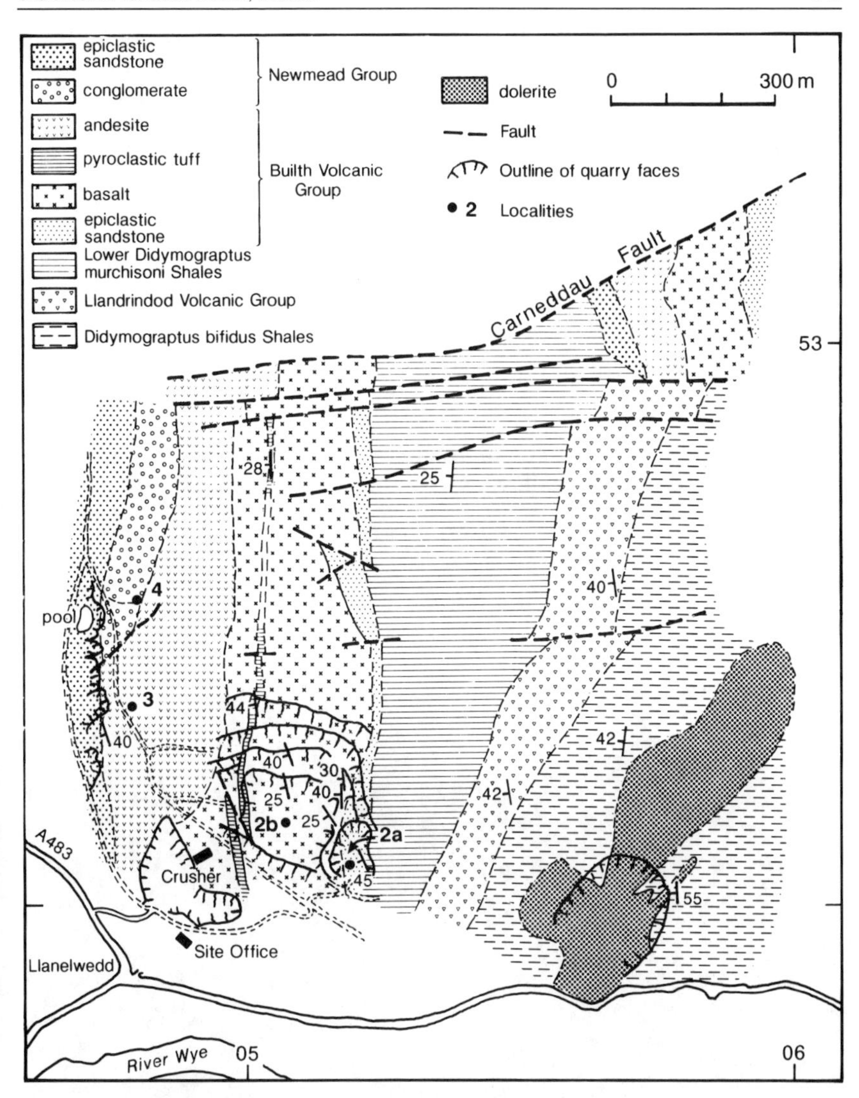

FIG. 5. Detailed geological map of the Llanelwedd Quarry area, showing the localities for Itinerary 1. This map is based on the mapping of R. A. Gayer, as well as the British Geological Survey 1:25 000 Llandrindod Wells Ordovician Inlier.

have been derived as a result of reworking of basaltic lavas emplaced in a shallow submarine environment. Loose blocks of hyaloclastite found on the quarry floor, although not seen *in situ*, support this environmental picture.

Certain bedding planes in the massive sandstones show tumescence structures up to 1 m across, which are cored by feldspar-phyric basalt. These basalts appear to represent minor intrusive bodies, emplaced into the epiclastic sediments while they were still wet, and highlight the coeval nature of sediment accumulation and igneous activity.

In the **north face of the quarry** the various epiclastic rocks are overlain sharply by grey-green, aphyric, massive basalts (Fig. 6). The basalts are distinguished by a strong jointing, the absence of the prominent spaced bedding planes seen in the epiclastic rocks, and by the presence of abundant calcite veins.

The **western face of the quarry** is defined by an obvious fault plane, showing prominent slickenside surfaces. The slickenfibres are principally of quartz or, more rarely, calcite. The fault, one of a suite of strike-slip faults traversing the area (Woodcock 1987), strikes approximately 165° and dips 40°E, while the slickenfibres plunge at a shallow angle, up to 20°, towards *c.* 180°.

Leave the quarry, turn to the west and **enter the lowest level of the main quarry**.

FIG. 6. Thickly-bedded, westward-dipping, feldspathic epiclastic sandstones overlain by well-jointed massive basalts of the Builth Volcanic Group, exposed at Locality 1.2A, Llanelwedd Quarry.

Locality 1.2b Centred on SO 0505 5215. From the entrance it is possible to gain an overview of the various lavas and pyroclastic rocks exposed on this level of the quarry. These are chiefly massive sheet-flows of basalt, up to 5 m thick, which dip westwards at approximately 40°, although in the west wall a prominent red-brown stained pyroclastic unit (the Felsite Agglomerate of Jones & Pugh 1949; the Pyroclastic Member of Metcalfe 1990) is exposed. In addition, numerous prominent quartz and/or calcite-filled fault fractures can be observed cutting through the basalts on the various quarry levels. The south-western face of the lowest level is defined by another prominent strike-slip fault striking N–S.

Proceed to the **eastern side of the quarry**. The lower parts of the quarry face are composed of massive, aphyric basalt (a continuation of the basalts from Locality **2a**), whilst the upper parts comprise a basaltic breccia, thought to represent a rubbly flow top to the underlying massive basalts. **Extreme care is required** as here the quarry face is extremely dangerous. In fact, the various lithologies are best seen in loose blocks on the quarry floor, while the relationship between the basalts and the basaltic breccias can be observed in the easternmost part of the **north face** of the quarry.

In this part of the quarry the metamorphic effects of a fossil hydrothermal system can be examined. Firstly, in the basalts thin veins (1–3 cm) and pods (up to 20–30 cm across) occur, composed principally of calcite surrounded by alteration haloes of bluish-green pumpellyite and cream prehnite, and intimately associated with black, shiny, friable masses of pyrobitumen which is > 85 wt% pure carbon. It is considered most likely that the pyrobitumen results from maturation of organic matter in the underlying Didymograptus bifidus Shales. Support for this comes firstly from the fact that leaves of shale in the lavas at Llanelwedd Quarry are associated commonly with pyrobitumen, and secondly from isotopic evidence, in that the pyrobitumen (with $\Delta^{13}C_{PDB}$ of −29.4%) has lighter ^{13}C values than the total organic carbon of the underlying shales, which have a $\Delta^{13}C_{PDB}$ value of −25.9%. Calcite from the veins releases a hydrogen sulphide aroma when crushed. The calcite-bearing veins can be traced up into the overlying basaltic breccias, which are themselves heavily impregnated by pumpellyite (indicated by the strong bluish-green colouration of the breccias). It has been argued by Metcalfe (1990) that organic-rich fluids responsible for the alteration were channelled along the relatively permeable breccias and were forced downwards into the more massive, less permeable underlying basalts as a result of over-pressuring. The heat source responsible for the hydrothermal activity was most probably provided by an intrusion associated with the volcanic activity during Llanvirn times. On the basis of various lines of evidence, Metcalfe (1990) has suggested that the heat source was most probably the silicified dolerite intrusion now exposed in Gelli Cadwgan Quarry, a short distance to the east. In both the basalts and the breccias small (*c.* 1–2 mm) cream-coloured spherules of prehnite are present, although these are thought to relate to the regional metamorphic effects in view of the fact that prehnite with this form is found in altered basic igneous rocks throughout the inlier.

Further to the west, along the north face of the quarry, the contact between the massive rubbly basalt breccias and the overlying massive basalts is exposed. This contact is offset slightly by a prominent fault, which is actually an extension of the strike-slip fault that defines the western face of the quarry at Locality **2a**. The offset of the breccia/lava contact indicates a sinistral sense of movement on the fault.

A dominant characteristic of the lavas comprising the **northern quarry face** is their highly amygdaloidal and feldspar-phyric nature. Smaller amygdales in the lavas are typically infilled with dark green chlorite, whilst larger amygdales usually have a rim of chlorite surrounding a core of calcite (or more rarely quartz). Feldspar phenocrysts are typically of albitic composition as a result of alteration, although compositions in the andesine range are preserved infrequently. The phenocrysts may reach up to 10 mm in length, although more usually are 5–7 mm; they are euhedral and commonly contain enclaves of chlorite. In thin section these phenocrysts are seen also to be partially replaced by calcite, prehnite and pumpellyite. Concentrations of the phenocrysts vary within flows, but commonly a marked break between phenocryst-free margins and phenocryst-rich centres is seen. The proportion of phenocrysts in the flow centres, however, is too high to be a direct crystallization product, implying concentration of these crystals at some stage. Chemical analyses of these flow centres show pronounced positive europium anomalies, related to the high feldspar contents, but the absence of a corresponding negative europium anomaly in the aphyric margins negates the possibility that excess phenocrysts were derived from these parts of the flows. This implies crystallization and concentration of these phenocrysts prior to magma eruption, most probably in a high-level magma chamber. This evidence also indicates that the various flows of this character exposed in the quarry are, in fact, compound in nature.

Turning attention to the **western side of this quarry level**, the southern part of the western face is defined by a further fault (orientated approximately 175° and dipping 60°E with near horizontal slickensides) belonging to the stike-slip duplex of Woodcock (1987b). The northern part of the face provides excellent exposures of a prominent pyroclastic flow deposit (Fig. 7).

This pyroclastic deposit is 9 m thick, and is composed of two flow units, each approximately 4.5 m thick. At the top is a thin (*c.* 5 cm), fine-grained tuff. The lower of the pyroclastic units has a fine-grained, friable, chlorite-rich base some 0.25 m thick, which passes gradationally upwards into the main facies of the unit. The unit contains prominent basaltic scoria clasts which both coarsen (from 4 mm to 16 mm) and become more abundant upwards; these scoria clasts are typically flattened, which may be taken to indicate that the unit was hot when it was emplaced. Matrix-supported lithic clasts are also present in the unit but in contrast these become finer upwards, from 20 mm to 4 mm. The clasts are heterolithic, being predominantly of igneous origin, including basalt, gabbro, rhyolite, micrograpite and microgranophyre. In the upper half of the unit, prominent ovoid siliceous nodules with chlorite rims are present. The nodules generally have long axes parallel to the flow margins and show an increase in size upwards. These nodules are thought to have been generated as a result of silicification of basaltic scoria clasts (R. Metcalfe and R. J. Sloan, unpublished data). The unit is thought to have been deposited from a hot pyroclastic flow.

In contrast to the lower unit, the upper unit has a much higher proportion of lithic clasts compared with scoria clasts. The lithics, which are again heterolithic and matrix-supported, coarsen upwards from 10 cm at the base to 80 cm near the top. The scoria clasts, however, remain constant in size at between 4 and 10 mm, and are, in contrast with the lower unit, only slightly flattened. This unit is also considered to have been deposited from

FIG. 7. Pyroclastic flow-unit exposed at the northern end of the western face of Llanelwedd Quarry (Locality 1.2b).

a pyroclastic flow, although perhaps not as hot as that responsible for the lower unit.

Leave the quarry at the south-west corner and proceed westwards, passing one of the principal crushers on the left. Continue westwards, passing a turn to the right leading to the second level of the quarry. After 250 m, at a major right-hand bend, strike off to the north-north-west, across open ground, passing through a gate adjacent to a small oak tree.

Locality 1.3 (SO 0477 5233) is **approximately 20 m north-north-west of the gate**. Numerous small exposures adjacent to the path are composed of a fine-grained, grey platey-jointed igneous rock containing scattered microphenocrysts. This rock contains approximately 58 wt% SiO_2 and hence is an andesite. In thin section the andesite can be seen to contain colourless, tabular microphenocrysts of albite (0.25–3 mm in length) and equant to tabular turbid, colourless to pale green, non-pleochroic clinopyroxene crystals (0.1–2 mm across). The microphenocrysts are set in a matrix of albite associated with chlorite and titanite. Unfortunately, from the exposures available it is not possible to determine whether this andesite is a lava flow or whether it represents a high-level intrusion.

Follow the path north-westwards, passing, after some 20 m, coarse feldspar-rich epiclastic sandstones, eventually reaching a fence. Turn northwards, following a path which parallels the fence. Further epiclastic sandstones, this time showing prominent cross-bedding, are **exposed to the west of the fence**. Continue along the path until the point where it passes through a gate. Here take a minor path which heads upslope to the east-south-east.

Locality 1.4 (SO 0479 5254). After 100 m, small scattered outcrops are composed of coarse conglomerates belonging to the Newmead Group. The boulders, which are typically well rounded, reach up to 30 cm in diameter, and are predominantly of basaltic lava. These conglomerates comprise the 'Boulder Bed' of Jones & Pugh (1949), thought to represent a beach deposit. Retrace the path downslope, pass through the gate, and descend to the track. Turn to the south and follow the track downslope.

Locality 1.5 (SO 0470 5255 to SO 0470 5225), is a **series of quarries** in epiclastic sandstones of the Newmead Group which overlie the conglomerates of Locality 1.4. The sandstones are commonly structureless, with only a broadly-spaced bedding. Under the microscope the sandstones are seen to be lithic-rich, clast supported, and to be associated with a minor crystal (albite) component. The lithic clasts are predominantly of plagioclase-rich basalt along with rare rhyolitic lithics, associated with interstitial calcite, yellow-brown to dark brown chlorite, titanite and rare pore-filling albite. The clasts are sub-angular to well rounded, and are generally equant. Like the sandstones of Locality 1.2a, below the lava sequence, these are thought to have been derived by erosion of a basalt lava pile in a subaqueous environment. The presence of sparse orthid brachiopods in these sandstones testifies to a marine environment. Although an unconformity separates the various lavas and pyroclastic rocks of the Builth Volcanic Group from the epiclastic sandstones and conglomerates of the Newmead Group this is not thought to be of a profound nature, and neither is it thought to represent a significant time break. Unconformities are common features in volcanic sequences and reflect the unstable nature of such environments. An environmental picture therefore emerges of the development of a volcanic pile in a shallow submarine environment which was subjected to minor, sporadic, contemporaneous instabilities and constant erosion. This erosion produced thick accumulations of basaltic sands, pebbles and boulders, now forming the various epiclastic sandstones and conglomerates of the Builth Volcanic Group and of the Newmead Group.

Continue down the track to the south which leads back to the entrance to the working quarry.

ITINERARY 2. CARNEDDAU

This itinerary details a west-to-east traverse through a part of the Builth Volcanic Group and older volcanic rocks some 3.5 km north of Builth Wells (Fig. 3). Total walking distance for this itinerary is approximately 4 km.

The itinerary starts at Tan-lan (SO 0569 5474), reached by following a narrow unclassified road which joins with the A483 Builth Wells–Llandrindod Wells road at SO 0472 5418, 2 km north of Llanelwedd (see Fig. 3). **The road is too narrow for coaches** but cars and minibuses can proceed as far as Tan-lan where there is **limited parking** available.

Locality 2.1 (SO 0570 5470). **Crags in the vicinity of Tan-lan** (Fig. 8) are composed of dolerite, which is resistant in comparison with the relatively weak sedimentary rocks into which it is intruded. Good exposures are provided by **a small quarry** approximately 40 m south-east of Tan-lan. Here intrusive contacts between shale and dolerite are exposed, which are broadly concordant, implying a sill-like form to the intrusion. However, the intrusion can be traced eastwards for some distance, cutting through the stratigraphy, and here appears to be dyke-like. The concordant contacts exposed in the quarry are interpreted as being sill-like apophyses off the margin of a dyke-like body. Such apophyses most probably result from magma intruding into sediments which were not completely lithified. The shales and siltstones strike approximately 180–200° and dip at 35–40° north-west, and have been extensively bleached up to 1 m away from the contact due to

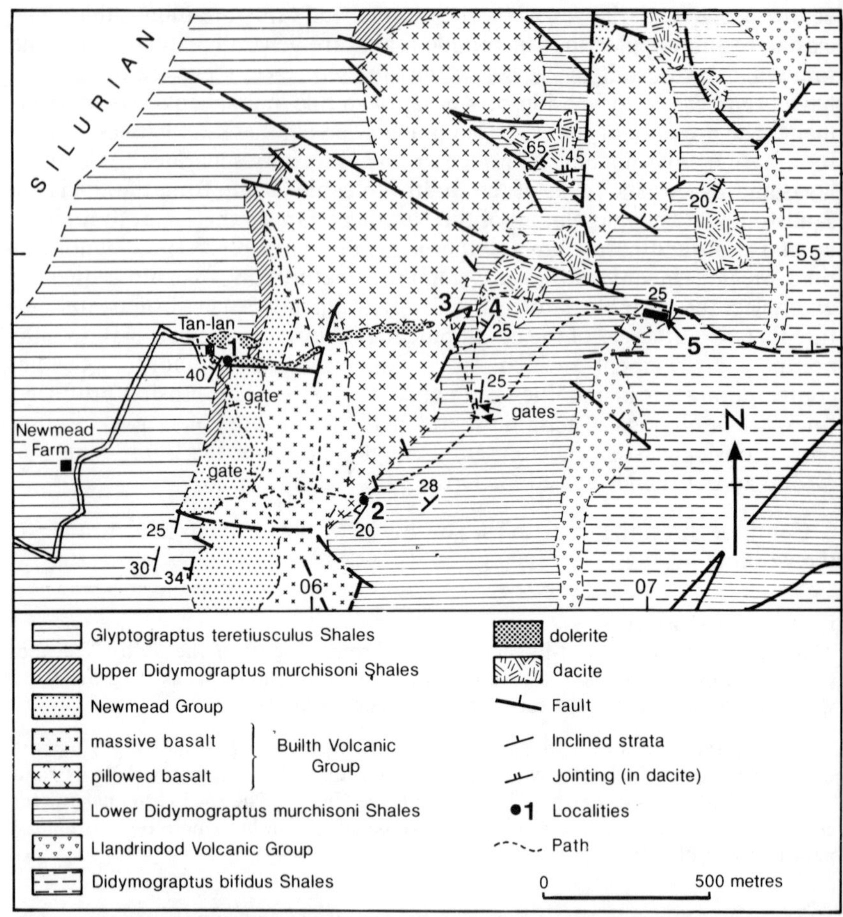

FIG. 8. Detailed geological map of the north Carneddau area, showing localities for Itinerary 2. Based on the British Geological Survey 1:25 000 sheet Llandrindod Wells Ordovician Inlier.

thermal metamorphism. Bedding planes in the shales contain scattered small inarticulate brachiopods. As the intrusion can be traced through lavas of the Builth Volcanic Group and overlying sedimentary rocks, it is clearly younger than the main Builth Volcanic Group magmatic episode. Follow the path leading upslope to the south-east, passing through a first gate after *c.* 100 m and a second after *c.* 300 m. The track then turns eastwards, in the vicinity of crags composed of feldspar-porphyritic basalt. These basalts are lateral equivalents of those exposed in the quarries at Llanelwedd (see Itinerary 1, Locality **2b**). Continue due east, passing over the crest of the hill.

Locality 2.2 (SO 0621 5432) is a **prominent set of crags** which face south-east and open out onto a small, poorly drained hollow. Along these crags are three prominent clusters of small trees which are at the level of the contact between pillowed and brecciated porphyritic basalts (best seen at the next locality) and underlying sandstones. The contact is concordant, with bedding in the sandstones striking approximately 225° and dipping at about 20° to the north-west. The sandstones comprise both thickly-bedded, massive units and thinner, cross-bedded units, as well as minor coarse-grained horizons with scattered, matrix-supported pebbles up to 2 cm across. Traverse north-east for 250 m, to a low col and pass through a gate. Proceed for a further 200 m to the north-north-west, to Locality 2.3.

Locality 2.3 (SO 0640 5474). In **steep crags** feldspar-porphyritic basalts are exposed, which are laterally equivalent to those at Locality 2.2 and at Llanelwedd Quarry. Here, however, the lavas are pillowed, with individual pillows reaching 50 cm across. Broken pillow lava fragments form an interpillow breccia. Like the basalts exposed further south these lavas are highly feldspar-porphyritic, with individual crystals reaching up to 6 mm in length. The feldspars are now albites, although at the time of crystallization they were undoubtedly more calcic in composition. The matrix is composed of scattered lath-shaped albite crystals (0.1–0.2 mm in length) showing well-developed dendritic terminations reflecting quenching of the lavas in water. The feldspars are set in brown or yellow-green chlorite which presumably has replaced primary volcanic glass. Prehnite, white mica and, more rarely, calcite are also present, partially replacing the feldspar phenocrysts. Rims to pillows are commonly amygdaloidal, with individual amygdules variably comprising quartz, chlorite, prehnite and calcite. The presence of pillow structures confirms that these basalts were emplaced in a subaqueous environment. As mentioned above, matrix feldspars show evidence of quenching and hence were crystallizing at the time of eruption; in contrast, however, the phenocrystic feldspars are tabular with square terminations and thus were almost certainly present in the lava *before* eruption and quenching, supporting the evidence from Llanelwedd Quarry for crystallization and concentration of these phenocrysts in a high-level magma chamber. Return south-eastwards, through a gate in a N–S-trending fence, and then proceed northwards for 275 m to a **prominent rocky knoll.**

Locality 2.4 (SO 0651 5488). Outcrops on this knoll comprise buff-coloured, platey-jointed dacite containing scattered tabular feldspar microphenocrysts (1–3 mm in length) set in a fine-grained siliceous matrix. The dacite is thought to be intrusive and the irregular jointing, in places folded, is considered to be a primary cooling-related phenomenon. Proceed 500 m to the south-east, passing down a NW–SE-oriented valley to examine **exposures in crags** on the south-west side of the valley.

Locality 2.5 (SO 0690 5465). Crags at this locality expose a sequence of rhyolitic ash-flow tuffs interbedded with epiclastic sandstones. The beds strike approximately north–south and dip about 25° west. Two major tuff units are recognized. A lower rhyolitic tuff unit is up to 25 m thick, and contains pumices (showing spectacular splayed ends), rhyolitic lithics, crystals, and rare shards set in a finer siliceous matrix. This unit is coarse at the base, with individual pumices and lithics up to 3 cm across, but passes upwards into fine-grained flinty rhyolitic tuffs, while the upper unit is generally fine-grained throughout. These rhyolitic tuffs underlie Lower Didymograptus murchisoni Shales, and hence probably belong to the Llandrindod Volcanic Group, which represents a volcanic episode earlier than the more basic lavas and tuffs of the Builth Volcanic Group. Return via the two gates to Locality 2.2 and to the track which leads back to Tan-lan.

Acknowledgements. Thanks are due to Dr R. A. Gayer for providing field maps which form the basis of Figure 5, and also for commenting on an early draft of the manuscript. Mr Lawrence Tomlinson, the Manager of Llanelwedd Quarry, is thanked for his interest, and also for permitting access.

11. THE ORDOVICIAN AND LLANDOVERY IN THE LLANWRTYD WELLS TO LLYN BRIANNE AREA

by A. H. MACKIE

Maps	*Topographical*	1:50 000	Sheet 147 Elan Valley and Builth Wells
		1:10 000	Sheets SN74, SN75, SN84, SN85
	Geological:	1:250 000	Sheet 52°N–04°W Mid Wales and Marches
		1:100 000	Special Sheet, Central Wales Mining Field

THE Ordovician and Silurian sequences in this area comprise a variety of very low-grade metasedimentary rocks, with both intrusive and extrusive igneous rocks near the base of the succession. The sediments document a transition from a marine shelf-slope environment via a glacio-eustatically shallowed shelf environment to a deeper water, slope to basinal environment. The area was subjected to a metamorphic event, probably during the late Silurian to early Devonian, with grades ranging up to lowest greenschist facies.

The area is interesting structurally because it spans two regional structural features – the Tywi and the Central Wales lineaments (see Fig. 2). Both these features were important controls on sedimentation during the Ordovician and Silurian (Woodcock & Smallwood 1987; Smallwood 1986*a*, *b*; Smith 1987*b*).

Important early work on the stratigraphy and sedimentary rocks was carried out by K. A. Davies (1933; Davies & Platt 1933). J. H. Davies (1981) revised sedimentological and structural interpretations, proposing a novel thrust-based model. More recent work by Mackie (1987), Smallwood (1986*a*, 1986*b*), Mackie & Smallwood (1987) and Woodcock & Smallwood (1987) has further refined the stratigraphy, sedimentological models and structural geology.

The stratigraphy used in this chapter (Fig. 3) is that of Mackie (1987), Smallwood (1986*b*) and Mackie & Smallwood (1987), and incorporates the stratigraphy for the Llanwrtyd Wells area derived from J. H. Davies (1981).

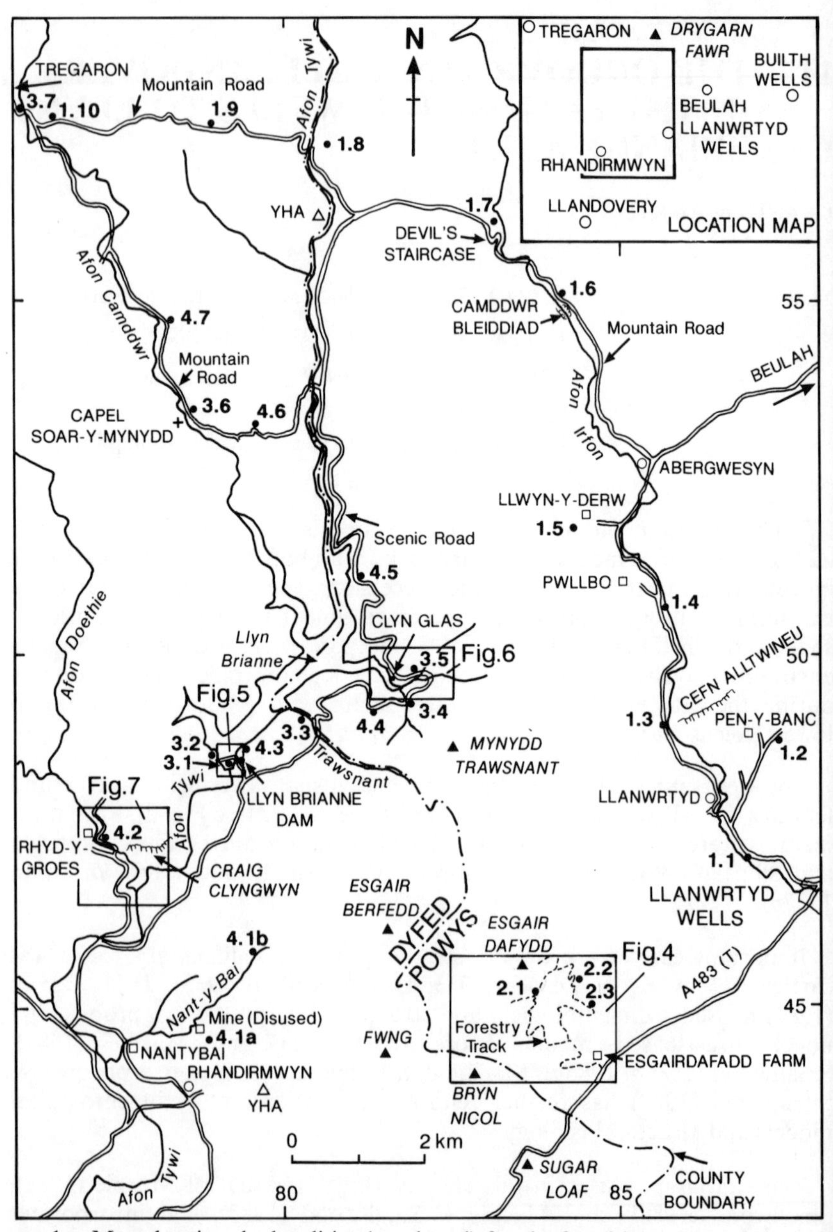

FIG. 1. Map showing the localities (numbered) for the four itineraries and areas of detailed maps in Figs. 4 to 7. Regional geographical setting shown inset at top right.

OVERVIEW OF SEDIMENTOLOGY

The succession is dominated by mudstones and fine siltstones, punctuated with some coarse and very coarse sediments particularly near the base (Ashgill). The Llandovery succession generally coarsens upwards from mudstones at the base to very coarse sandstones at the top.

The sediments are thought to have been deposited on the south-east margin of an elongate, NE–SW aligned trough (Mackie 1987; Mackie & Smallwood 1987). Tectonics exerted a considerable control upon sedimentation both locally (fault scarps and slumps) and more regionally (uplift to the south probably generated large amounts of sediment in late Llandovery times). The Tywi Axis (a fault-cored, anticlinal structure) probably acted as a shelf-slope demarcation line during the late Ashgill and early Llandovery. The effects of this can be clearly seen on a geological map of the area (Fig. 2). A glacio-eustatic sea-level fall had a marked effect on sedimentation in the late Ashgill.

OVERVIEW OF STRUCTURAL GEOLOGY

The structure is dominated by NE–SW trending, SE-verging, asymmetrical, sub-cylindrical and periclinal folds. These structures occur on all scales (km to cm) and represent a bulk shortening of about 20–30 per cent. The largest scale fold structures are the Tywi Axis (which bounds the area to the south-east) and the Central Wales Synclinorium (forming the north-west margin). Both of these probably overlie basement fault belts that were involved closely in their development (Woodcock 1987*a*; Smith 1987*b*). Intermediate scale folds (3–4 km wavelength) probably initiated around major arenite bodies in the mud-dominated sediment pile.

There is a moderate to well developed cleavage associated with most folds, dependent on lithology. Although not readily apparent in the field, this cleavage generally transects the fold axes by about 5–15° in a clockwise sense.

Although faults are not common at the surface, both normal and reverse types have been recorded. Strike-slip motion has occurred on some fracture surfaces but has not produced any major offsets.

ITINERARIES

The more accessible outcrops in the area are arranged into four separate itineraries, on logistic and thematic bases. Itinerary 1 is a stratigraphical overview in a north-westward traverse from Llanwrtyd Wells, requiring about 4.5 hours. Itinerary 2 covers the detailed sedimentology of the early to mid Ashgill sediments south-west of Llanwrtyd, and takes about 3 hours. Itinerary 3 covers the sedimentology of the late Ashgill and Llandovery sequence around and to the north of Llyn Brianne, requiring about 5 hours. Itinerary 4 concentrates on the structural geology of the area between Nant-y-bai and the Camddwr Valley, north of Llyn Brianne; this itinerary takes about 4 hours. Itineraries 3 and 4 follow broadly the same route and

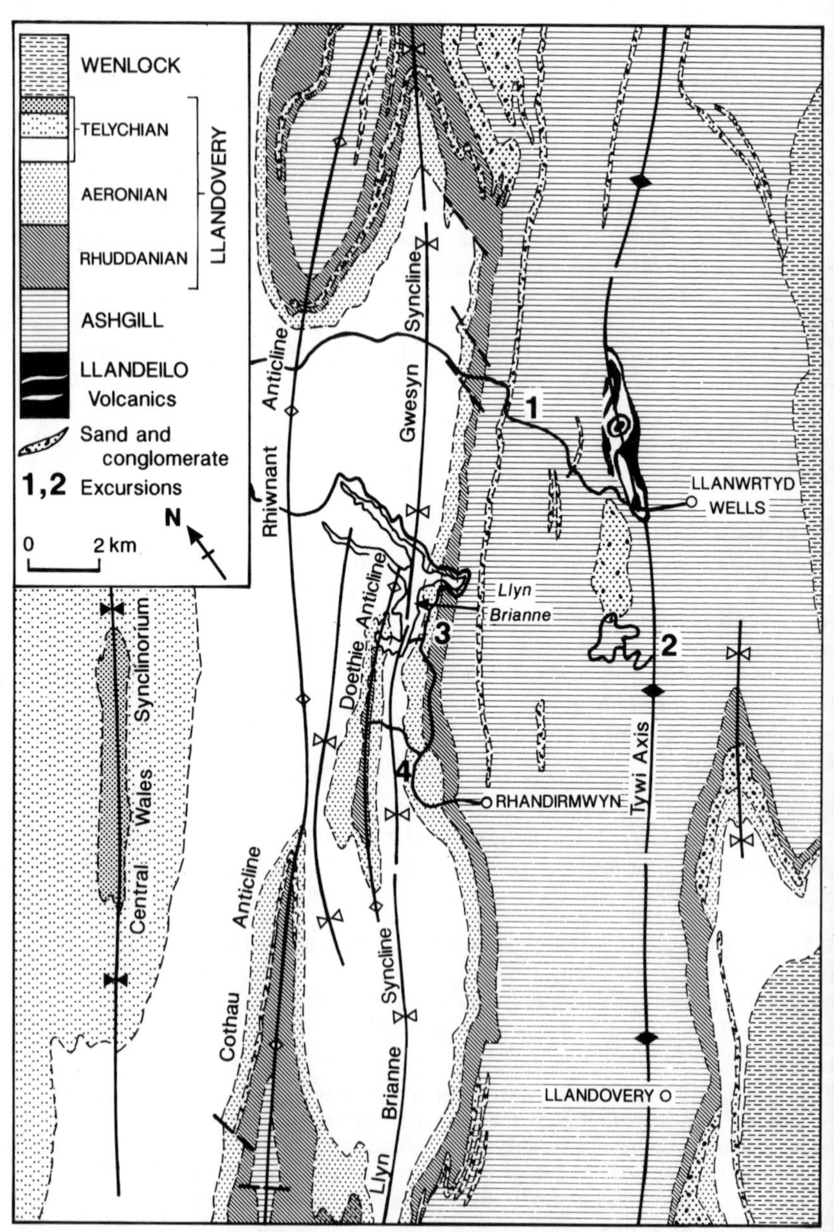

FIG. 2. Geological map of Llanwrtyd Wells/Llyn Brianne (after Stamp & Wooldridge 1923; Davies 1933; Davies & Platt 1933; Smallwood 1986*b* and Mackie 1987). Map also shows major structural features and location of itineraries.

could be followed in parallel to make a full day's excursion. The variations in metamorphic grade have been determined using laboratory techniques (illite crystallinity) and are not readily discernible in the field.

The access roads to most localities are **not suitable for large coaches.** This applies particularly to the mountain road from Abergwesyn to Tregaron. In order to preserve geological coherence, all but itinerary 2 transgress the county boundary of Powys with adjoining Dyfed.

ITINERARY 1: Stratigraphical overview

This itinerary begins at the town of Llanwrtyd Wells on the main A483 road (SN 8785 4670), and follows the Tregaron mountain road (Dolecoed Road) heading west from the town. Refer to Figs. 1, 2, and 3.

1.1. Dole-y-coed Park. Basaltic rocks occur in the core of the Tywi Axis at SN 8675 4715, approximately 1.2 km from the village. Good exposures of fine-grained, medium-grey weathering, pale to dark green basalts crop out **in the River Irfon section** approximately 100 m downstream from the footbridge. They show feldspar phenocrysts (0.5–3 mm) and vesicles 2–20 mm

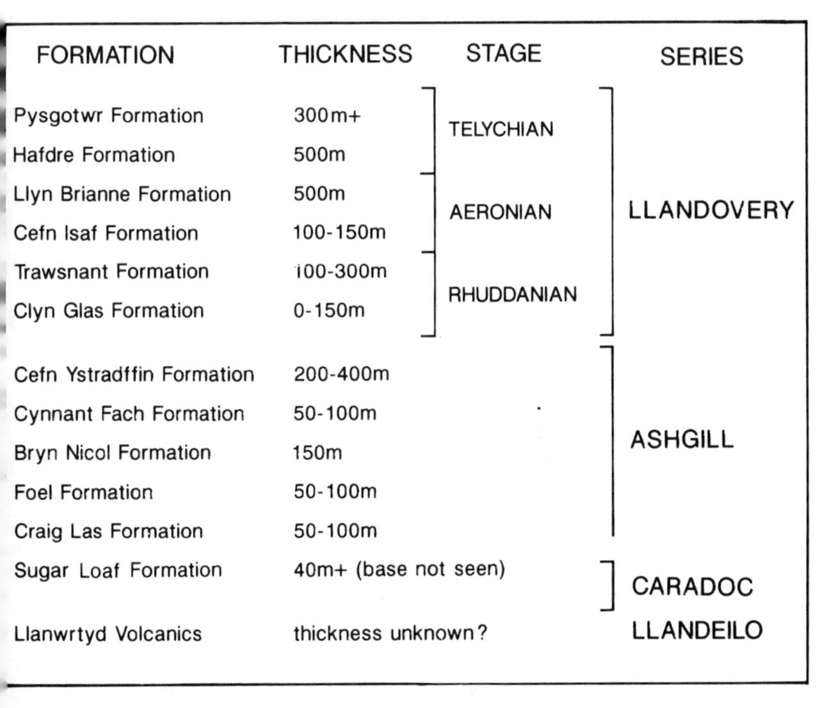

FORMATION	THICKNESS	STAGE	SERIES
Pysgotwr Formation	300m+	TELYCHIAN	LLANDOVERY
Hafdre Formation	500m		
Llyn Brianne Formation	500m	AERONIAN	
Cefn Isaf Formation	100-150m		
Trawsnant Formation	100-300m	RHUDDANIAN	
Clyn Glas Formation	0-150m		
Cefn Ystradffin Formation	200-400m		ASHGILL
Cynnant Fach Formation	50-100m		
Bryn Nicol Formation	150m		
Foel Formation	50-100m		
Craig Las Formation	50-100m		
Sugar Loaf Formation	40m+ (base not seen)		CARADOC
Llanwrtyd Volcanics	thickness unknown?		LLANDEILO

FIG. 3. Stratigraphical summary table.

in diameter. Pillow structures up to 1 m long have been reported (J. H. Davies 1981), implying their extrusive nature. At this locality, strongly developed jointing obscures some detail but some concentric (relict pillow) structures are discernible. These lavas are thought to be of late Llandeilo age (J. H. Davies 1981) and thus represent the oldest rocks seen in this area.

1.2. Garn Dwad. Drive a further half kilometre westwards to SN 8655 4760 where a track leads off towards Pen-y-Banc Farm. Although the track is driveable right up to the exposures, it is narrow and locally very steep with restricted turning space. It is thus **recommended that vehicles are left at the main road.** Follow this track, taking the left fork at Kilsby. About 600 m from the main road, the track forks again: take the right fork. For about **the next 400 m** there are sporadic exposures of interbedded, iron-stained, very dark grey mudstones and ripple cross-laminated siltstones (up to 10 cm thick). These rocks are thought to correlate with the Sugarloaf Formation of Smallwood (1986*b*) and would thus be of Caradoc age.

Where the main track leads off down to the north-west (Pen-y-Banc Farm), follow the right fork up a steep, stony track to a **small summit on the western shoulder of Garn Dwad** (approx. 100 m from the fork). Here there are exposures of a buff weathering, dark green, medium to coarse-grained, igneous rock. The origin of these rocks is unknown but K. A. Davies (1933) implies that they are related to the extrusive basaltic rocks seen further south and that they have a faulted eastern margin. Thin section examination reveals assemblages consistent with prehnite-pumpellyite facies metamorphism.

1.3. Alltwineu. Return to the main road and drive north-westwards for 2 km along the Irfon Valley. **Park at a picnic area** on the left side of the road (SN 8562 4919). Cross the road and follow the line of the hedge down to **the river** where there are good exposures of medium to coarse-grained sandstones of the Foel Formation showing extensive quartz veining. This unit is dominated by mudstones but is also characterized by dark-grey to buff coloured, coarse, moderately-well sorted sandstone units up to *c.* 1.6 m thick. Sedimentary structures include parallel, ripple- and trough-cross laminae.

The line of the outcrop can be followed to a **prominent crag** to the north-east (Cefn Alltwineu) which is composed of this unit. This particular sand packet is lenticular on a 1–2 km scale.

1.4. Penybont Uchaf. Return to the road and drive north-west along the Irfon Valley to a bridge at SN 8565 5060. **In the river bed**, under the bridge, there are good exposures of folded, cleaved, laminated mudstones of the Cefn Ystradffin Formation of late Ashgill age. Bedding is on a 2–10 cm scale, with dark and light grey laminae.

1.5. Nant Rhydgoch. Continue north-west along the road until, at SN 8500 5175, a track leads off to the west towards Llwyn-derw. It should be possible to **park a vehicle off the track.** Follow the footpath marked on the 1 : 50 000 map leading west from Llwyn-derw to the valley of Nant Rhyd-goch. The **lower reaches of the stream** section here expose very dark grey, fine-grained, well cleaved mudstones with thin silts and fine- to coarse-grained sandstones up to 0.5 m thick. These sediments represent the top of the Cefn Ystradffin Formation.

Near the confluence of Nant-y-Brain and Nant Rhyd-goch (SN 8430 5165) there are **stream-bed exposures** of finely laminated dark and medium grey mudstones showing well-developed soft-sediment deformation structures, bioturbation and microlaminated (sub-mm scale) layers. These represent the Trawsnant Formation (Rhuddanian) which comprise the bulk of the hillside north-west of Llwyn-derw.

This valley is bounded to the north-west by **a prominent ridge** which is capped by the pale-grey weathering, silty mudstones of the Cefn Isaf Formation (Rhuddanian-Aeronian).

1.6. Camddwr Bleiddiad. Return to the vehicle and continue north-east along the Irfon Valley towards Abergwesyn. Turn left at Abergwesyn onto a road signposted to Tregaron. Follow the road into the Irfon Gorge, stopping at **Camddwr Bleiddiad** (SN 8415 5500) to admire the view. The steep north-east side of the valley is **capped with crags** of folded, pale-grey weathering silty mudstones; the Llyn Brianne Formation (Aeronian–Telychian). The line of crags on the north-east valley wall coincides with a major synclinal hinge zone that can be traced south-westwards to beyond Llyn Brianne.

Walk down to the stream section at this point to observe the lithology in detail and the spectacular, **narrow rocky gorge** of the Irfon. Note the pale-grey weathering, mud-dominated lithology with fine silt layers up to 2 cm thick (Llyn Brianne Formation).

1.7. Pen y Cnwc. Continue to the north-west along the valley floor to the Devil's Staircase (SN 8310 5590), where the road climbs steeply out of the Irfon Valley through pine forests. At this point a major anticlinal structure brings older (Upper Ashgill and Lower Llandovery) rocks close to the surface (see Fig. 2). This is a periclinal fold which plunges at about 15–20° to the south-west. Around Pen y Cnwc (SN 8250 5630), the relatively steep plunge can be determined from **roadside exposures** where the bedding-cleavage intersection lineation plunges to the south-west.

1.8. Bryn Mawr. The road gradually loses height as it passes into the Tywi Valley near Nant-yr-hwch. At a junction (SN 8110 5625), carry straight on towards Tregaron and head north-west along the Tywi Valley. Pause at about SN 8060 5670 to observe well-defined fold structures in the north-east valley side. These folds have wavelengths of about 0.5 km and occur in rocks of the Llyn Brianne Formation. Notice that the folds are NE–SW-trending and SE-verging (i.e. anticlines display steep to overturned south-eastern limbs and more gently dipping north-western limbs). This fold style is typical throughout the area.

Continue westwards, climbing again steeply out of the Tywi Valley towards the high moors. **Near Bryn Mawr** (SN 7960 5735), low grassy outcrops expose open fold culminations **on either side of the road.** They are NE–SW-trending and gently plunging. At this point, measurements of cleavage orientation and cleavage-bedding intersection lineations have indicated that the cleavage transects the fold hinges.

1.9. Nant y Gerwyn. Stop at a **junction with a rough forestry track** (SN 7830 5760) to examine roadside exposures of the Hafdre Formation (Telychian). This formation differs from the underlying Llyn Brianne Formation in the presence of distinct silt to fine-sandstone layers, which weather-out from the background mudstone to form tabular fragments. Sand layers are typically 1–10 cm thick in this area and commonly show well-developed, parallel and ripple-cross lamination, basal bioturbation and tool-mark structures. The way-up of the bedding can be determined readily using these structures, although anomalous 'inversions' are often caused by local down-slope creep in the overlying soils.

Bedding dips in this area reveal open folding of approximately 100–200 m wavelength. Owing to this folding, a traverse to the west along this section of the road to SN 7785 5760 encounters both the Hafdre and Llyn Brianne Formations.

1.10. Cefn Cerrig. Continue westwards to SN 7660 5760, where a small stream passes under the road. On the **north side of the road** are low craggy exposures of the Pysgotwr Formation (Telychian). Close examination reveals a thickly-bedded (20–50 cm), coarse-grained, poorly sorted sandstone with a high mud component in the matrix. This mud component causes the development of a cleavage in the sandstone. Interbedded with the thick sandstone layers are units of interbedded mud and fine-sandstone similar to those seen in the underlying Hafdre Formation.

ITINERARY 2: early to mid Ashgill shelf-slope sedimentation

Most of the Ashgill localities are exposures alongside forestry tracks. The forestry area is **not accessible by vehicle (without permission)**, but the localities can be visited on foot. **Permission for access must be obtained** from the Forestry Commission Offices in Llandovery. The forest can be entered via a well-made service track from the main A483 road just south of Esgairdafadd Farm at SN 8465 4415. Follow the track up the valley to a gate which marks the entrance to the plantation, where a vehicle can be **parked on the verge.** Most localities lie within 2–3 km of the forest entrance (Fig. 4).

2.1. Esgair Dafydd Quarry. The Bryn Nicol Formation can be examined at the type locality (Mackie & Smallwood 1987) in **a quarry** at SN 8370 4520. More recently, however, the quarry has been largely destroyed by blasting. The site may be occupied by a roadstone dump but there should be some relict, useful exposure along the north-east edge. Most of the loose rock in the dump is representative of the Bryn Nicol Formation.

The lithology includes interbedded conglomerates, sandstones and mudstones. The conglomerate beds are usually 10–50 cm thick, normally graded, matrix-supported and with pebbles up to 15cm in diameter. Both inverse- and normally-graded bed types occur.

Sand units are 10–20 cm thick, with parallel lamination and normal grading. Most of these sand beds are interpreted as Bouma T_a. Thinner, finer-grained sandstones show more internal structure and rare, nearly

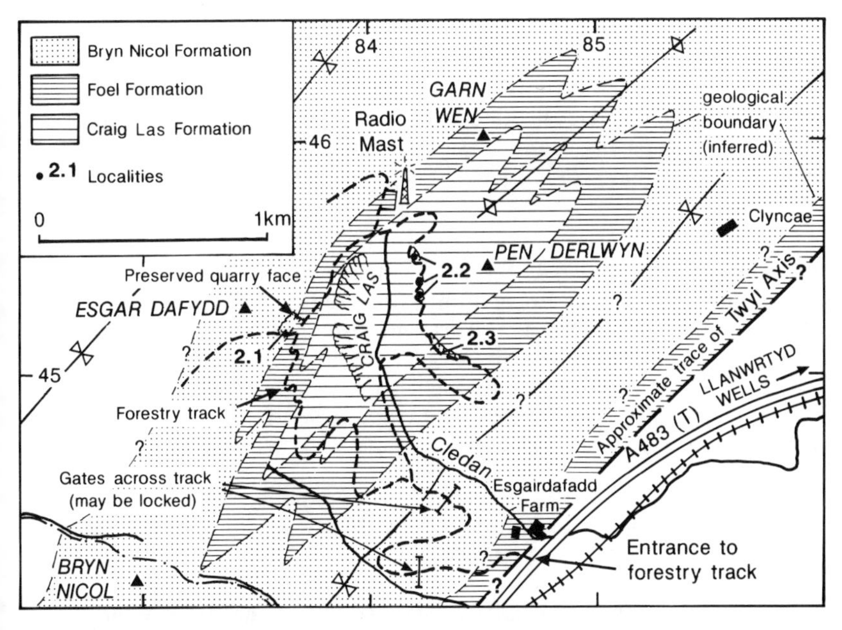

FIG. 4. Sketch map of Esgair Dafydd forest localities.

complete Bouma sequences (T_{abde}) can be seen. The mud facies is developed only locally but typically has silty, fine sand or, rarely, pebbly bases in units 5–10 cm thick. In contrast with other coarse facies in this area, the Bryn Nicol Formation is rich in faunal debris, including brachiopods, corals, gastropods, bryozoa, and crinoid ossicles.

The sedimentological features of the coarse units lead to an interpretation as the products of rapid deposition or lags on steep slopes in canyons or channels. The fossil debris indicates that the flows were probably sourced from relatively shallow water, probably on the shelf. The graded T_{abde} units represent the deposits of waning, moderate density, turbulent flows, and the mud facies probably resulted from dilute turbidity currents.

The environment was probably a coarse clastic wedge in a slope environment, sourced from relatively shallow water, possibly by powerful storm return flows on the shelf, or by the back-cutting of slump scars into the tectonically active shelf edge. The overlying Cynnant Fach Formation appears to represent a similar environment, but contains generally finer sediment and was thus probably formed slightly further from a shoreline or delta-front source.

2.2. West of Pen Derlwyn. Walk to the north-east along the forest track taking the right fork after about 0.5 km. Examine the silty mudstones of the Craig Las Formation in a **well-exposed track cutting** at SN 8422 4554. Note the typical beds 5–10 cm thick with thin, diffuse, rarely parallel laminated silty bases. These units correspond to T_{de} units of Stow & Piper (1984) and are interpreted as fine-grained turbidites deposited in an outer shelf or upper slope environment. Interpretation is on the basis of sedimentological features and proximity to the contemporaneous Tywi Axis (a shelf-slope break).

2.3. South of Pen Derlwyn. A sharp contact between the older Craig Las Formation and the Foel Formation is exposed at SN 8436 4512 on the north-east side of Cwm Cledan. This unit also contains a high proportion of mud but is characterized by tabular, fine-grained sandstone beds 5–30 cm thick, which are especially well displayed at SN 8441 4502. Some sand beds show clear grading, parallel laminae and muddy tops and correspond to Bouma T_{abe} or T_{ade} units.

This facies was probably deposited in a similar environment to the underlying Craig Las Formation, but with periods of lowered sea-level introducing coarser sediment periodically.

ITINERARY **3: late Ashgill and Llandovery slope and basinal sedimentation.** Most of the late Ashgill and Llandovery localities are well-exposed close to public roads and are readily accessible. Begin this itinerary at the Llyn Brianne dam site (SN 7930 4830) where there is ample parking. Refer to Figs. 5 and 6.

3.1. Llyn Brianne dam site (east). This is the type locality of the Cefn Isaf Formation (Rhuddanian-Aeronian). Here the formation is exposed superbly in **quarry faces on the eastern side of the dam.** The background lithology shows very little variation throughout the area and is widespread throughout mid Wales. It consists of medium–dark grey silt-mudstone units (weathering to pale grey) typically 1–20 cm thick. The silt layers are usually very thin (1–10 mm) and grade very rapidly into mudstone corresponding to T_{de} units of the Stow & Piper (1984) model. The dark grey, organic-rich microlaminated layers common in the underlying Trawsnant Formation are virtually absent.

Close examination of the **cliff exposures** reveals soft-sediment normal and reverse faulting, soft-sediment folding, and slump scars (0.1–1 m scale; Fig. 5).

Also exposed at the dam site is a rare coarse facies, consisting of conglomerates and coarse sandstones, locally rich in shelly debris. This facies occurs as discrete, semi-continuous beds, small-scale lenses (0.1–1 m scale) and as large slumped beds (some with relict bedding) and mega-balls (Fig. 5). The facies has not been traced for more than 100 m along strike.

Typical coarse-sand units show basal scours with gravel lags and may grade upwards into fine sandstones. The conglomerates are generally matrix-supported with clast sizes ranging from 0.5–10 cm. Clasts are subangular to rounded and consist of quartz-vein fragments with subsidiary quartzites and calcareous mudstones. Dark grey mud 'rip-up' clasts are also common. Shelly debris includes brachiopods, crinoids, trilobites and corals.

The silt-mudstone component is interpreted as the deposit of dilute turbidity currents, well away from primary sediment transport paths. Sedimentation was probably extremely rapid, as indicated by the lack of hemipelagite layers which would have had insufficient time to develop between successive turbidity currents. The presence of soft-sediment deformation is consistent with a slope setting.

The coarse facies shows a variety of forms – bedded, lensoid and massive, slumped. The bedded units probably result from erosive, high density turbidity currents (Lowe 1982) that flowed in channels down the mud covered slope. The lensoid bodies are probably smaller scale, genetic precursors of the above.

The slumps and mega-balls imply a different sedimentation process. They are interpreted as debris-flow products and were probably deposited *en masse*

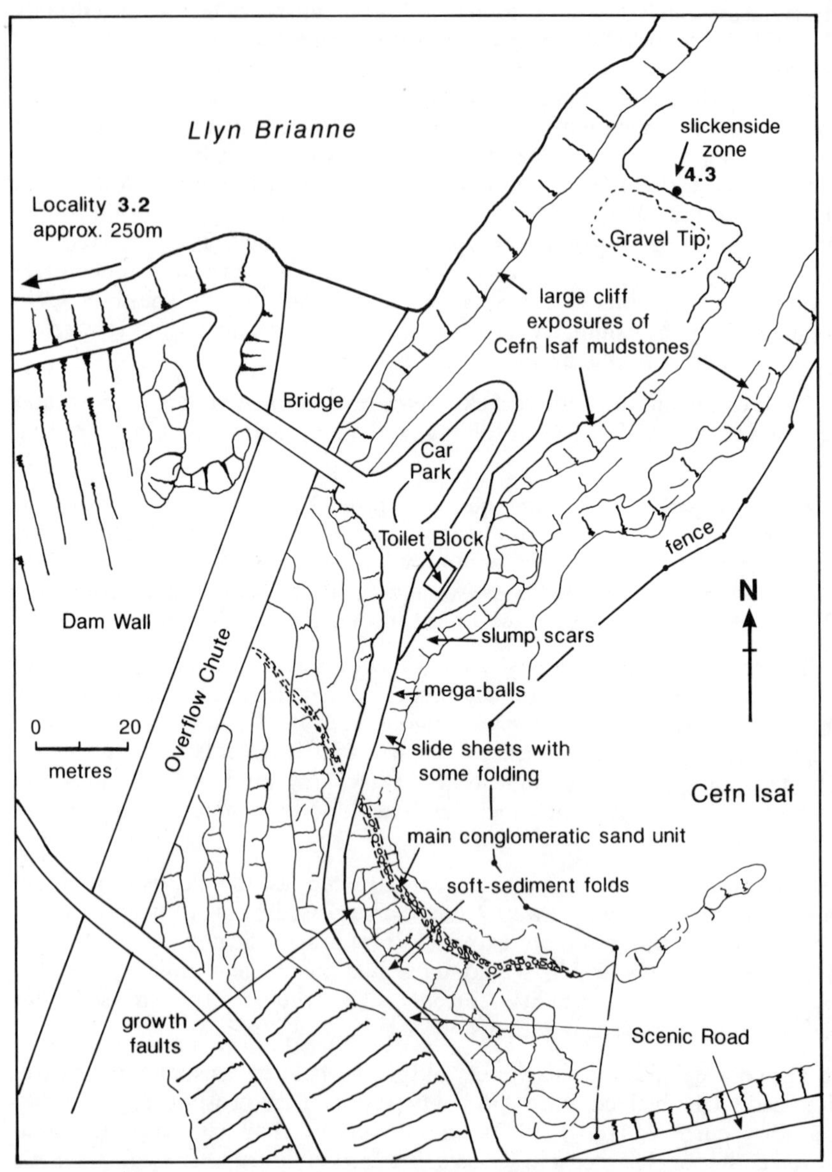

FIG. 5. Sketch map of Llyn Brianne dam site showing location of features discussed in the text.

after originating as reworked, coarse-facies turbidites. Semi-continuous masses with relict bedding may represent discrete slide-sheets. The mega-balls resulted from overturn, driven by density contrasts after relatively denser conglomerates and sands were dumped onto very water-rich silt-muds.

In the case of all the coarse facies, sedimentation (or final emplacement) was probably induced by a sudden change of slope. This may have been local (slump scar) or a more regional base of slope.

3.2. Llyn Brianne dam (west). Walk across the dam wall to examine the type locality of the Llyn Brianne Formation (Aeronian to Telychian).

The formation differs from the underlying Cefn Isaf Formation in having generally thicker bedding units, a higher component of silt and fine sand, better-developed ripple lamination, and an apparent lack of soft-sediment deformation structures. This is particularly well illustrated in the **cliff exposures** between SN 7895 4851 and SN 7872 4870.

In this area, the lithology consists of silt to fine sand/mud units, compositionally similar to the Cefn Isaf Formation. These units are 2–5 cm thick, parallel and ripple cross-laminated, and grade upwards rapidly into medium–dark grey, essentially massive mudstones (5–10 cm thick).

Some silt/fine sand layers show shelly and gravel lag bases which are often erosive. These graded, laminated, units correspond to the T_0 to T_8 units of Stow & Piper (1984).

The Llyn Brianne Formation is thought to represent a similar sedimentary environment to that of the Cefn Isaf Formation. The generally higher proportion of silt/fine sand indicates a location closer to a main sediment transport path, and the lack of soft sediment faulting/folding suggests deposition on a more level substrate. The Llyn Brianne Formation was probably deposited in an outer fan/basin plain setting.

3.3. Trawsnant. From the dam site, drive approximately 1.5 km to the north-east to a point where the stream Trawsnant crosses the road (SN 8050 4893). Examine the **stream-bed exposures downstream from the road** within 300 m. This is the type locality of the Trawsnant Formation (Rhuddanian).

A particularly good polished surface is exposed in the stream bed below a point where a new fence (in May 1990) is attached to the easternmost silver-birch tree on the south bank. Here, the smoothly polished and wet rock surfaces show finely and very finely laminated dark/pale grey mudstones with rare silt layers. Lamination is on a scale of 1–2 cm. The pale layers are composed of very fine silt and mud and are essentially unstructured. The darker layers are themselves finely laminated (sub-millimetre scale). A variety of sedimentary features is preserved, including soft-sediment faulting, chaotic slump folding, and soft-sediment injection structures. Some beds also show extensive bioturbation (*Chondrites*).

On the basis of a high proportion of well-laminated, organic-rich mudstones with a terrigenous component and soft-sediment deformation, this facies is thought to have been deposited on a quiet slope fairly close to the basin margin.

3.4. Mynydd Trawsnant. Drive north along the scenic road to SN 8180 4930 where a stream draining Mynydd Trawsnant passes under the road. Examine the Nant y Ffin Member of the Cefn Ystradffin Formation (Upper Ashgill) in **roadside cuttings.**

The mudstone is massive, dark blue-grey and commonly shows a purple-indigo surface staining (probably manganese) or a rusty film, reflecting the pyrite content. In fresh exposures, master-bedding surfaces can be seen at 10–30 cm intervals. There are some thin silt layers and bedding-parallel phosphatic concretions. Sedimentary structures are rare but some silt layers show parallel and ripple-cross lamination.

The formation also contains several lenticular, coarse sand and conglomerate bodies. If time allows, one of these can be examined by walking approximately 250 m up the **stream section.** Here, a medium to coarse, parallel-laminated sandstone unit is exposed **on the west bank** of the stream, adjacent to a prominent rowan tree. The unit is about 6 m thick here and can be traced along strike, uphill away from the stream where it thins rapidly, showing its lenticularity. The sand body is not visible on the east bank of the stream.

The lower member is generally inaccessible from the road but is represented at Locality 1.3. This lower member (the Cwm Henog Member) is finely laminated on a 2–5 mm scale and locally shows extensive mottling (*Chondrites*) or slumping.

The laminated facies is interpreted as a reflection of seasonal factors: the dark layers resulting from periods of greater organic activity (spring/summer) and the paler layers reflecting a higher proportion of clastic input (winter storm activity).

The more massive facies (in the roadside cutting) reflects a different origin. These units, especially those with silty bases, are interpreted as T_{de} units of Stow & Piper's (1984) silt turbidite model and probably represent the products of dilute turbidity currents. The environment was probably a slope setting, well away from the main sediment transport paths. The higher incidence of silt/mud units close to the top of the formation reflects the onset of a glacio-eustatically lowered sea-level. The coarse facies in both members probably represents deposits of high density turbidity flows, probably in channels cutting into the slope.

3.5. Clyn Glas. Walk or drive further north-east along the scenic road to SN 8208 4970, where there is **adequate space to park** off the road. At this point, **a stream (Nant Gwrach)** drains the valley to the north-east. This is the type locality of the locally developed, but significant, Clyn Glas Formation (Hirnantian to Rhuddanian). The outcrop pattern has been mapped out along strike and is highly lenticular (Fig. 5).

The lithology is dominated by dark–medium grey mudstones with interbedded siltstones and sandstones. The proportion of coarser material decreases gradually upwards as the unit passes into the Trawsnant Formation (at SN 8160 4978).

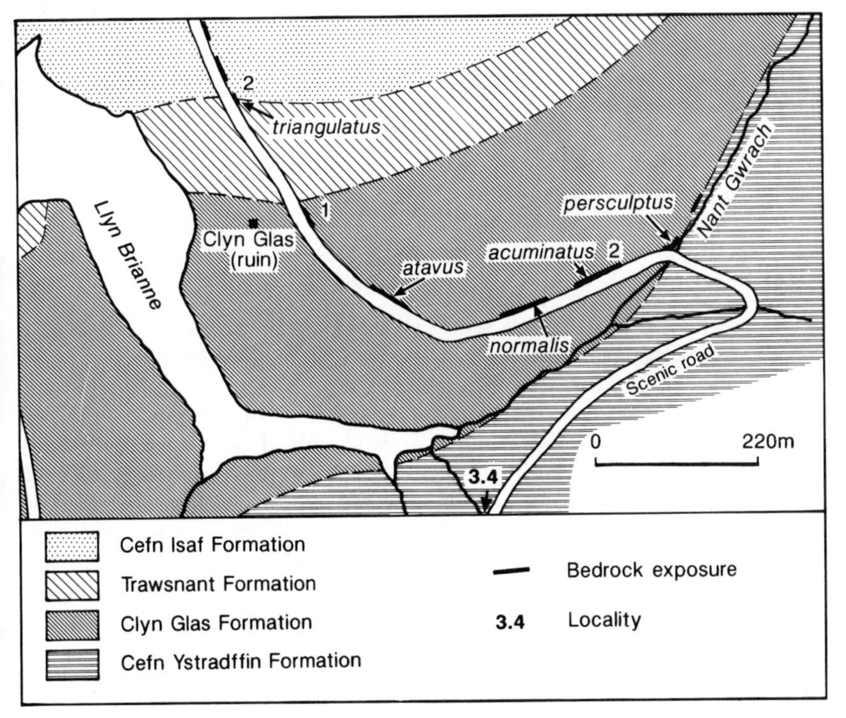

FIG. 6. Biostratigraphical geological sketch map of the Clyn Glas area showing graptolite sample points and major lithological boundaries. 1 = transitional facies to overlying Trawsnant Formation; 2 = shelly debris and flaggy sandstones. Bedrock exposures shown in solid black.

Within 100–200 m to the west of the stream, **along the road cut**, there are several shelly and micaceous beds (5–10 cm thick) which contain fragments of brachiopods, gastropods, crinoids and bryozoa. Higher in the formation, palaeocurrent directions derived from ripple cross-lamination indicate on-shore (i.e. east-north-east) directed currents.

The Clyn Glas road section has yielded graptolites that enable a well-constrained biostratigraphy to be applied to the base of the Llandovery in this area. Figure 6 shows the localities where graptolite assemblages enable zones to be identified. Full biostratigraphical details of the area are available in Mackie (1987), Smallwood (1986*b*) and Mackie & Smallwood (1987).

The lenticular outcrop pattern, relatively high proportion of clastic material, shelly debris, and indication of on-shore directed palaeocurrents leads to an interesting environmental model. The formation was probably deposited in a series of negative relief features on a slope (large-scale channel scours or slump scars). The high proportion of clastic and shelly material represents a lowered sea-level which would have effectively moved the sediment source closer. This is supported by the on-shore directed palaeocurrents, which probably represent the movement of regular deep-tidal and storm-driven currents. The outcrop pattern represents large-scale scouring of the shelf mud-apron, which would have been facilitated during a period of lowered sea-level.

3.6. Capel Soar-y-mynydd. Continue along the eastern side of the reservoir, cross the bridge at SN 8051 5377 and head west for about 2.5 km.

Examine the Hafdre Formation (Telychian) near the **southern end of a roadside exposure** near Capel Soar-y-mynydd (SN 7846 5358). In an unbroken 150 m section on the eastern side of the road, fine-grained sandstones are interbedded with medium–dark grey mudstones. Sand units are typically 2–10 cm thick, and are well sorted with well-developed parallel and ripple cross-lamination. On the bases of some beds are well-developed scours, flute casts, tool marks and bioturbation trails. Bedding thicknesses are remarkably constant and evidence of soft-sediment deformation has not been found. These beds correspond to the T_{cd} sequences of Bouma (1962). Towards the northern end of the section, the rocks grade into the less sandy lithology of the underlying Llyn Brianne Formation.

On the basis of sedimentary structures, the high proportion and maturity of the sand component, the lack of coarse material, and the continuity of bed thicknesses, deposition from distal, medium density turbidity currents is favoured as the sedimentation process. This was probably a mid–outer fan or basin plain environment (cf. Mutti & Ricci-Lucci 1978). The mudstone component was probably deposited either by separate, more dilute turbidity currents or the waning stages of the currents responsible for the sandstone beds.

3.7. Nant-y-Maen. Follow the road north-westward alongside Afon Camddwr for about 6 km to a road junction at SN 7625 5755. Head towards the telephone box at SN 7625 5770 where **a vehicle can be parked.** A few metres west of the telephone box, a thick bed of the Pysgotwr Formation (Telychian) forms **a small waterfall** in the stream (Nant-y-Maen). On the northern side of the road, an exposure at SN 7620 5765 shows thick (up to 0.5 to 1 m), medium grey, very coarse-grained, poorly sorted sandstones with a high proportion of mud in the matrix. There is normal grading near the base of some units and parallel lamination near the top of the beds. Some beds show mud rip-up clasts.

In this area, the thick, coarse sand beds can be traced along strike for considerable distances (> 1 km) and form mappable, NE–SW-trending ridge features across the hills. This facies is interbedded with a sand–mud lithology very similar to that of the Hafdre Formation. The two interbedded facies types (sand/mud and coarse sand) are a key to the environmental setting. The thick, coarse sandstones are thought to be the deposits of high density turbidity currents (T_a units of Middleton & Hampton 1973). The thickness and lateral continuity of these beds indicates a probable outer-fan lobe origin (Smith 1987*a*). The interbedded sand/mud facies represents background, more dilute turbidity current activity.

ITINERARY 4: structural features

4.1. Nant-y-bai. Take the road north-west from Rhandirmwyn (SN 7850 4370) for about 1.5 km to Nant-y-bai (SN 7755 4460). At this point the main road curves to the west and a forestry service track joins from the east. There may be a **locked gate preventing vehicular access** to this site. **Permission for access** can be obtained from the Economic Forestry Group regional office in Carmarthen. Walk up the track, following the stream to the open, abandoned mine site. This is the old Nant-y-mwyn lead–zinc mine.

From SN 7850 4470, look north-east along the Nant-y-Bai valley. This major feature represents the course of a basin-down normal fault. The current displacement is approximately 100 m, downthrown to the north-west. Although offsets can only be demonstrated by mapping outcrop patterns, the marked valley feature, presence of a major mine site, and abundant quartz veining are secondary evidence of a fault. If time allows, walk up the valley on the **west bank of the stream** to SN 8010 4655 where there is a large exposure of net-veined, brecciated sandstone and mudstone.

4.2. Craig Clyngwyn/Rhyd-y-Groes. Return to the main road and drive north-west from Nant-y-bai to SN 7730 4590 where a side road to the west leads to a bridge across Afon Tywi. Cross the bridge and continue for about 1.5 km along a narrow, tarmac road until you reach a bailey bridge at SN 7760 4720 (Fig. 7). **Leave the vehicle** on the south side of the river as the **land to the north is private.**

You can continue on foot as there is **a public footpath.** From the bailey bridge, walk about 400 m westwards, following the line of the fence on the upslope side of the road. At the point where the fence turns at 90° to head directly upslope, a prominent sheep track leads towards the base of some crags. Follow this path up to the crags where it passes through **a low gully.** Examine the relatively flat-lying surfaces either side of this gully where there are numerous, very well developed, open, close and chevron folds (cm scale) in interbedded silt/sand and mudstones.

Notice also that there is a sudden change in the course of the river below this point (where the Afon Doethie turns sharply to the north-east at an elbow bend). This coincides with the trace of a significant fault zone which is associated with the major, faulted Doethie Anticline. The fault is aligned NE–SW and is also responsible for a series of valley features between SN 7780 4870 and SN 7440 4475. This is a composite fault zone with a predominant reverse displacement of about 100 m (down to the south-east). Mapping shows that the displacement diminishes gradually to the north-east and south-west of here.

There are subsidiary normal and reverse faults associated with this zone (Fig. 7). A complex group of outcrops on the **south side of the valley** (SN 7700 4720) exposes a series of truncated, reclined folds and low-angle thrust faults with NW–SE displacement vectors indicated by slickenlines. These can be examined by returning to the bailey bridge and walking along a rough track for about 600 m to the west. **Please obtain permission** from the RSPB warden before entering this area (tel. 0550 6228). The thrust zone is exposed on the thickly-wooded lower slopes due south of Rhyd-y-Groes.

Figure 7 shows that this is a structurally complex area with folds, normal faults, and steep and low-angle reverse faults closely juxtaposed. This is thought to relate to the major Doethie Anticline/Fault, which may have a component of lateral as well as reverse displacement.

4.3. Llyn Brianne dam site. Return to the bailey bridge and drive back to rejoin the main road at Afon Tywi. Turn left and head north to the Llyn Brianne **dam site car park** at SN 7940 4846. Walk about 50 m north from the edge of the car parking area to a vertical, approximately E–W-trending **quarry face on the ground level terrace** at SN 7945 4811 (see Fig. 5).

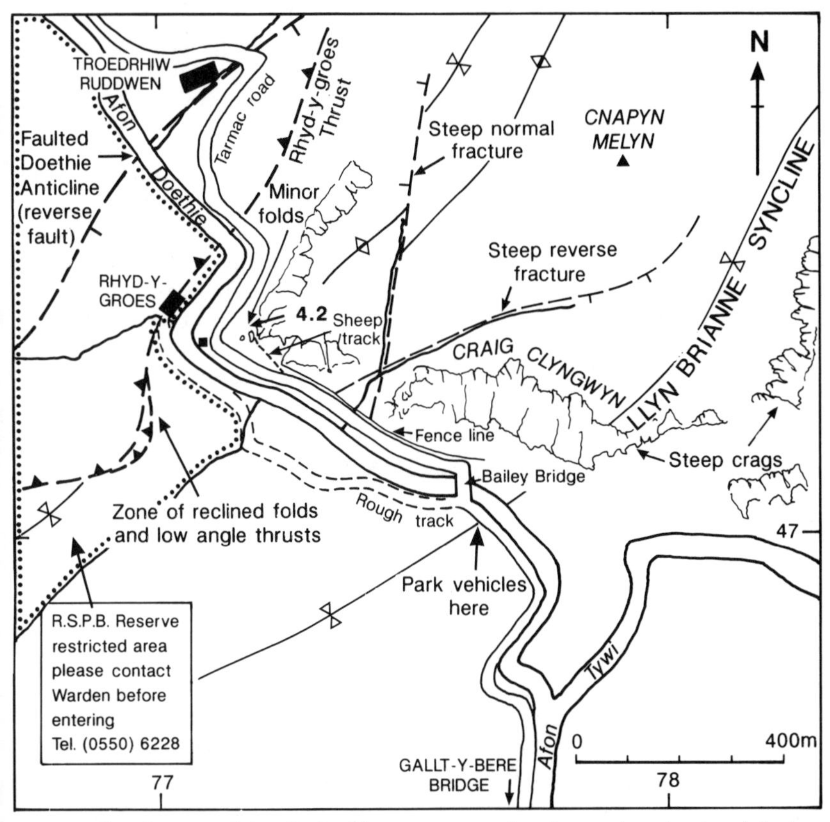

FIG. 7. Sketch map of the Craig Clyngwyn area showing major structural features and location of well-developed minor folds.

Here, in the centre of the quarry face a 10 cm thick, bedding-parallel fracture dips at about 36° to the north-west. This is a zone of concentrated slickencryst development, with fibres consisting of quartz and carbonate with iron staining. Lineations plunge to the north-west but there is no indication of the scale of movement as the lithology is so monotonous.

This feature is thought to have formed early in the deformation sequence because other, similar features show the slickencrysts to be offset by cleavage. Also, measurements elsewhere have shown that the lineations are not perpendicular to mean fold axes. This implies that they do not result from flexural slip during folding. These zones are thought to have formed by bedding-parallel slip during pore-fluid, pressure-aided gliding of sediment sheets prior to the major deformation (cf. Fitches *et al.* 1986).

4.4. Mynydd Trawsnant. Lithology plays a major role in controlling the style of deformation in this area. Drive north-east **along the scenic road** to SN 8135 4925. This is a well-exposed section through intensely cleaved Upper Ashgill mudstones with rare silts. Notice that minor (cm scale) folds are absent in this mud-dominated lithology.

A small stream crosses the road at the eastern end of the section. Follow the line of the fence along the top of the road cut to the first double-braced fence post. Closely examine **the rock face** directly below this where an anticlinal closure can be seen. Note that the fold is south-east verging and that the south-east limb is cut by a reverse fault. This structure reflects the structural style throughout the area. Faults are not common, but where found are often closely associated with the hinges of anticlines on various scales (the Doethie Anticline and fault zone is another example).

4.5. Nant Craflwyn. Continue northward along the scenic road to SN 8100 5100 where a relatively open anticline is **exposed by the road side.** Note the well-developed cleavage which appears approximately axial planar to the fold.

Further folds are exposed along **the cliff face** between SN 8080 5200 and SN 8070 5245.

4.6. Bryn Mawr. Drive north to the bridge at SN 8051 5377. Cross the bridge and continue heading west for about 1.5 km to the brow of the hill. Drive or walk downhill from here (heading west) and note the repeated changes in lithology from silt/mud to sand/mud lithologies caused by folding of approximately 0.25 to 0.5 km wavelength.

This folding pattern is repeated **along the road** as far as Locality 4.7. The folds occur along the boundary between the older Llyn Brianne Formation (mud/silt dominated) and the Hafdre Formation (mud/sand) and are thought to reflect the competence contrast along the boundary between the different lithologies.

4.7. Nant-y-Graig. Stop at SN 7830 5470 where there is an **elbow bend in the road.** Note that the stream (Afon Camddwr) also bends at right angles here.

Examine **the rock face on the north side of the road** approximately 3–5 m west of the minor stream. Here a thrusted anticline is exposed, very similar to that described at Locality 4.4. A sandier lithology is thrust over more mud-dominated sediments.

Look closely at the axial region of the fold. There is a prominent bedding surface 2.5 m above ground level. A kink-band on this surface is clearly oblique to the axis of the fold. The axes of these kinks are steeply plunging and the kink-band clearly post-dates the fold, indicating a late origin in the deformation sequence.

Acknowledgements. I thank Nigel Woodcock and Stuart Neilsen Smallwood for help and advice given both during the preparation of this chapter and during the period when the original work was done. I also thank PRO-IV Systems Ltd. for allowing me the use of word-processing and printing facilities during the preparation of the manuscript. Finally, my thanks go to Elizabeth Dainty who acted as my field companion during May 1990 and provided many useful comments and constructive criticisms of the itineraries.

12. THE SILURIAN OF THE NEWBRIDGE–BUILTH–EPPYNT AREA

by M. G. BASSETT

Maps *Topographical:* 1 : 50 000 Sheet 147 Elan Valley and Builth Wells
1 : 25 000 Sheets SN 94, SO 04, SO 05

THIS excursion is designed as a north to south traverse across an ascending sequence of Llandovery–Wenlock–Ludlow–Přídolí strata down the Wye valley from Newbridge on Wye to Builth Wells and continuing onto the northern flanks of Mynydd Eppynt. From Newbridge to Builth the rocks form a much-faulted, shallow, and in places overturned asymmetrical synform which pitches gently to the south-west and whose eastern flanks wrap around a core of Ordovician volcanics and sediments of the Carneddau Range (Jones 1947, plate 1). To the south of Builth the younger beds continue in a relatively uncomplicated succession upwards onto the Eppynt, striking E–W with only minor flexures and faults and maintaining a gently southerly dipping to sub-horizontal disposition (Straw 1937, plate 29; Fig. 9 herein).

The area as a whole is one in which a number of significant contributions to Silurian geology were first made. Murchison (1839) gave the first descriptive account of the region, followed by other general accounts through the nineteenth century, and then in 1900 Elles used the Wenlock succession between Newbridge and Builth as the basis for defining a standard graptolite biostratigraphy that is still essentially valid world-wide. Similarly, Wood (1900) used the sequence in support of her initial definitions of Ludlow graptolite biozones. This is also the region in which Straw (1937) first produced a refined classification of Ludlow Series strata based on detailed lithostratigraphical subdivision and bio-stratigraphical content, to break away from the long-used standard recog-nition of 'Lower Ludlow', 'Aymestry Limestone' and 'Upper Ludlow' divisions that had been recognised almost everywhere in Wales and the Welsh Borderland since the early nineteenth century. In the same publication, Straw demonstrated convincingly for the first time the role of sedimentary slumping in the emplacement of Ludlow deposits, and this role was later amplified fully by Jones (1947) for the Wenlock strata north of Builth.

Reference to slumping gives an immediate pointer to the palaeogeographical setting of the Silurian sediments described in this excursion. Background sedimentation was largely as hemipelagic deposition in the Welsh Basin, but the repeated incursion of slumped units indicates clearly the proximity to the slope environments of the trough–shelf margin. Many of the shelly faunas are clearly allochthonous and were transported by the mass movement of sediment from the shelf areas to the east and south-east.

ITINERARY

The itinerary is described as starting from Builth Wells and is arranged in approximately stratigraphical order from the Llandovery upwards. Many of the roads in the area are narrow and **unsuitable for large coaches**. Localities 1–4 are also **not suitable for large parties**, but are better restricted to small specialist groups. Localities 5–9 concentrate on the Ludlow succession but are of more general interest and accessibility; together they are particularly suitable for a half-day excursion, and provide some beautiful scenic views in addition to the geology. These localities also link conveniently with those of Excursion 13 to give an extended examination of the Ludlow Series on both sides of the Wye Valley.

1. River Ithon: Dol-fawr – Craig Fain. From Builth take the A470 road north-north-westwards towards Newbridge on Wye, and turn off to the north-east immediately south of Pont a'r Ithon (SO 0195 5725) along the farm track to Dol-fawr (Fig. 1). **Parking for small vehicles** is available at

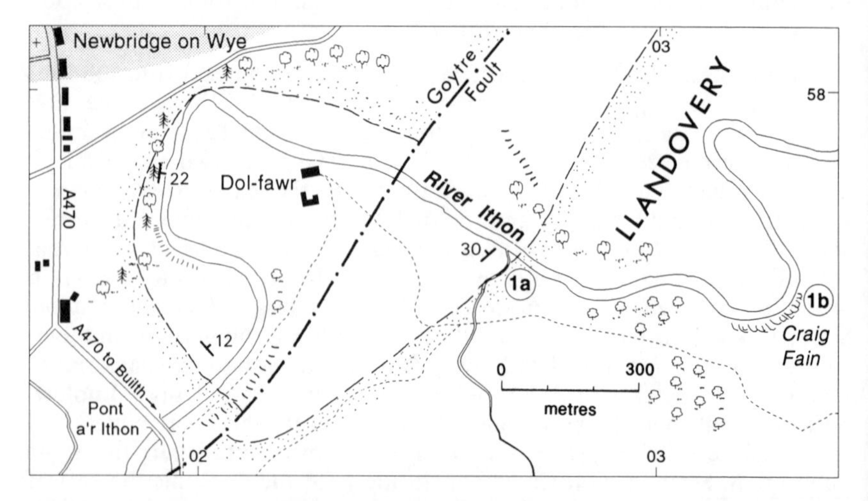

FIG. 1. Location and simplified geology of the River Ithon between Dol-fawr and Craig Fain (Locality 1). Fine stippling indicates the top of the Llandovery beds. Grid lines are 1 km apart.

the farm, where **permission must be sought** to visit the exposures. Walk eastwards along the south bank of the River Ithon.

Wenlock shales and mudstones in this region are preserved within a faulted envelope of upper Llandovery Pale Shales (Fig. 1: Jones 1947, plate 1). The Llandovery–Wenlock boundary is exposed in a **high cliff about 70 m above the mouth of a tributary** entering the Ithon from the south (Locality **1a**; SO 0265 5764); dips are to the north-west at 30–35°. The Llandovery sediments are distinctive 'striped' mudstones, with closely and uniformly alternating bands of thin black and pale grey lithologies. They are mostly hemipelagic muds, with the grey beds showing marked mottling (bioturbation) and oxygenation because of biogenic activity; the trace fossil *Chondrites* is common. There are some thicker bands of dark mudstone. About 1.5 m below the top of this unit, the beds yield abundant graptolites of the uppermost Llandovery *crenulata* Biozone, including *Monoclimacis crenulata*, *Monograptus priodon* and *Retiolites geinitzianus* (Jones 1947, p.9; Siveter *et al.* 1989, p. 78).

Above the 'striped' beds are conformable, darker grey-green to blue mudstones, occasionally bioturbated and with pale laminae. Within centimetres of the base there are graptolite faunas indicative of the basal Wenlock *centrifugus* Biozone, including *Cyrtograptus insectus*, *C. centrifugus*, *Monoclimacis vomerina vomerina*, *Retiolites geinitzianus*, *R. angustidens* and *Monograptus priodon*. There is then an upward transition into blue-grey laminated shales with a *murchisoni* Biozone fauna, followed by *riccartonensis* and *rigidus* Biozone faunas along the river towards Dol-fawr.

Further east along the Ithon, the **cliff section at Craig Fain** (Locality **1b**; Fig. 1; SO 0325 5752) preserves a thick succession of upper Llandovery Pale Shales, with greenish sandy mudstones and siltstones near the base, passing up into more massive beds into which 'striped' shales are interspersed increasingly. Graptolites suggest that the beds are entirely within the *crenulata* Biozone and only a short distance stratigraphically below the sequence at Locality **1a**.

2. River Wye. The lower banks and bed of the River Wye, southwards and south-eastwards from Newbridge and the confluence with the Ithon, contain numerous exposures of Silurian rocks (see Jones 1947, plate 1), but they are generally accessible only at periods of **very low water**. Access can be gained to either bank via the numerous small roads leading off the A470 or the minor road to the west, but **permission must be asked** at the appropriate farm for right of way to the river bank. **Large vehicles are not allowed** and this part of the itinerary is definitely not appropriate for large parties. It is suitable only for **small specialist groups** interested in details of the graptolite succession and in slumped sediments.

A representative traverse can be made along the northern banks by approaching from **Pen-min-cae** (SO 0065 5435) after leaving the main A470 at SO 0238 5439.

The major Goytre Fault trends NE–SW across the river at SO 0010 5445, throwing to the south-west within upper Llandovery Pale Shales. Southwards from here, **at the sharp river bend** just above Goytre Wood (SO 0015 5319), basal Wenlock graptolitic mudrocks and calcareous flags of the *centrifugus/murchisoni* Biozone follow above the Llandovery; the beds dip and young downstream at about 20–40°. Apart from centrifugids, the faunas include *Monograptus priodon* and *Monoclimacis vomerina vomerina*. Succeeding *riccartonensis* Biozone faunas then occupy the stretch of river almost to the point opposite the **south-east end of the small island** above the old ford (approximately SO 0525 5416 to 0540 5410). *M. riccartonensis* itself is present, accompanied still by *M. vomerina vomerina*. Some metres above the ford, however, these beds are cut off by the SE-throwing Gaufron Fault, which may have some lateral movement (Jones 1947, p.29), but whose main effect here is a downthrow of some 115 metres against mid Wenlock *ellesae* Biozone strata; the latter now dip and young north-westwards, passing successively downstream into older rocks of the *flexilis* and *rigidus* biozones. The eponymous cyrtograptids *ellesae* and *flexilis* are present accompanied upward by a fairly diverse fauna including *Pristiograptus dubius*, *Monoclimacis flumendosae* and *Monograptus flemingii*.

Within another few hundred metres, at SO 0120 5365, a further SE-throwing fault (the Dol yr Erw Fault of Jones 1947) throws lower Wenlock thinly-bedded *rigidus* sediments against a major unit of slumped mudstones of probably early Ludlow age (?*nilssoni* Biozone); the chaotic slumped masses are well displayed in the **high north bank** of the river through a stratigraphical thickness of some 90 metres. They pass down into Ludlow graptolitic shales, which in turn are faulted on the **sharp river bend** due south of Glan Gwy (at SO 0200 5340) against another slump sheet of mid Wenlock age (*ellesae* Biozone).

A descending bedded sequence of *ellesae* and *flexilis* Biozone shales, mudstones and calcareous flags then extends down river to some **30 m above the railway bridge**, where a further slump sheet crops out within the *flexilis* Biozone. This is the oldest slump unit present in the Builth district (Jones 1947, p.14), about 75 m thick and occupying a lateral distance of some 180 metres; its base is particularly well exposed on the underlying bedded strata. From this point downstream, a descending sequence of *rigidus* and *riccartonensis* Biozone rocks dips north-westwards at 30–40°, with beds assigned to the *murchisoni* Biozone cropping out at the base at

SO 0290 5250. Here the Silurian sits unconformably on Ordovician volcanics **in the river adjacent to Pen-ddol rocks**. The presence of shelly Llandovery sandstones mapped in by Jones (1947, plate 1) at this point between the Ordovician and Wenlock is difficult to verify.

3. Trecoed. From the A470 road adjacent to the Wye, either drive via Builth and northwards towards Llandrindod Wells on the A483, or take the minor turning heading north-eastwards from Cŵm-bach (SO 0294 5391). At the cross-roads at SO 0440 5472, take the small road to the north-east via Castle Crab to Trecoed (Fig. 2); **small vehicles** can be **parked at Trecoed Farm, with permission** of the owners.

To the south-east of the farm, in an **old shallow quarry** adjacent to the sharp bend (Locality **3a**; SO 0528 5516; Fig. 2), beds of the Trecoed Formation are coarse, rubbly, micaceous sandy mudstones, quartzitic

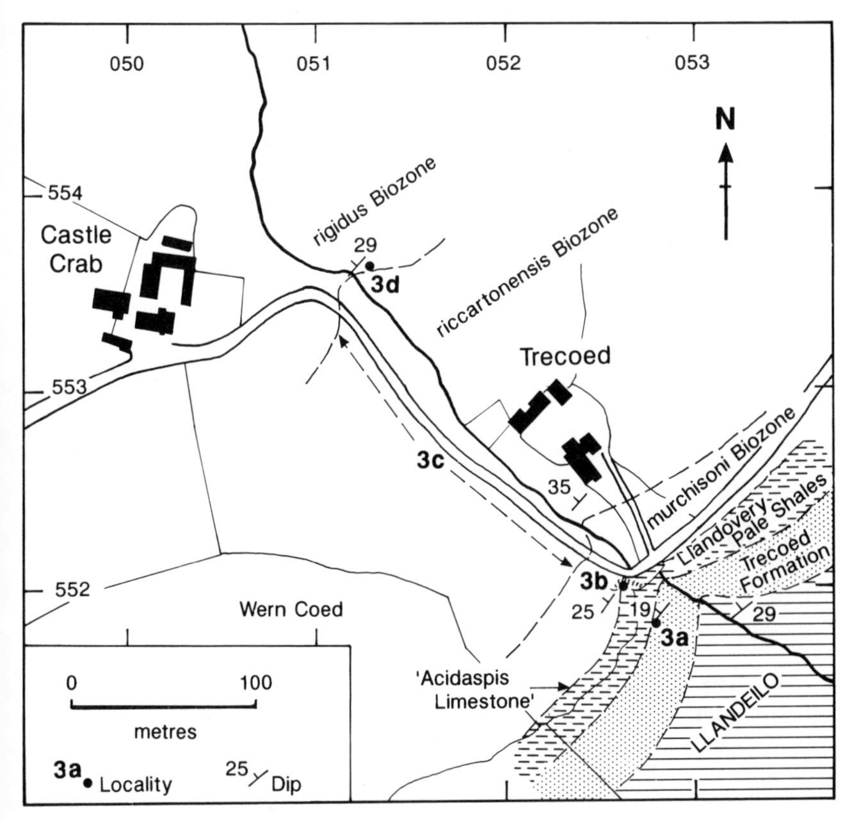

FIG. 2. Location and geology of the Trecoed section (Locality 3). After Siveter *et al.* 1989, fig. 75.

sandstones, and dark mudrocks. This is one of the few localities in the region to preserve transgressive arenitic Llandovery sediments with probable autochthonous faunas, which include the brachiopods *Pentamerus oblongus* and *Stricklandia laevis*. The fossils suggest an early Upper Llandovery age (C_4, Telychian) and a mid–outer platform setting with a mixed *Pentamerus–Stricklandia* Community (Ziegler *et al.* 1968, p.770). Although a contact is not exposed, the Trecoed beds clearly sit unconformably on Ordovician (Llandeilo) sediments, which crop out as dark flaggy shales and mudstones in the **stream banks** only a few metres to the east. The Llandeilo faunas include trilobites (*Ogyginus corndensis*, *Ogygiocarella debuchii*) and sparse graptolites (cf. *Diplograptus foliaceus*) (Baker *et al.* 1979, p.71).

Above the Trecoed Formation at Locality **3b among trees** immediately adjacent to the sharp road bend (SO 0526 5518) are Llandovery Pale Shales and then the Acidaspis Limestone of Jones (1947), which is a calcareous and rubbly mudstone containing small, fragmentary shelly fossils including the trilobite *Leonaspis* cf. *hughesi*, cardiolid bivalves and small, mostly indeterminate brachiopods. Sparse graptolites from the **adjacent Trecoed stream and track** (Elles 1900) suggest that the Acidaspis Limestone is of early Wenlock age (*centrifugus*/*murchisoni* Biozone); Jones (1947, p.5) considered that it follows unconformably above the Llandovery sandstones and shales, but the evidence for this is equivocal (Hurst *et al.* 1978, p.205; Siveter *et al.* 1989, p.78).

From Locality **3b** north-westwards along the farm road for about 200 m (Locality **3c**) and in the adjacent stream there are intermittent exposures of grey-green mudstones of 'Wenlock Shale' type similar to the Coalbrookdale Formation of platform areas across the Welsh Borderland. Hurst *et al.* (1978, p.205) report the presence of a brachiopod community containing *Visbyella trewerna* only 0.5 m above the Acidaspis Limestone, indicative of an offshore, low energy environment. Graptolites are not uncommon through the shaly sequence, including *riccartonensis*, *rigidus* and *flexilis* Biozone faunas towards and beyond Castle Crab. The **small old quarry** 100 m west of Castle Crab (Fig. 2; Locality **3d**; SO 0511 5534) yields a *rigidus* fauna including *Cyrtograptus rigidus rigidus*, *M. flemingii*, *M. antennularius* and *P. dubius*.

4. Irfon Bridge, Builth Wells. From Trecoed return southwards along the A483 to Builth and continue through the town on this road, turning to the west after crossing the River Wye. Take the minor road turning south at Irfon Bridge (SO 0319 5101) along the east bank of the River Irfon and stop after about 180 m. There is a fairly continuous section across the Wenlock–Ludlow boundary **in the river bank** (Locality **4**; SO 0322 5084 to 0322 5082). Access is via steps cut into the rock (Fig. 3).

This is a classic section, described in some detail by Elles (1900, pp.380–4, Fig. 1), Wood (1900, pp.431–3, Fig. 5), and Jones (1947, pp.15–16). It extends from well down in the *ellesae* Biozone close to Irfon Bridge itself, well up into the *nilssoni–scanicus* Biozone around the meander of Cae Beris (SO 029 506). Those wishing to trace the full graptolite succession in some detail should consult the publications quoted above; the purpose here is to give a general guide to the section across the Wenlock–Ludlow boundary.

The succession identified in Fig. 3 comprises mostly dark mudrocks with calcareous bands and nodules, dipping to the south-south-east at 35–45°. Faunas of the *lundgreni* Biozone are rich and diverse, including the eponymous species together with *Pristiograptus dubius dubius*, *Monograptus flemingii*, *Monoclimacis flumendosae*, and *Pristiograptus pseudodubius*. Accompanying shelly faunas include orthoconic nautiloids, cardiolid bivalves, and phacopid trilobites.

Close to the south end of this part of the section, **adjacent to the concrete steps** shown in Fig. 3, hard flaggy siltstones and pot-holed limestones form a

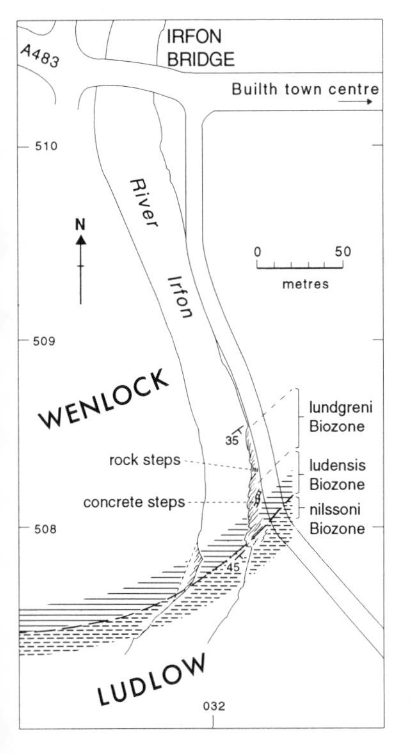

FIG. 3. Wenlock-Ludlow boundary section in the River Irfon (Locality 4). After Siveter *et al.* 1989, fig. 76.

prominent reef across the river, representing the 18 m thick slump unit identified by Jones (1947, p.16). Graptolites here are indicative of the topmost *ludensis* Biozone of the uppermost Wenlock (Siveter *et al.* 1989, p.79). Also present is the dalmanitid trilobite *Delops*, which is known from similar calcareous beds at this horizon in north Wales, the Long Mountain of Powys, Cumbria, Scania and Gotland in Sweden, and the Prague Basin of Czechoslovakia.

The *ludensis* siltstones are succeeded by softer, dark, brown-weathering, laminated shales containing a basal Ludlow *nilssoni* Biozone fauna. The Wenlock–Ludlow boundary is about **18 m south of the concrete steps** (SO 0322 5079) close to the point where the river swings away from the road. The lowest *nilssoni* faunas here include *Cucullograptus progenitor*, *Colonograptus colonus* and *C. compactus* accompanied by *Pristiograptus dubius*, with *Saetograptus varians* and *S. leintwardinensis incipiens* appearing in slightly higher beds.

5. Llangammarch Wells. From Irfon Bridge continue westwards on the A483 for 8 km to Garth and turn south-west onto the minor road at SN 9500 4935 to Llangammarch Wells. Locality **5** is a **small old quarry** off the minor road leading east out of the village, some 250 m south-east of the church (SN 9374 4720). **Small vehicles** can be parked on the **road adjacent to the quarry**, but **larger vehicles should be left in the village itself**.

The quarry exposes dark, occasionally calcareous, flaggy hemipelagic shales that are almost vertical to locally overturned. Rich graptolite faunas are indicative mostly of the *nilssoni* Biozone, with the better specimens being found in the slightly weathered scree material on the quarry floor. Species include *Neodiversograptus nilssoni*, *Pristiograptus dubius*, *Bohemograptus bohemicus*, *Monograptus auctus*, *M. ludensis*, *Saetograptus* cf. *colonus colonus*, *S. c. compactus* and *S. varians varians*. Synrhabdosomes of the latter species are not uncommon.

The rare occurrence of specimens of *Lobograptus progenitor* in the scree material (Bassett in Baker *et al.* 1979, p.74; Siveter *et al.* 1989, p.79) shows that the *scanicus* Biozone is also present here, although its precise location in the outcrop remains uncertain.

Accompanying faunas in thin shelly bands include orthoconic nautiloids, cardiolid bivalves, hyoliths, gastropods and brachiopods, of which *Howellella* cf. *elegans* and *Microsphaeridiorhynchus* sp. are the most common. Also present is a rare, branching, non-calcareous alga, *Powysia bassetti* (Fig. 4; Edwards 1977); the best preserved specimens have hold-fast structures, indicating that the alga was benthic in habit and was thus probably washed into the area together with the allochthonous

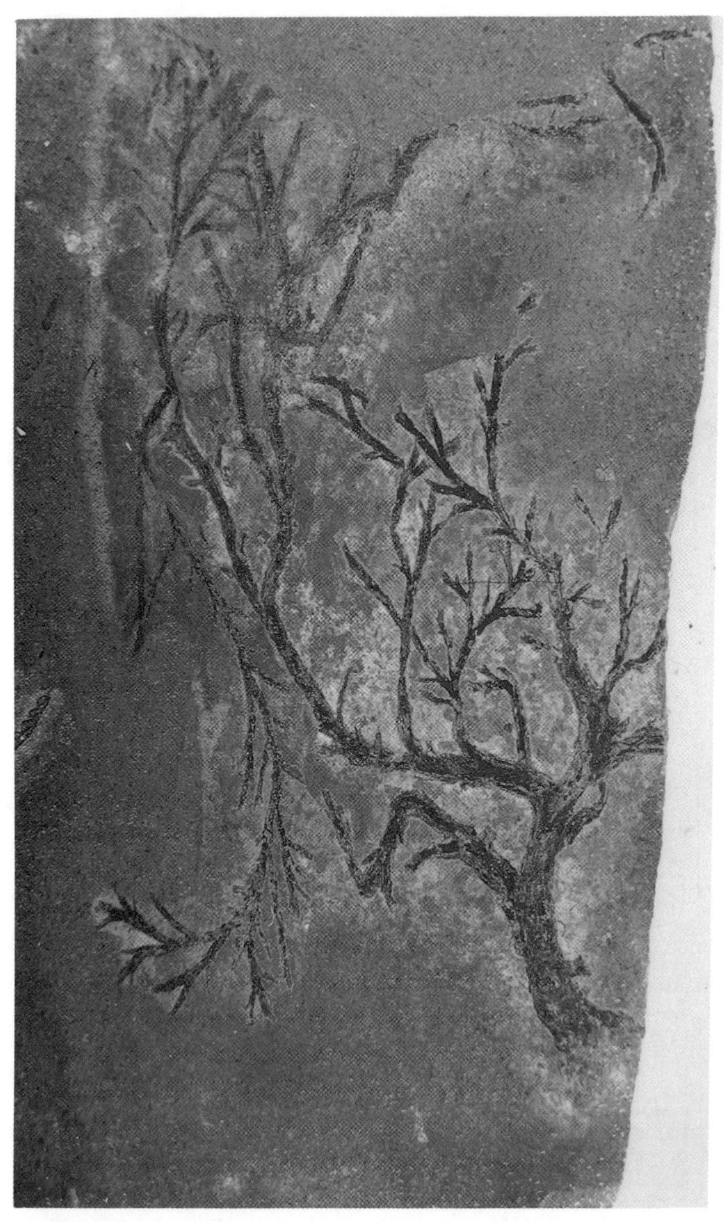

FIG. 4. *Powysia bassetti* Edwards from Llangammarch Quarry (Locality 5) (approx. ×2).

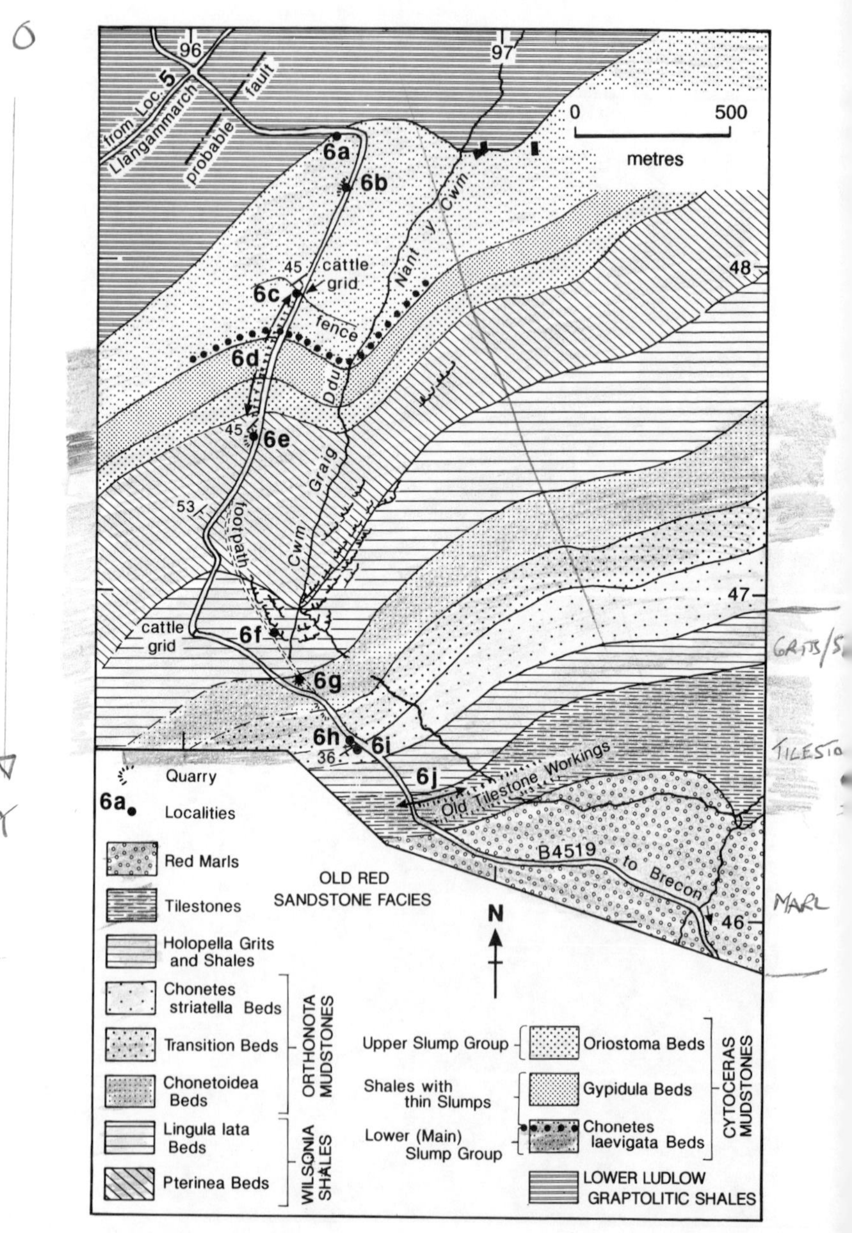

FIG. 5. Geological succession at Cwm Graig Ddu (Locality 6). Modified after Straw (1953) and various sources.

benthic shelly faunas. The graptolites, nautiloids and cardiolids, by contrast, represent probable indigenous elements of the pelagic fauna.

6. **Cwm Graig Ddu.** Continue eastwards from Locality **5** to join the B4519 road at SN 9700 4859 and turn right (south-east) towards the **road section running up the western side of Cwm Graig Ddu** (Locality **6**; Fig. 5; SN 964 479 to SN 963 467). **Vehicles can be parked** at a number of points along the road.

Cwm Graig Ddu is a fine example of a glaciated valley. The geology was described in detail by Straw (1953), who introduced a lithostratigraphical subdivision to reflect the prominent presence of slumped units in the sequence; in Fig. 5 these are related to the more widespread divisions of the Ludlow Series that he established for the Builth area in 1937. Additional accounts of the sedimentology and palaeontology have been given by Bailey & Woodcock (1976), Bassett (in Baker *et al.* 1979), and Siveter *et al.* (1989). The guide described here is essentially taken from the latter comprehensive account, with only a few modifications. The faunal succession upward through the Ludlow of Cwm Graig Ddu is summarised in Fig. 6.

The top of the lower Ludlow graptolitic shale unit seen at Llangammarch crops out **near the sharp bend** in the road at Locality **6a** (SN 9650 4838), where it dips south-east at 50°. The succeeding 'Cyrtoceras Mudstones', a sequence of blue-grey shales and rubbly mudstones, are subdivided into three units, the lowermost of which, the 'Lower (Main) Slump Group', is 350 m thick and comprises several superimposed slump sheets that can be examined at two localities. In the roadside quarry 200 m south of the bend (Locality **6b**; SO 9648 4820), the slumped and contorted nature of the laminations in the massive, somewhat calcareous mudstones is well displayed; the slump sheets entered this area from the south-east, having moved down slope across the basin margin, which at that time lay along the Church Stretton Lineament. They belong to the series of slumps that were probably triggered by seismic activity along the fault zone during mid Ludlow times. The slumps should be examined carefully, since apparently balled-up masses or 'pillows' of sediment can easily be confused with them; these are the result of ellipsoidal weathering and can be recognised, especially when wet, by continuous laminations passing uninterrupted through them. Fossils are not common here, and are usually small, and confined to thin sandy laminae; they are commonly weathered out to leave ochreous moulds. Ostracodes (*Hemsiella 'maccoyana'*), pelmatozoan ossicles and small brachiopods (*Mesopholidostrophia laevigata* and *Dayia navicula*) are the most common.

The upper part of the 'Main Slump Group' crops out in exposures **alongside the cattle grid** (SN 9634 4790), at the beginning of the continuous

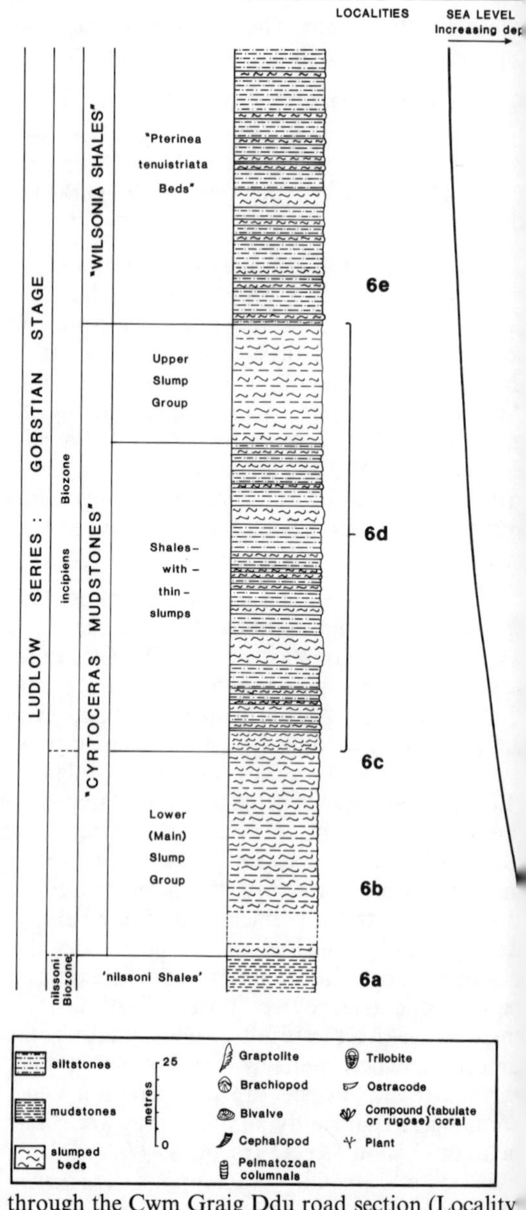

FIG. 6. Lithological succession through the Cwm Graig Ddu road section (Locality 6) with a generalized sea-level curve and ranges of selected fossils. Group A = allochthonous faunas introduced with slumps. Group B = autochthonous faunas and floras. After Siveter *et al.* (1989, fig. 79).

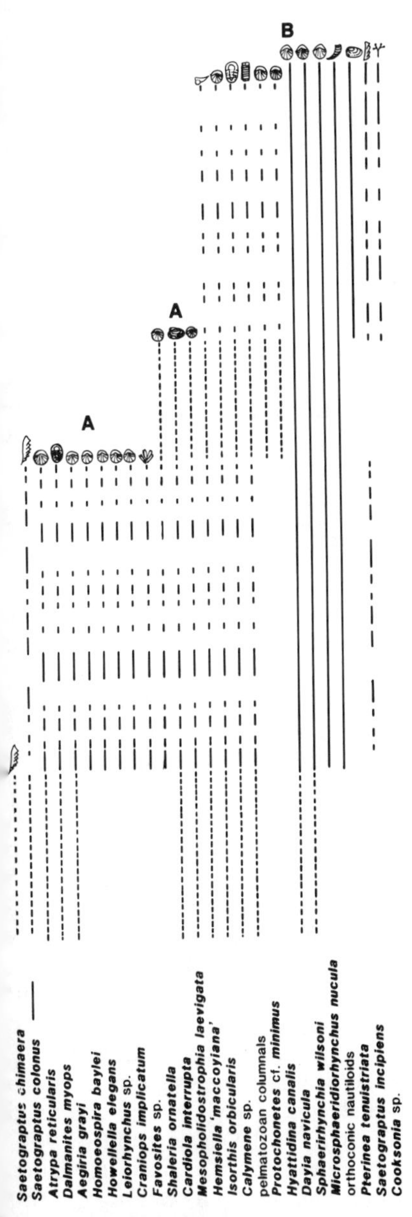

FIG. 6. *Continued.*

roadside section (Locality **6c**) (this is also a convenient place to leave vehicles). The sediments and fauna here are similar to those at the previous locality, but fossils are more abundant, tending to occur in thin layers. They include the brachiopods *M. laevigata*, *D. navicula* and *Isorthis orbicularis*, and the ostracode *H. 'maccoyana'*. Graptolites are rare in this part of the sequence and are generally poorly preserved. Straw (1953) reported *Saetograptus colunus* and *S. chimaera*, suggesting the *scanicus* Biozone. The Main Slump Group continues to crop out up the hill for **some 90 m to the south of the cattle grid**, and over the next 180 m (SN 9631 4782 – SN 9625 4764) the succeeding 'Shales-with-Thin-Slumps' are exposed (Locality **6d**). The thin slumps are richly fossiliferous mudstone units that range from a few centimetres up to a metre in thickness. Most are single slumps which contain a much richer fauna than the underlying 'Main Slumps'. They again include the brachiopods *M. laevigata*, *D. navicula* (abundant) and *I. orbicularis*, but in addition there are others appearing for the first time, such as *Homoeospira baylei*, *Shaleria ornatella*, *Leptostrophia filosa* and *Howellella elegans*. Straw (1953) noticed that some bands have a preponderance of a particular species. Other fossils include bryozoans, bivalves, ostracodes (*Hemsiella 'maccoyana'*), trilobites (*Dalmanites myops*), corals and pelmatozoan columnals. The intercalated, undisturbed finely flaggy grey siltstones have a sparse pelagic fauna of small bivalves, orthoconic nautiloids and graptolites (*Saetograptus incipiens*) indicating the *incipiens* Biozone. The contrast between the abundance and composition of the faunas of the slumps and those of the intercalated beds in the whole of this section is striking. The former are evidently allochthonous, having been introduced from shallower waters to the south-east by the fluid mudflow-type thin slumps, whilst the latter represent the indigenous fauna inhabiting deeper water, edge-of-slope environments.

The uppermost unit of the 'Cyrtoceras Mudstones', the 'Upper Slump Group' **crops out over the next 70 m**, between 270 m and 340 m (SN 9625 4764 – SN 9624 4758) south of the cattle grid as a continuation of Locality **6d** through continuous exposures. It is some 40 m thick and lithologically resembles the 'Lower Slump Group', but is somewhat darker in colour. The fauna is similar to underlying beds, but *Microsphaeridiorhynchus nucula* now becomes abundant. *I. orbicularis* and *D. navicula* are also abundant, as are pelmatozoan columnals and the ostracode *H. 'maccoyana'*. The lower division of the 'Wilsonia Shales', the 'Pterinea Beds', occupies the remainder of the road section. They are best examined in the **roadside quarry** (Locality **6e**) 450 m south of the cattle grid (SN 9647 4721), where the typical dark, blue-grey, rather hard, thinly laminated flaggy siltstones and a typical thin slump can be seen. Slumps in this division are usually less than 30 cm thick, and rarely over 60 cm. They contain rich allochthonous shelly faunas; the brachiopod *M. nucula* is especially abundant, and *I. orbicularis*, *D. navicula*, *Whitfieldella* sp., and *Sphaerirhynchia wilsoni* are common. Other fossils, including pelmatozoan columnals, bryozoans and beyrichiacean ostracodes also occur. In the intervening flaggy siltstones, extensive smooth bedding planes are exposed in the quarry which are strewn with orientated orthoconic nautiloids and bivalves; this is suggestive of active bottom currents. This distinctive facies becomes commoner higher in the succession, at the expense of the slump units. Some bedding planes are crowded with the bivalve *Pterinea tenuistriata*, for which this quarry is the type locality. Some fragmentary graptolites have been found here, including *Saetograptus incipiens*, indicating the *incipiens* Biozone. The most interesting fossils recorded from the quarry are fragments of fossil plants, that include the rhyniophyte *Cooksonia* (Fig. 7), characterised by tiny Y-shaped axes with terminal sporangia; these are some of the earliest land plants known (Edwards *et al.* 1979). They appear as dark carbonaceous remains, a few millimetres long and usually under a millimetre in diameter.

In order to examine the highest part of the succession, it is necessary to visit some **small quarries** to the south of the main section, **at the head of the**

FIG. 7. *Cooksonia* sp. from the Pterinea Beds of Cwm Graig Ddu (Locality 6e) (×4). Note the smooth, forked axis with compressed terminal sporangia on the two left hand branches.

valley (Localities **6f, g).** The upper division of the 'Wilsonia Shales', the 'Lingula lata Beds' crop out in a quarry (Locality **6f**; SN 9627 4687) alongside the east side of a footpath that leaves the road about 200 m south of Locality **6e**. The 'Lingula lata Beds' are some 240 m thick, and comprise more coarsely laminated shales and silts than the *Pterinea* Beds. They contain the brachiopods *S. wilsoni*, *M. nucula*, *W. didyma*, *D. navicula*, *Orbiculoidea rugata* and *Lingula lata*. *Saetograptus leintwardinensis* also occurs here, showing that these beds belong to that biozone. Another **quarry** (Locality **6g**; SN 9635 4673) on the **west side of the track**, shortly before it rejoins the main road and 150 m from the previous exposure, preserves 7.5 m of rather soft, greenish flaggy shales which Straw (1953) regarded as probably transitional between the 'Lingula lata Beds' and the overlying 'Chonetoidea grayi Beds'; *D. navicula* occurs in abundance in thin bands, along with other brachiopods such as *Aegiria grayi*, and *Shaleria ornatella* and the bivalve *Fuchsella amygdalina*. The 'Chonetoidea grayi Beds', lowest division of the 'Orthonota Mudstones' are exposed in a **roadside quarry** (Locality **6h**; SN 9649 4658) 145 m to the south, where they comprise dark, rather hard, greenish and bluish flaggy shales and mudstones. They yield the brachiopods *A. grayi*, *Protochonetes ludloviensis*, *D. navicula*, *S. ornatella*, *Atrypa reticularis* and *I. orbicularis*, together with trilobites (*Encrinurus* sp.) and ostracodes (*Neobeyrichia lauensis*). Some **28 m south of this quarry** is another (Locality **6i**; SN 9653 4654) in blue flaggy siltstones of the 'Transition Beds' yielding the brachiopods *P. ludloviensis*, *M. nucula*, *D. navicula*, *Salopina* cf. *lunata* and the bivalve *F. amygdalina*. *A. grayi* and *N. lauensis* are both absent, indicating correlation of this level with the Lower Whitcliffe Formation of the Ludlow area.

The whole succession represents generally upward shallowing conditions, but the highest Ludlow rocks of this district, the 'Chonetes striatella Beds' within the 'Orthonota Mudstones', together with the 'Holopella Grits and Shales', all of shallowest water origin, are not exposed here. The Přídolí age Tilestones are now poorly exposed, but can be traced along old workings on either side of the road (**6j**) as debris of highly micaceous yellow and green sandy siltstones. They are transitional upwards into Old Red Sandstone marl facies higher up the hillside along the B4519 road.

7. Llanddewi'r Cwm. Continue south-eastwards along the B4519 for about 7 km towards Upper Chapel. The route is entirely across Lower Old Red Sandstone red marl facies, cropping out at numerous points along the road-side; they represent typical alluvial plain deposition of silts, muds and sands and they contain occasional, rare fish fragments. Note that the road from Cwm Graig Ddu to Upper Chapel is **sometimes closed for military manoeuvres; on no account should you enter the area when the red range-warning flags are flying**.

At SO 0064 4117 take the B4520 road back towards Builth. Locality **7** is 8 km to the north at Llanddewi'r Cwm, at the point where the road bridge crosses the Duhonw River (Fig. 8; SO 0362 4885). Exposures here are entirely in the Pterinea Beds, described by Straw in 1937. **Parking is available on the roadside** north of the bridge.

This is one of the most convincing localities in the Builth area for the demonstration of slumping as a sedimentary mechanism, and the exposures here were where O. T. Jones demonstrated to Straw (1937, p.447) that submarine sliding as opposed to tectonic thrusting was responsible for the features of contorted bedding. The rock face on the north side of the road **at the sharp bend** over the bridge (Locality **7a**; Fig. 8) preserves an impressive example of a slump fold showing incipient shearing, which breaks the continuity of the distinctive lamination in the unit.

In the 12 m high **river-cliff section to the west of the bridge** (Locality **7b**) there are three distinct slumped horizons separated by thin (7.5–15 cm), undisturbed, flat-lying interbeds. The relationships dispel any notions of tectonism; instead the lowermost shales of the undisturbed units are sandy and fill hollows in irregular erosion surfaces on top of the slumps. Successive laminations overlap until the hollows are filled, and the higher laminae are then continuous, providing undoubted evidence that they were deposited between episodes of slumping.

8. Waun Hirwaun. From Llanddewi'r Cwm, return southwards along the B4520 road for 3 km and turn off to the south-east at SO 0270 4661 along the minor road through Pen-y-garreg (Fig. 9); **this road is narrow** but is metalled and is suitable for **small vehicles**.

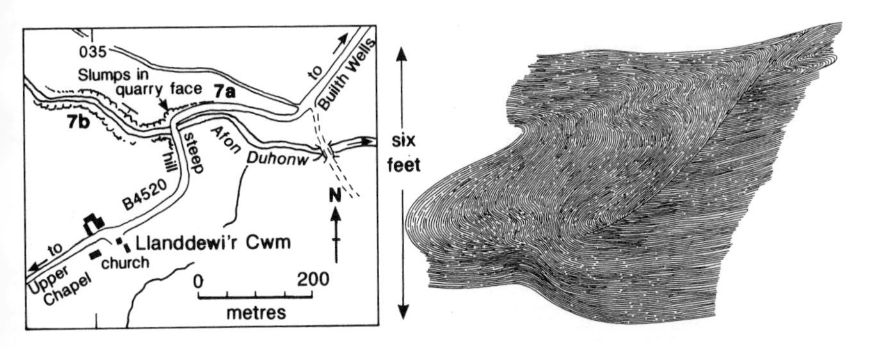

FIG 8. Locality map (left) for Llanddewi'r Cwm (Locality 7), with reproduction (right) of Straw's original sketch (1937, fig. 5) of the sheared slump-fold at Locality 7a.

After the farm the road climbs steeply up the flanks of Mynydd Eppynt, first through upper Ludlow Chonetes striatella Beds, and then at an altitude of about 350 m onto the Holopella Grits and Shales (Fig. 9). There are crags in both units along the roadside. Continue up onto the summit area where the topography flattens out across the Eppynt plateau at about 400 m. There are wonderful views in all directions across Wales and adjacent England.

Fig. 9 shows that a small segment of Old Red Sandstone marls is let down into the Holopella Beds just before the road veers to the south-east at SO 0437 4583. The rounded **hill top of Banc y Celyn** to the north is at 472 m in Ludlow Transition and Chonetes striatella Beds, with Holopella beds in the slope down to the road across Waun Hirwaun. **Park at the roadside** near the bend and walk across the bracken moorland for about 100 m to

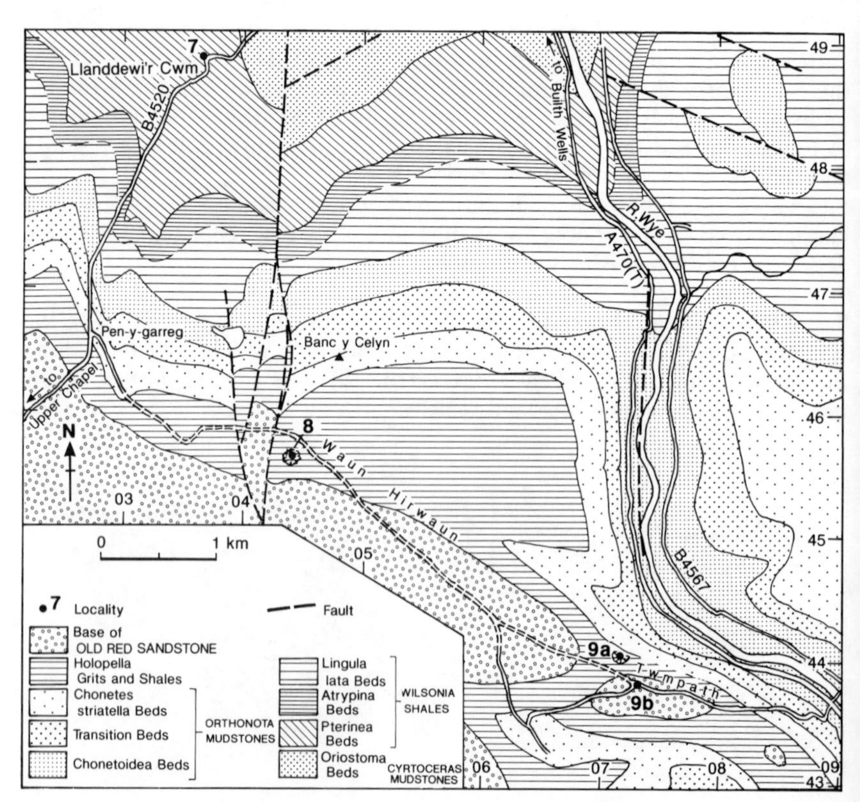

FIG. 9. Geology of the Ludlow succession between Llanddewi'r Cwm and the Wye Valley on the north flanks of Mynydd Eppynt south of Builth Wells, with localities 7, 8 and 9 of this excursion. After Straw 1937, plate 29.

the south, where Locality **8** is a **shallow remnant of a former small quarry** at SO 0440 4575.

The sparse exposure comprises sub-horizontal or gently southerly dipping thin shales, siltstones and sandstones. Fossils are not common, but include the typical fauna of the Holopella beds, including species of *Loxonema*, *Cucullella ovata*, *Salopina lunata* and *Protochonetes ludloviensis*. As elsewhere, the environment is clearly restricted, close-inshore marine and marginal into the non-marine Old Red Sandstone facies forming the bulk of Mynydd Eppynt immediately to the south.

9. Viewpoint: Twmpath. Continue south-eastwards along Waun Hirwaun via Cefn-hirwaun to the **road bend and junction** at the top of the Twmpath (SO 0729 4385). From here (Locality **9a**) the road descends steeply from the Old Red Sandstone down the Ludlow succession (Fig. 9) to the A470 road and River Wye at Erwood Bridge. Exposures through this sequence down the Twmpath are described at Localities **6a–c of Excursion 13** in this volume.

At a number of points down the Twmpath from Locality **6a** there are **superb views** to the north along the Wye Valley towards Builth, and across the river to the distinctive ridged topography of **Aberedw Rocks** described in Excursion 13 (see also cover photograph). It is worth emphasizing here the historical importance of the view and the Aberedw exposures, because it was here that Roderick Impey Murchison, the founder of the Silurian System, was to state that on his first field excursion to Wales in 1831 he recognized for the first time the important junction between the marine Ludlow and the Old Red Sandstone. In his own words:

> I took notes from Dr Buckland [of Oxford] of all that he knew of the slaty rocks, or grauwacke, as it was then called, which lay beneath the Old Red Sandstone, and the relations of which I was determined to begin to unravel; and I recollect that he then told me that he thought I would find a good illustration of the succession or passage on the banks of the Wye near Builth . . .
>
> . . . Travelling from Brecon to Builth by the Herefordshire road, the gorge in which the Wye flows first developed what I had not till then seen. Low, terrace-shaped ridges of grey rock dipping slightly to the south east appeared on the opposite bank of the Wye, and seemed to rise out quite conformably from beneath the Old Red of Herefordshire. Boating across the river at Cavansham Ferry, I rushed up these ridges [i.e. Aberedw Rocks], and to my inexpressible joy found them replete with transition fossils, afterwards identified with those at Ludlow. Here then was a key, and if I would only follow this out on the strike of the beds to the north east the case would be good.

Murchison thus clearly recognized the importance of using faunas in correlation and mapping, and indeed in that first field season he was able to map out the base of the Old Red Sandstone up to Ludlow, thus establishing the framework for his subsequent studies of the Silurian.

Part of the transition sequence can be examined on the flanks of the Twmpath itself (Fig. 9), including a further exposure of Holopella Grits and Shales in the **crags and old quarry** at Locality **9b** (SO 0713 4405), which can be reached by traversing the hillside north-westwards from **9a**; lithologies and faunas are similar to those at Locality 8.

Acknowledgements. I am especially grateful to Dr Jonathan Harris (via Dr R. B. Rickards) for allowing me to quote details of graptolite faunas freely from his unpublished Ph.D. thesis (University of Cambridge 1987); this work forms an essential reference for anyone studying the Wenlock sequence in the Builth area.

13. THE SILURIAN OF THE WYE VALLEY SOUTH OF BUILTH

by L. CHERNS

Maps *Topographical*: 1:50000 Sheet 147 Elan Valley and Builth Wells
1:25000 Sheets SO 04, SO 05

THE Wye valley south of Builth comprises a sequence of marine Silurian rocks that strike roughly from north-east to south-west and dip gently to the south-east, overlain conformably to the south by rocks of Old Red Sandstone facies. It is a classic area for Silurian geology, since it was near Erwood in the Wye valley that Murchison in 1831 recognized the transition down from the Old Red Sandstone into older fossiliferous rocks that were subsequently to be included in his 'Silurian System' (Murchison 1839). Rocks in this excursion area are on the northern limb of a broad synform which has a Caledonoid NE–SW trend, and which to the south passes into a complementary antiform, the 'Brecon Anticlinal area'. Minor open folds are parasitic on these broader regional structures. During Silurian times, the Builth region lay close to the offshore shelf margin on the southern side of the Anglo–Welsh trough. The shelf edge was controlled structurally by the Welsh Borderland Fault System, defined by the Tywi, Pontesford and Church Stretton lineaments (Woodcock and Gibbons 1988). Intermittent tectonic activity during Silurian times affected sedimentation by controlling the position of sedimentary troughs and topographic 'highs', and by triggering downslope slumping of sediment. Background trough sedimentation was largely of low energy, hemipelagic muds, among which coarser sediments were introduced episodically by turbidity currents and through storm disturbances. In less rapidly subsiding, shelf environments, deposition of coarser silts and sands that had shelly benthic assemblages and active burrowing faunas was interrupted by storm influxes of laminated sheets of sediment. Rock exposures are fairly limited across much of the area, which was mapped by Straw (1937) and Kirk (1947, unpublished Ph.D. thesis, University of Cambridge; 1951b).

This excursion demonstrates the upper Silurian succession (Figs. 1, 2), which is mostly in offshore shelf to trough mudstone and siltstone facies. The occurrence of slumps (Locality 1; see also localities 6 and 7 of Excursion 12) that represent contemporaneous sliding among early

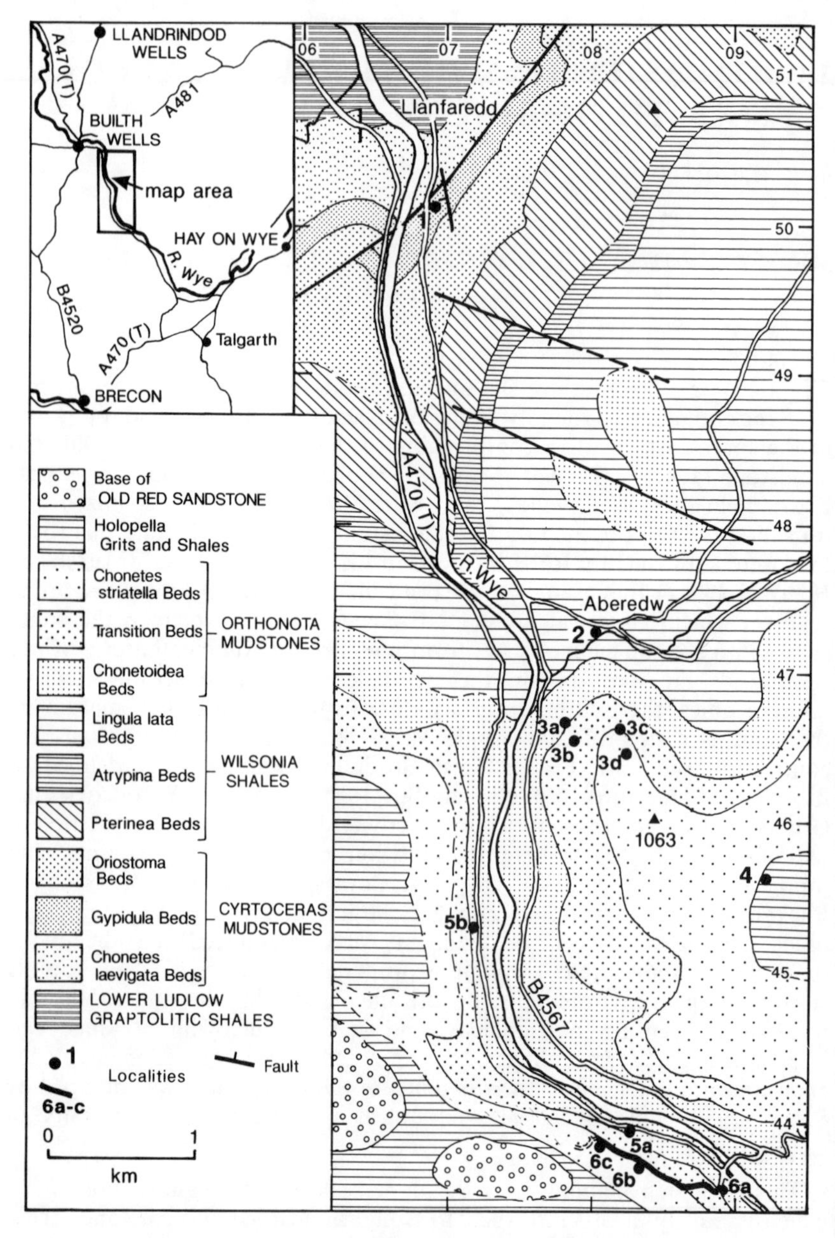

FIG. 1. Geological map (after Straw 1937) of the Wye valley south of Builth and the itinerary for this excursion.

Ludlow (Gorstian) sequences indicates the palaeogeographical position in the sloping and unstable shelf edge region. From late Ludlow (Ludfordian) times a transition up from sparsely fossiliferous and graptolitic, laminated siltstone trough facies (Locality 2) into more calcareous and coarser siltstone facies with benthic skeletal faunas, i.e. shelf facies (although faunas are impoverished cf. Ludlow type area; localities 3–6), indicates increased basin circulation and late Silurian regression.

The excursion starts from Llanfaredd, approximately 3 km east of Builth and on the east side of the Wye valley on the B4567 (turn south off the A481 at SO 062 517). Localities are arranged to demonstrate the succession up-sequence. Alternatively, this excursion can be linked with excursion 12, which finishes at Locality 6 on the west side of the Wye valley; the itinerary would then be followed in reverse order.

ITINERARY

1. Llanfaredd road cutting (SO 0685 5015 – 0687 5025). This section is on the east side of the B4567, 0.5 km south of Llanfaredd. **Park opposite the road cutting** where the road is reasonably wide.

The bank, some 80 m long and up to 10 m high, is in very thinly bedded, buff-olive siltstones from the Cyrtoceras Mudstones and *Saetograptus incipiens* Biozone (Gorstian Stage; Fig. 2). Four slump units (2–5 m) which show undulating and disturbed bedding occur among normally bedded units (0.5–2 m) of similar siltstones. The slumps display plastic deformation of bedding lamination, and have abrupt contacts with under- and overlying, normally bedded units, although with no indication of irregular upper erosion surfaces (cf. comparable slumps at Llanddewi'r Cwm, SO 0356 4883 – 0363 4885; see Excursion 12, Locality 7; Straw 1937). The beds in undisturbed units show a fine and rather shaly, flat lamination.

Between the second and third slump units, bands of shelly limestone (<20 cm) which weather to decalcified rottenstone form prominent marker horizons. These coquinas are lenticular, although fairly persistent laterally across the width of the outcrop, and comprise tightly packed accumulations of broken and disarticulated skeletal material that includes articulate brachiopods (*Gypidula* sp., *Dayia navicula*, *Atrypa reticularis*, *Shagamella minor*), bryozoans, corals and pelmatozoans. They represent current-swept accumulations of epibenthos, resulting from storm disturbance of bottom waters.

2. Aberedw (SO 0805 4728). **Park by the small inn** (Seven Stars) in Aberedw and walk down to the river via the footpath south from the road at SO 0806 4732.

STAGE	LOCAL STAGE	GRAPTOLITE ZONATION	LUDLOW ANTICLINE	BUILTH Straw 1937	BRECON Kirk 1949
LUDFORDIAN	WHITCLIFFIAN	no younger graptolite zones known in Britain	Upper Whitcliffe Formation	*Holopella conica* Beds	*Chonetes* Flags
				Chonetes striatella Beds	
		Bohemograptus proliferation	Lower Whitcliffe Formation	Transition Beds	Transition Flags
	LEINTWARDIN–IAN	*Saetograptus leintwardinensis*	Upper Leintwardine Formation	*Chonetoidea grayi* Beds	*Chonetoidea grayi* Flags
			Lower Leintwardine Formation	Upper *Lingula lata* Beds	Striped Flags
GORSTIAN	BRINGE–WOODIAN	*Pristiograptus tumescens*/ *Saetograptus incipiens*	Upper Bringewood Formation	Lower *Lingula lata* Beds	
			Lower Bringewood Formation	*Atrypina* Beds	
				Pterinea tenuistriata Beds	
	ELTONIAN		Upper Elton Formation		
		Lobograptus scanicus	Middle Elton Formation	*Cyrtoceras* Mudstones	*nilssoni-scanicus* Mudstones
		Neodiversograptus nilssoni	Lower Elton Formation	graptolitic shales	

FIG. 2. Correlation chart of the local stratigraphy against the standard Silurian stratigraphy in the Ludlow area and graptolite biozonation.

Rocks exposed in the **northern river bank and low cliffs** are very thinly bedded, dark laminated shaly siltstones of the Upper Lingula lata Beds and *S. leintwardinensis* Biozone (Lower Leintwardine Formation, Ludfordian Stage; Fig. 2). The beds dip generally to the south-south-east (e.g. 070/21°S) but here show a series of small folds with ENE–WSW axes. Tectonic wrinkling is shown by bedding plane lineations that are parallel to fold axes.

The thin (mm scale) lamination picks out small-scale current structures, such as very shallow round-crested ripples and erosional scours among dominantly flat-laminated, tabular beds. Isolated, narrow silt-filled burrow traces displayed on some bedding surfaces and in section are insufficient to destroy primary bedding lamination.

Skeletal fossils are very rare in these beds, which belong to the laminated siltstone facies of Cherns (1988). Moulds of ribbed rhynchonellide brachiopods (*Sphaerirhynchia wilsoni*) and of *D. navicula* valves which are scattered across some bedding surfaces appear to show weak current alignment, and there are a few very thin (<1.5 cm) and impersistent lenses of fragmental skeletal material which includes these articulate brachiopods. In addition, valves of the small inarticulate brachiopod *Lingula lata* and moulds of orthoconic nautiloid cephalopods occur among the siltstones. Wood (1900) reported the occurrence at this locality, if rare, of some bedding planes crowded with *L. lata*, or the graptolite *S. leintwardinensis*, or of the two together.

These beds represent hemipelagic muds and fine silts of low energy trough environments, where a paucity of bioturbation traces and sparse benthic skeletal faunas except apparently lingulides suggest that dysaerobic conditions may have been limiting (Cherns 1979, 1988). The faunal assemblage corresponds to Association D of Cherns (1988), and the *L. lata – S. leintwardinensis* Association of Cherns (in press).

3. Pont Shoni (SO 078 467) **and Aberedw Rocks** (Fig. 3). **Park in the lay-by** on the west side of the B4567 about 300 m south of the turning to Pont Shoni. Follow the footpath through Pont Shoni (**ask permission at the house**) and up the steep path to the south (starting between ash and sycamore trees) through the woods about 150–200 m to the **lower crags of Aberedw Rocks**.

The cliffs comprise very thin and thin bedded, dark olive-grey calcareous siltstones in thick to massive faces that weather into flaggy slabs. Many of the beds in the lower crags (Locality **3a**) have a homogeneous or slightly mottled texture which indicates bioturbational reworking, but others on weathering show a remnant faint bedding lamination. These lower beds represent the Chonetoidea grayi Beds and *S. leintwardinensis* Biozone (Upper Leintwardine Formation, Ludfordian Stage), and pass up without lithological change into the Transition Beds (Lower Whitcliffe Formation, Ludfordian Stage; Fig. 2). The sequence includes very thin (<5 cm) shelly limestone beds and lenses, bedding plane assemblages of fossils, and there are also scattered fossils within beds. By comparison with Aberedw, the sediments are coarser and more calcareous; skeletal fossils are fairly common and burrow mottling of sediments indicates the intense activity of *in situ* benthic faunas.

The skeletal fauna of the lower crags (i.e. Upper Leintwardine Formation) includes bedding plane assemblages of valves of the small brachiopod *Aegiria grayi.* The brachiopods *D. navicula*, *Microsphaeridiorhynchus nucula*, *Salopina lunata*, *Protochonetes ludloviensis*, *Atrypa reticularis* and the bivalve *Fuchsella amygdalina* are also quite common. Valves of the brachiopod *Shaleria ornatella* and the trilobites *Calymene puellaris* and *Encrinurus stubblefieldi* are characteristic Upper Leintwardine Formation fossils. Beyrichiacean ostracodes are reported to include the index neobeyrichiacean species *Neobeyrichia lauensis*, and poorly preserved specimens of the graptolite *S. leintwardinensis* are recorded from here (Straw 1937).

Higher crags in the hillside have shelly assemblages which lack *A. grayi* and have increased abundance of *P. ludloviensis*, *S. lunata* and *F. amygdalina.* These assemblages are typical of the Lower Whitcliffe Formation, or Transition Beds, which here are about 45 m thick. The boundary between the two divisions is essentially faunal but was taken by Straw (1937, p.421) at a line of calcareous concretions near the base of the crags at SO 0780 4662, north of the NE–SW dry channel. The passage up into the higher division is **well exposed in faces south of the channel head** (Locality **3b**), where very thin to thin (<30 cm), sheet-laminated siltstone units occur among calcareous siltstones that are mostly bioturbate in texture with little primary bedding lamination preserved. The laminated sheets typically show parallel lamination and wavy cross-lamination, whilst some thicker units have erosional bases and shallow ripples on the upper surfaces. They represent episodes of higher energy, storm disturbance and offshore transport of sediment.

Farther up the footpath onto Aberedw Rocks at **Pen-y-Gareg** (SO 0815 4670), another dry channel (Locality **3c**) towards the south-east exposes higher calcareous siltstones of the Transition Beds in its walls. Thin to medium sheet-laminated siltstone units (<50 cm) are interbedded among mostly rather homogeneous and poorly calcareous, unevenly splitting siltstones. Many of the laminated units are tabular and fairly laterally extensive, with wavy cross-lamination and parallel lamination. Some show cross-bedding and a few have rippled tops (<10 cm). Fossils are scattered in the bioturbated siltstones, but there are also some limited bedding plane

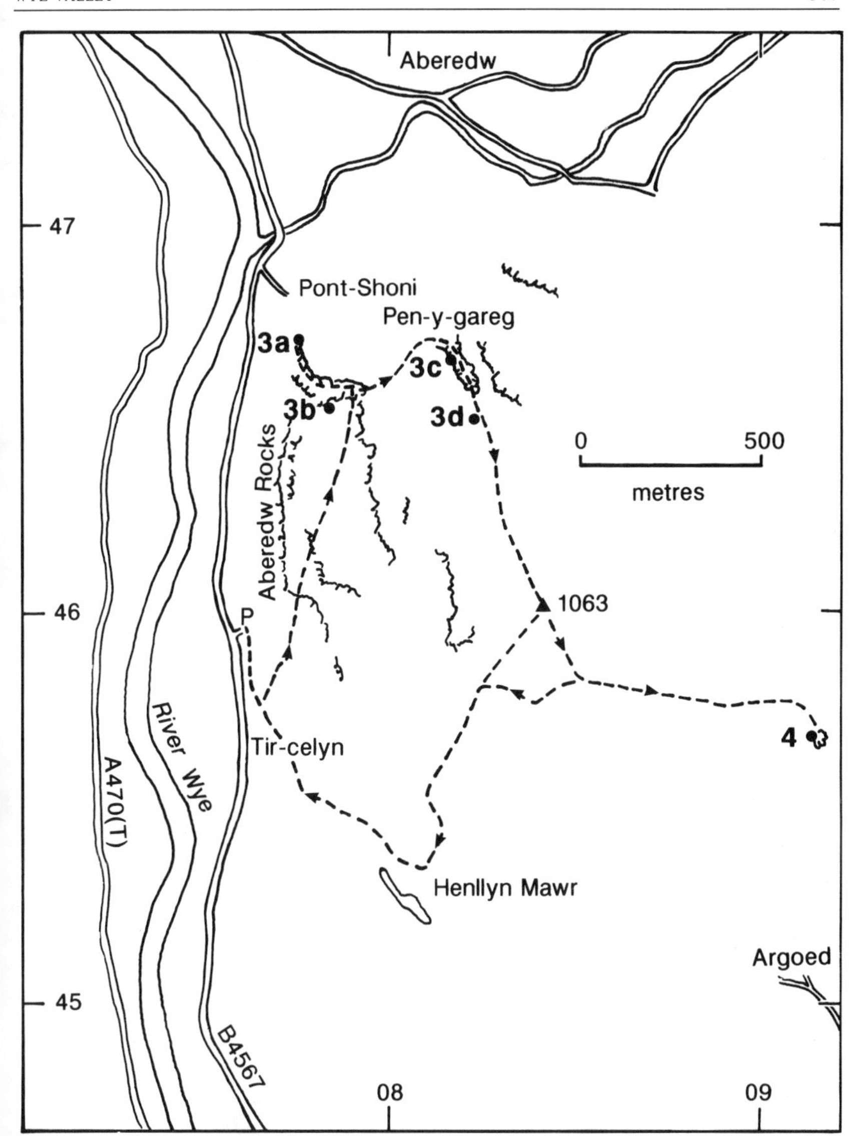

FIG. 3. The area of Aberedw rocks showing the excursion route for Locality 3.

assemblages of shells, notably of articulated bivalves. *F. amygdalina*, *M. nucula*, *P. ludloviensis*, *S. lunata*, *D. navicula*, the inarticulate brachiopod *Orbiculoidea rugata* and nautiloid cephalopods comprise common elements of the skeletal fauna, and there are chondritic burrow traces. The proportion of laminated and cross-bedded sheets increases, and the sediments become somewhat coarser into higher siltstones and fine sandstones. This sequence from Localities **3a–c** demonstrates increasing storm influence on sedimentation in shelf areas, perhaps linked to some regression. A coarse sandstone ('grit band'; 20–30 cm) **exposed in the south of the channel and the line of crags beyond** was taken by Straw (1937) as the base of the overlying C. striatella Beds (Upper Whitcliffe Formation, Ludfordian Stage; Fig. 2), here about 55 m thick.

The C. striatella Beds crop out beyond the channel in the **ascending line of crags** that dip gently north-west on the west of the path (**3d**). Sedimentary structures are not well displayed in weathered faces, but the flaggy siltstones and fine sandstones have a shelly fauna most commonly with *P. ludloviensis* and *S. lunata*, and *F. amygdalina*. *M. nucula* and nautiloid cephalopods are also fairly common. This unit forms the higher cliffs along the **western scarp of Aberedw** Rocks south to Henllyn Mawr (SO 081 452).

Alternatively, the upper Ludlow succession described above (and Locality **4**) can be reached along an attractive circuit walked from Tir-celyn (Fig. 3; Localities **3a–d**, about 3.8 km; **3a–d** and Locality **4**, about 5.3 km). Park across the cattle grid in the **disused quarry** north of the small road off the B4567 at 0758 4595 (turning signed to Bracken Lodge and Brynydd). The C. grayi Beds (equivalent to Locality **3a**) are exposed **in the quarry**, and in the lower **two lines of crags** crossed along the footpath running north-east from the road bend at the pylon (0765 4575). **Higher crags** expose the Transition Beds, and although these rocks are weathered, thin laminated sheets are prominent among the flaggy siltstones. Follow the footpath across to the dry channel south of Pont Shoni (Locality **3b**), and then follow the itinerary to Localities **3c** and **3d**. Walk across to the summit at 1063 ft (324 m) where the view across the Wye valley to the west shows the same upper Silurian sequence. To return directly to Tir-celyn from the summit, join the footpath south-west of the summit at the boundary stone and follow the track down via Henllyn Mawr (Fig. 3).

4. Argoed (SO 0915 4565). The small **disused quarry** approximately 500 m north of Argoed is reached from the walk described above, or directly from Tir-celyn along the footpaths shown on Fig. 3 (about 4.5 km return). To continue from Locality 3, join the footpath south of the summit at 1063 ft (324 m) and follow it eastwards for about 700 m across to the quarry (Fig. 3).

The younger, Holopella conica Beds (Upper Whitcliffe Formation; Fig. 2) are exposed in the core of the Argoed syncline (Straw 1937), where they have a minimum thickness of about 35 m. The quarry exposes 4–5 m of gently dipping, very thin to thin bedded siltstones and fine sandstones with shaly siltstone intercalations. The coarser beds include laminated sheets (<20 cm) with flat or wavy lamination and rippled tops. Several have basal lenses of skeletal sandy limestone, in which spired gastropods (*Loxonema* spp. = previously *Holopella*) are characteristic, as well as *P. ludloviensis* and *M. nucula*; such assemblages represent storm accumulations. Interbedded muddy and well bioturbated siltstones have a scattered skeletal fauna of *P. ludloviensis*, *S. lunata*, *M. nucula*, *F. amygdalina* and other bivalves, and phosphatic fragments of the tube-dwelling worm '*Serpulites*' *longissimus*. The gastropod-rich assemblages of storm accumulations, and relatively high proportion of bivalves among skeletal faunas of the H. conica Beds suggest more inshore shelf environments.

5. Tre-gaer Bridge (**5a**, SO 0815 4398; **5b**, SO 0765 4430). From Tir-celyn continue south along the B4567 to Erwood Bridge (SO 0895 4375) and cross the River Wye, then turn north-west along the A470(T) for approximately 0.8 km.

Locality **5a** is a **road cutting** some 50 m long (and <4 m high) on the south side of the A470(T) and east of Tre-gaer Bridge. **Park near Erwood Bridge** (or continue to Locality **5b**). The cutting of **5a** exposes C. grayi Beds (Upper Leintwardine Formation, Ludfordian Stage; Fig. 2; see also Locality **3a**) in thin and very thin bedded, irregularly splitting, hard olive-grey calcareous siltstones that dip gently west-south-west (154/3°W). Weathered flags show a faint lamination, but fresher rocks have a homogeneous or mottled texture that indicates bioturbational reworking.

Fossils are quite common, mostly scattered through beds but with a few more concentrated assemblages in calcareous bands. *D. navicula*, *A. grayi*, *M. nucula* and *F. amygdalina* are all common. *E. stubblefieldi* and beyrichiacean ostracodes are characteristic of this unit, and the fairly diverse fauna also includes nautiloid cephalopods and gastropods. The calcareous siltstone facies, with mostly well-bioturbated sediments and fairly common skeletal benthos, indicates deposition in shelf environments.

Similar beds crop out at Locality **5b**, the **old quarry** south of the A470(T) and west of Tre-gaer Bridge (SO 0765 4430; about 0.6 km north-west of Locality **5a**). **Park in the lay-by** beside the quarry. The quarry is accessible though somewhat overgrown.

6. Twmpath, west of Ynys Wye (**6a**, SO 0875 4355 – **6b**, SO 0825 4360, (**6c**, SO 0730 4385)). Return south-east along the A470(T) to about 30 m beyond Erwood Bridge, to the turning south to Ynys Wye. The road (Twmpath) ascends the western side of the Wye valley and affords a fine view across to Aberedw Rocks (Locality 3; see cover illustration). The **crags exposed to the south of this road** provide a fairly continuous traverse up sequence through from the Transition Beds into the H. conica Beds (Lower and Upper Whitcliffe formations, Ludfordian Stage; Fig. 2; see also Localities 3 and 4). **Park at the roadside** near the localities.

A **disused shallow quarry** south of the road at Ynys Wye (Locality **6a**, SO 0875 4355) is in thinly bedded, flaggy siltstones of the Transition Beds. The siltstones are tough, dark olive-grey, somewhat calcareous, and they show only a faint lamination or a homogeneous to mottled texture. Skeletal fossils occur in some bedding plane assemblages but more typically are scattered through beds. *D. navicula* remains common as well as *P. ludloviensis*, *M. nucula* and the bivalve *F. amygdalina*.

From the **high crags south of the road** near the cattle grid (SO 0865 4355) up about 450 m to the **small quarry** south of the road (**6b**, SO 0824 4362) there are **roadside cuttings** through thinly bedded hard siltstones which crop out as NW–SE-striking lines of crags, and dip gently (5–10°) south-west. A crude lamination shows more clearly in weathered rocks, but there are also some sheet-laminated siltstones that are prominent in outcrop; in these latter beds the bedding fabric is mostly even and flat. Other siltstones show a mottled texture which indicates intensive biogenic reworking. These beds are more shelly, with both scattered skeletal material within beds and also bedding surfaces with concentrations of shells. As at Aberedw Rocks (Locality 3), the laminated siltstone units represent deposition of sheets of silt sediment across shelf environments following storm events. The common fauna includes *M. nucula*, *P. ludloviensis*, *F. amygdalina* (numerous in some bands) and the inarticulate brachiopod *O. rugata*. *D. navicula* occurs among the fauna of siltstones up to around SO 085 436 ('Y' in Ynys Wye) but is absent from higher beds; these latter belong to the higher, C. striatella Beds, where *P. ludloviensis* becomes increasingly common.

The H. conica Beds were previously accessible in old quarries (SO 0740 4405) north-west of the sharp bend in the road, but **small bank exposures nearby** (Locality **6c**, SO 0730 4385) show the typically thinly bedded, flaggy olive siltstones and fine sandstones with muddy intercalations. High-spired gastropods (*Loxonema* spp.) are common especially in the coarser lithologies, whilst valves of *P. ludloviensis* and *S. lunata* are found more usually on bedding surfaces of finer and muddier siltstones.

14. THE OLD RED SANDSTONE OF THE BRECON BEACONS TO BLACK MOUNTAIN AREA

by J. ALMOND, B. P. J. WILLIAMS *and* N. H. WOODCOCK

Maps	*Topographical:*	1:50 000	Sheet 160 Brecon Beacons
		1:25 000	Sheets 1037 (SN 83/93) Mynydd Bwlch-y-groes, 1060 (SN 62/72) Llandeilo and Llangadog, 1061 (SN 82/92) Sennybridge, 1062 (SO 02/12) Brecon, 1085 (SO 01/11) Brynmawr and Pontsticill.
	Geological:	1:250 000	Bristol Channel
		1:50 000	Sheet 231 Merthyr Tydfil

THE southern third of Powys, the district of Brecknock, is dominated by the 'Old Red Sandstone', referred to subsequently here as the ORS. This sequence of mostly red-brown shallow marine to non-marine clastic sediments contrasts both with the dark grey Silurian marine sequence below and with the light grey Carboniferous Limestone above. In this area it ranges in age from Přídolí (late Silurian) to Famennian (latest Devonian), but encloses the major mid Devonian unconformity present in most southern British sequences (Fig. 1). This itinerary includes representative localities in both the Lower and the Upper ORS, working broadly up the sequence and from west to east. The first two localities are in the county of Dyfed, but are included to demonstrate an important part of the ORS sequence, present but poorly exposed in Powys. In particular the ORS of the spectacular Sawdde Gorge section is detailed here for the first time; it did not feature in the companion guide to Dyfed (M. G. Bassett 1982b).

STRATIGRAPHY AND GEOLOGICAL SETTING

The formal lithostratigraphy of the ORS in south Wales is still in a state of flux. The upper components of the scheme used here (Fig. 1) are essentially those of the Geological Society's Devonian correlation chart (House *et al.* 1977). The lowest unit of that chart, the Red Marls, has since been subdivided by Almond (1983) into three formations. The lowest, the Tilestones Formation, is equivalent to the Downton Castle Sandstone of the West Midlands and the Forest of Dean. The overlying Gwynfe Formation correlates

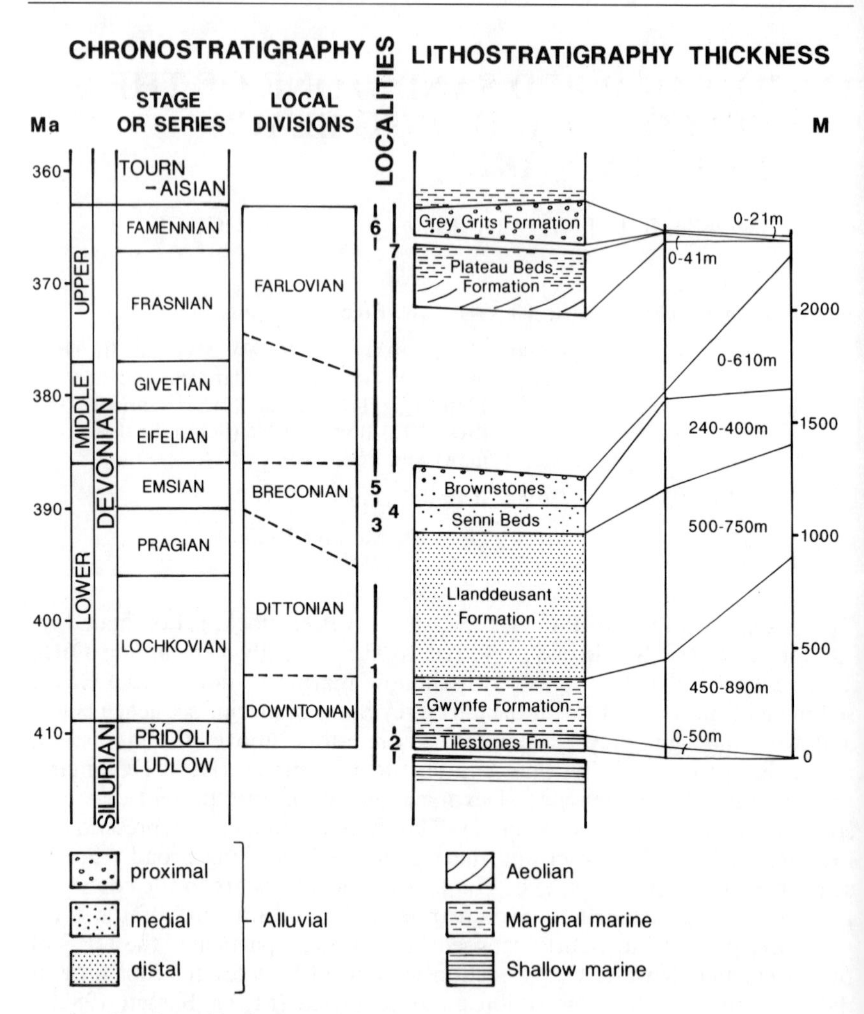

FIG. 1. Stratigraphical sequence in the itinerary area.

with the Temeside plus Ledbury Formations or the Raglan Marl Group, and the Llanddeusant Formation is roughly equivalent to the Ditton Group or St Maughan's Group.

The internationally defined Stage names for the Devonian are still not easily applied in Wales, and use of the local 'Stage' names persists; Downtownian, Dittonian and Breconian in the Lower ORS and Farlovian in the Upper.

The ORS preserved in southern Powys was deposited entirely on the site of the Lower Palaeozoic Midland Platform (review by Woodcock & Bassett, this volume). However, extensive deposition of ORS on the site of the old Welsh Basin to the north-west is suggested by the Přídolí remnants of Clun Forest, by the recycled ORS debris incorporated in the upper Lower ORS in Shropshire and South Wales (Allen 1962), and by the high burial metamorphic grade in the basinal Silurian (e.g. Roberts & Merriman 1985; Awan & Woodcock 1991). This former cover to the basin was eroded soon after its accumulation, as the Acadian (late Caledonian) shortening of the Welsh Basin culminated in late early to mid Devonian time.

The Lower ORS records the change from marine to non-marine environments as the Welsh Basin first filled up with sediment, then started to recycle its eroding contents south-eastward onto the Midland Platform down a major river system. Marine shelf conditions during Ludlow time were progressively replaced by marginal marine conditions during the Přídolí and early Lochkovian (Almond 1983). First the Tilestones Formation represents a shallow marine embayment or brackish lagoon and shoreface environment. At this stage there was probably still open marine water in the basin to the north-west. Then the Gwynfe Formation represents the gradual establishment of inter- and supra-tidal mudflats cut by channels which were still dominantly tidally influenced. However increasing frequency of south-east directed palaeocurrents suggests that the former NW-dipping basin slope was being reversed, and colonized by SE-flowing rivers. Common airfall tuffs record sporadic volcanism up to this time from an unknown source.

By mid Lochkovian time, in the Llanddeusant Formation, any tidal influence was finally lost. From then through Pragian and Emsian time an extensive alluvial plain was traversed by low-sinuosity channels of major SE-flowing streams. A gradual increase in the proportion of sandstone to siltstone through the Llanddeusant Formation, the Senni Beds and the Brownstones records the south-eastward migration of proximal over medial and distal alluvial facies. This progradation of the system may have been forced by the south-eastward advance of a fluvial fall-line at the Acadian deformation front, from a position within the basin to its final position along the Pontesford Lineament (Allen 1974; Tunbridge 1981; Woodcock & Bassett, this volume).

By mid Devonian time the culmination of the Acadian Orogeny ensured that even the Midland Platform was uplifted, tilted a few degrees southward, and was being eroded by its own streams. Sediment was fed southward to alluvial depositional systems such as those now represented in North Devon (Allen 1979; Tunbridge 1986). The resulting unconformity in south Wales spans some 15 Ma. Only in Frasnian time did uplift wane sufficiently to allow renewed deposition in south Wales. The Plateau Beds Formation represents a first pulse of alluvial, aeolian and marginal marine deposition. A non-depositional interlude was followed by accumulation of the alluvial Grey Grits Formation before the major transgression which marked the final replacement of the ORS by marine Carboniferous sediments.

ITINERARY

The itinerary (Fig. 2) begins at the spectacular gorge of the Afon Sawdde where it crosses the foothills of the Black Mountain south-east of Llangadog (Locality **1**). Here a section unique in central Wales and the Welsh Borderland exposes a near continuous sequence through the Tilestones, Gwynfe and lower Llanddeusant Formations. The lower part of the same sequence is then examined some 13 km along strike to the north-west (Locality **2**), east of Llandovery. The higher parts of the Lower Devonian sequence are seen in three localities (**3**, **4**, **5**) south-west of Brecon,

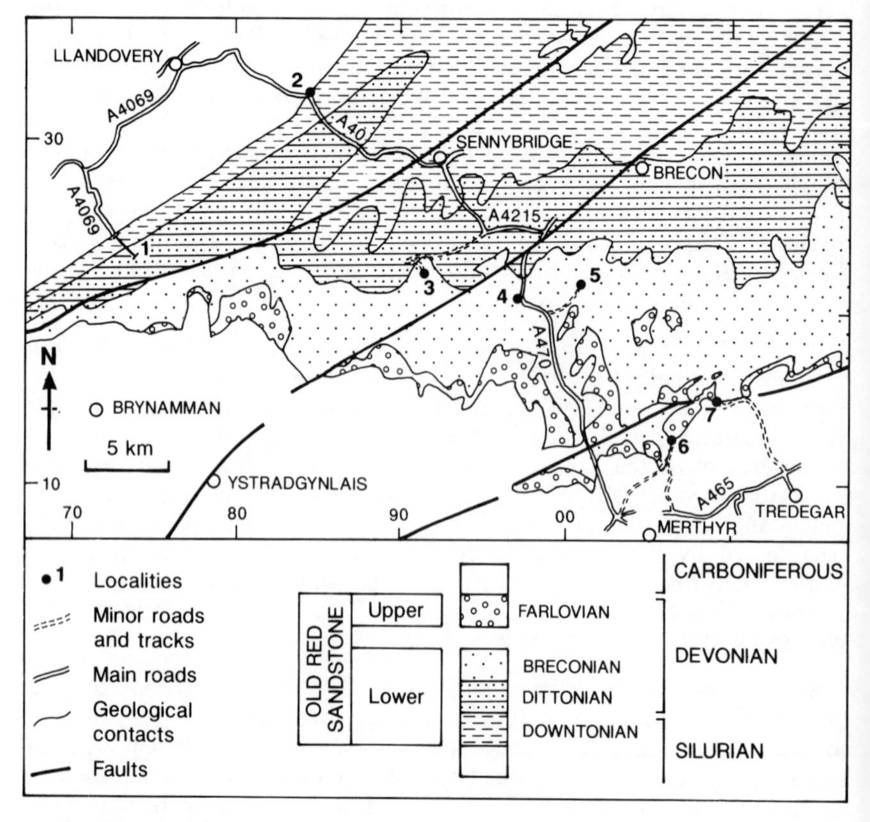

FIG. 2. Location map for the itinerary with outline geology.

and the itinerary is completed by two Upper Devonian localities (**6**, **7**) north of Merthyr Tydfil.

The whole excursion involves about 100 km of driving, and will take two days to complete fully. Localities **1** and **5** each require a minimum of half a day. An abbreviated itinerary giving a satisfactory sample of the Old Red sequence within a day could comprise Localities **1** or **2** with **3** and **7**.

The first locality is reached by taking the A4069 southwards from Llandovery and Llangadog or northwards from Gwaun-Cae-Gurwen and Brynamman. The lower ORS is well exposed from Pont-ar-llechau (SN 7282 2452) southwards to Pont Newydd (SN 7367 2336). **Parking along the road is not advised** and vehicles should be left at the carparks close to the Three Horse Shoes Inn (Pont-ar-llechau) or at SN 7310 2390 and the section traversed on foot.

1. Afon Sawdde (SN 7282 2452 to SN 7367 2336). Some 1250 m of ORS are **exposed in the gorge**, and therefore specific sub-localities have been highlighted as showing features of particular interest. The strata dip steeply and young south-east and are described in that direction. **Care should be taken at all times** as the gorge is narrow and steep in many places and the river often deep and fast flowing.

Locality 1A (SN 7282 2452), about 4 m **north of the bridge at Pont-ar-llechau**, is at the base of the Tilestones Formation. The contact with the underlying Lower Roman Camp Beds is more or less well exposed depending on the state of bank erosion. There is a distinct change in lithology from blue calcareous micaceous siltstones below to yellowish grey to green, highly micaceous sandstones and siltstones above. Although there is no angular discordance, the contact is regionally unconformable. The underlying Silurian rocks have been described by Bassett (1982b).

The Tilestones Formation here is 18.5 m thick (Fig. 3), and dominated by grey-green micaceous sandstones with a restricted nearshore shallow marine fauna (Stamp 1923; Straw 1930; Potter & Price 1965). The formation is best examined on the east bank of the gorge or alternatively (**with permission**) in the **disused quarry** behind the Three Horse Shoes Inn (SN 7279 2446).

The basal 3 m forms the Capel Horeb Member, of dull green to yellow chloritic, occasionally carbonaceous mudstones and micaceous siltstones with a few thin-bedded, erosively based, highly micaceous fine-grained sandstones. The sandstones contain abundant scattered intraformational mudclasts and faecal pellets, and are interpreted as storm deposits in a brackish water lagoon or protected shallow marine embayment (Almond 1983). The sediments have yielded specimens of *Modiolopsis* sp. and *Cornulites* Sp.

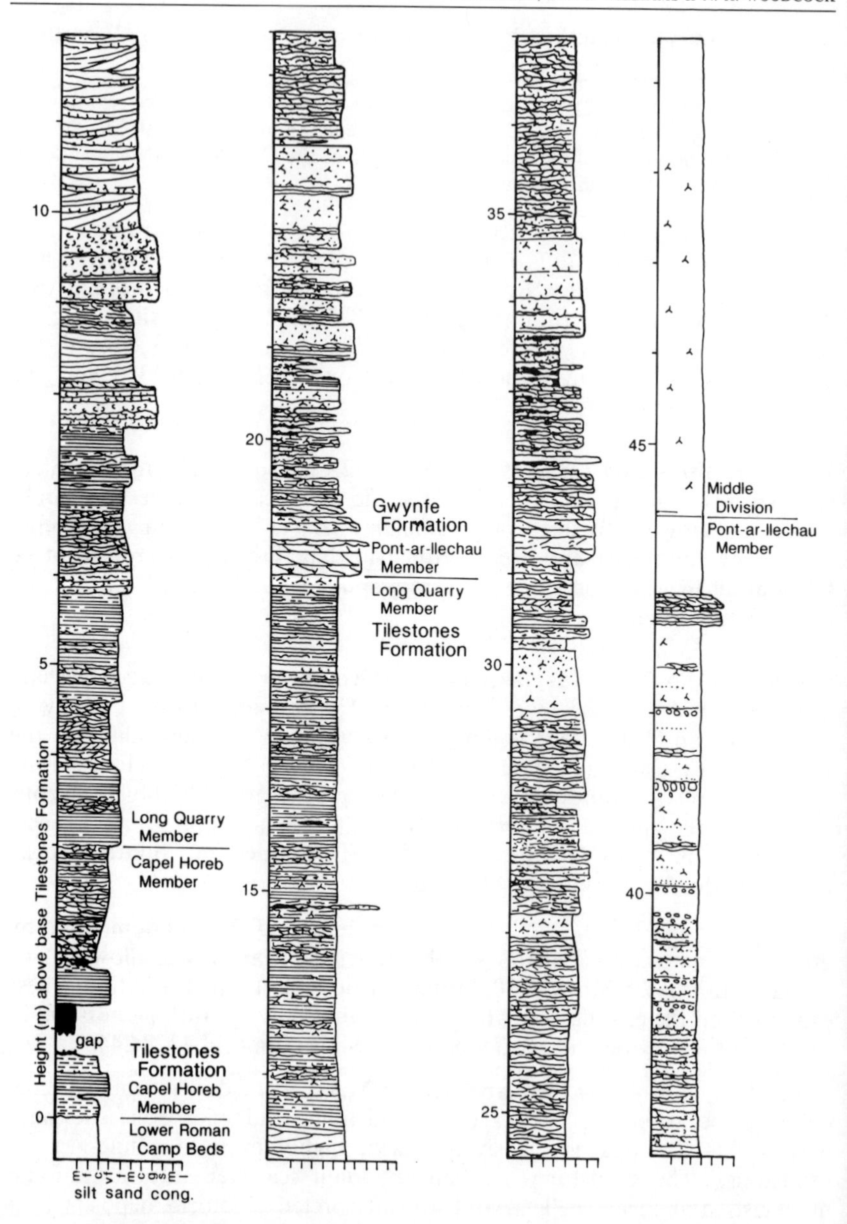

FIG. 3. Lithological log of the Tilestones Formation and the Pont-ar-llechau Member of the Gwynfe Formation in the Afon Sawdde (Localities 1A and 1B, SN 7282 2452 to SN 7286 2449). For key see Fig. 5.

The conformably overlying Long Quarry Member (15.5 m) is a stacked sequence of grey to greyish green, highly micaceous lithic sandstones. The sands are medium bedded and have a horizontal to low-angle lamination defined by concentrations of shell fragments, intraformational mudclasts, and rare granules of quartz and fragments of acid volcanic rock. Parallel and ripple cross-lamination is common, and trough cross-bedding occurs in thicker bedded sandstones in the middle of the member. The sandstones have yielded *Modiolopsis* sp., *Cornulites* sp., *Orthoceras* sp., *Lingula* sp. and abundant gastropods. King (1934) recorded a cornua like that of *Hemicyclaspis* sp. *Skolithos* burrows are common within the upper parts of beds and increase in abundance towards the top of the member. The Long Quarry Member is interpreted as a transgressive sequence of lower to middle shoreface sediments, deposited on a high energy coastline (Almond 1983).

At **Locality 1B** (SN 7286 2449), about 7 m **south of the bridge**, the Tilestones Formation is overlain conformably by the Gwynfe Formation. The basal 27 m of this formation is the Pont-ar-llechau Member (Fig. 3), distinguished by very fine to medium grained, highly quartzitic sandstones interbedded with thin green siltstones and mudstones.

The basal 5 m shows an upwards coarsening sequence of stacked thin to medium bedded sandstones with erosive bases locally scattered with quartz granules, mudstone clasts and shell fragments. The sandstones are interbedded with thin, dull yellowish-green, lenticular bedded siltstones and mudstones. The sandstones show low-angle to horizontal lamination, but often appear massive due to intense bioturbation. *Skolithos* burrows are locally preserved. A bedding plane towards the top of the basal sequence exposes large *Fucusopsis* type burrows. This basal sequence of the Pont-ar-llechau Member is interpreted as the infill of a shallow lagoon by washover sheet sandstones (Almond 1983).

The next 15 m of the member comprises a heterolithic sequence of ripple cross-laminated and wavy bedded very fine-grained sandstones and siltstones interbedded with mudstones. Within this interval occur two fining-upwards sequences, from thin-bedded quartzitic sandstones to wavy and lenticular bedded siltstones and mudstones which are interpreted as small tidal channels (Almond 1983). The quartzitic sandstones and associated mudstones contain a nearshore shallow marine fauna (Stamp 1923; Straw 1930; Potter & Price 1965) of *Modiolopsis* cf. *complanata*, *Lingula cornea*, *Cymbularia carina*, *Platyschisma helicites* and *Orthoceras* sp. A solitary cornua of *Cyathaspis banksi* has also been recorded (Almond 1983).

The upper 6 m of the Pont-ar-llechau Member consists of dull red, massive bedded mudstones and very fine-grained (locally gritty and pebbly) ripple cross-laminated sandstones interbedded with wavy flaser and lenticular bedded siltstones. The lithologies are commonly intensely bioturbated and Modiolopsids and Lingulids have been recorded in life position. The sediments are interpreted as a stacked regressive sequence of sub- to intertidal and inter- to supratidal mudstones, the latter with incipient calcrete profiles (Almond 1983).

The 'Middle division' of the Gwynfe Formation begins about 44 m above its base, at SN 7292 2440 and comprises the 680 m of strata to SN 7327 2383. It is characterized by a high (4:1) mudstone to sandstone ratio, with bright red mudstones and isolated grey to pink sandstones, typically 2 m thick. Thin airfall tuffs are common and calcrete profiles are well developed. The interval is poorly fossiliferous, yielding a few specimens of *Modiolopsis* sp., *Pachytheca* and disseminated plant remains. Bioturbation is common, particularly the large *Beaconites antarcticus* burrows.

Locality 1C (SN 7308 2400) is typical of the lower to middle parts of the 'Middle division' of the Gwynfe Formation (Fig. 4a). Individual sandbodies range from 2 m to 3 m in thickness, and consist of fine to medium grained, locally pebbly, purplish-grey micaceous lithic sandstones. They overlie a basal erosion surface and show a fining upwards trend. Their lower parts consist of planar cross-bedded sandstones, often separated by thin mudstone drapes. These pass up into horizontal laminated and ripple cross-laminated sandstones interbedded with lenticular-bedded silt and mudstones. Dessication cracks and Skolithoform burrows are common within the upper parts of the sandbodies. The sandbodies pass gradationally up into massive-bedded bright red mudstones containing rootlets and abundant carbonate nodules. Several well-developed calcrete profiles may be developed between individual sandbodies. The sandbodies are interpreted as tidally influenced channels or low sinuosity ephemeral channels, deposited in the lower reaches of a muddy low-lying coastal plain (Almond 1983).

Locality 1D (SN 7325 2385) exposes the stratigraphically important Townsend Tuff Bed (Allen & Williams 1981). It occurs 695 m above the base of the Gwynfe Formation, in a **weathered recess on the west bank** of the Afon Sawdde. As in most of south-west Wales, three airfalls are present (Fig. 4b). Fall A (0.25 m) is a muddy dust tuff with a characteristic upper surface strewn with faecal pellets. Fall B (2.35 m) has fine-grained cream, purple and yellow crystal, and crystal-lithic tuff passing up into siliceous dust tuff. Fall C (0.30 m) overlies Fall B with a prominent erosion surface and comprises a coarse to medium grained crystal lithic tuff grading up to a dull red and green mottled dust tuff.

Another thick airfall (0.75 m) occurs 23 m above the Townsend Tuff Bed

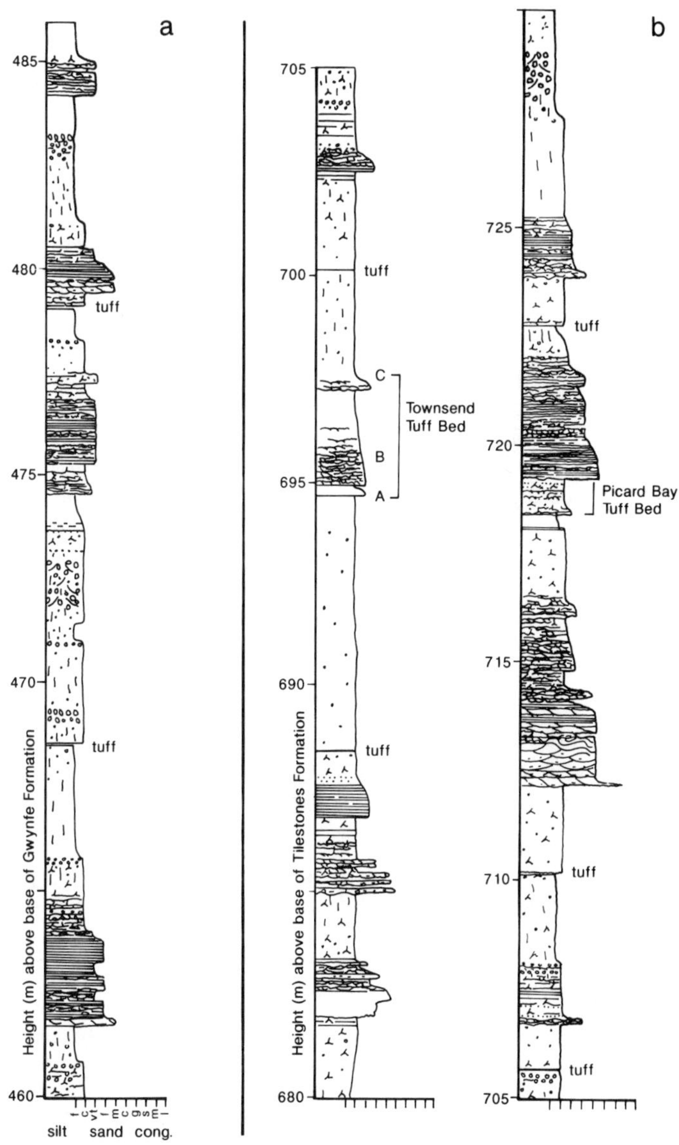

FIG. 4. Lithological logs of *a*) a typical part of the 'Middle division' of the Gwynfe Formation in the Afon Sawdde (Locality 1C, SN 7308 2400) and *b*) the part of the 'Middle division' of the Gwynfe Formation containing the Townsend and Pickard Bay Tuffs, Afon Sawdde (Locality 1D, SN 7325 2385). For key see Fig. 5.

and is equated with the Pickard Bay Tuff of south-west Wales (Allen & Williams 1981). It comprises two airfalls, one of fine-grained crystal lithic tuff and one of muddy dust tuff.

Between the Townsend and Pickard Bay Tuffs the red mudstone sequence contains one 3 m thick sand body and two thinner (<1 m) sandbodies comprising purple-grey to pink, coarse to fine-grained sandstone. The major sandbody overlies a sharp basal erosion surface and fines up from planar and trough cross-bedded sandstones first to horizontal and ripple cross-laminated sandstones then to wavy-bedded siltstones and mudstones, rarely containing Modiolopsids. The sandbody is interpreted as a low sinuosity tidally influenced channel and the thinner sands as small tidal gullies deposited on the lower reaches of a coastal plain (Almond 1983).

Locality 1E (SN 7330 2381 to SN 7332 2372) displays the 'Upper division' of the Gwynfe Formation, from about 23 m stratigraphically above the Pickard Bay Tuff to a point 10 m **south of Turkey Cottage**. The 'Upper division' is mudstone dominated (Fig. 5a), but with a coarsening upwards trend shown by the increasing frequency of sandbodies. A transition from bright red mudstones to dull red siltstones and mudstones occurs in the upper part of the division. Airfall tuffs are absent.

Five sandbodies occur in the 'Upper division', each consisting of light grey to purple, medium to fine grained, highly micaceous lithic sandstones. The sandbodies each overlie a scoured lower erosion surface and exhibit an upward-fining trend. Planar and trough cross-bedded sandstones pass up into horizontal and ripple cross-laminated sandstones and then into wavy-bedded heterolithic siltstones and mudstones. Several calcrete profiles may be present between each sandbody. The sandbodies are interpreted as low to moderate sinuosity channels, possibly exhibiting tidal influences (Almond 1983).

No precise equivalent of the Psammosteus Limestone is developed in this area, but well-developed calcrete profiles occur in both the lower and upper parts of the 'Upper division'. The lack of a more conspicuous limestone is ascribed to the continuous subsidence of this area, perhaps due to movements along the nearby Careg Cennen Disturbance (Cope 1979).

The Gwynfe Formation is taken to pass gradationally up into the Llanddeusant Formation at a point 10 m south of Turkey Cottage (SN 7332 2372). The lower 400 m of the Llanddeusant Formation is exposed dipping at 40–60° SSE between here and Pont Newydd (SN 7367 2336).

Locality 1F (SN 7340 2364) is a representative section through the Llanddeusant Formation beginning about 65 m stratigraphically above its base (Fig. 5b). It consists of medium to very fine grained, grey, pink or red

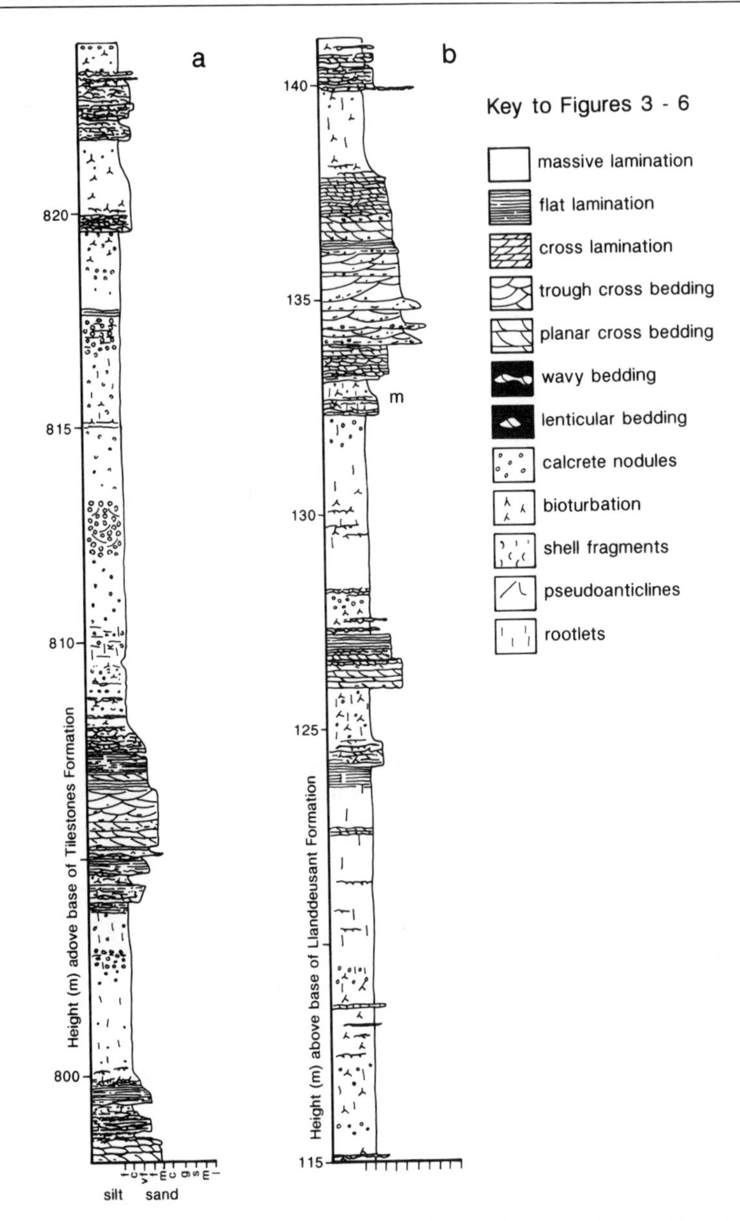

FIG. 5. Lithological logs of typical parts of *a*) the 'Upper Division' of the Gwynfe Formation in the Afon Sawdde (Locality 1E, SN 7330 2381 to SN 7332 2372) and *b*) the Llanddeusant Formation in the Afon Sawdde (Locality 1F, SN 7340 2364).

sandstones. The sandbodies have a fining-upwards trend similar to those in the 'Upper division' of the Gwynfe Formation, but with some lateral accretion surfaces also developed. The sequence is interpreted as moderate to high sinuosity fluvial channels interbedded with overbank floodplain siltstones and mudstones (Almond 1983).

The Llanddeusant Formation is exposed intermittently **south** of **Pont Newydd** and is probably about 700 m in total thickness. **Near Pont Aber** (between SN 7380 2280 and 7380 2255) a zone of fracture and irregular dips marks the WSW-striking Careg Cennen Disturbance. South-east of this, the Senni Beds overlie the Llanddeusant Formation. They can be examined in the **Nant Sawdde Fechan** (SN 7565 2190) on the lower slopes of the Black Mountain, but are better displayed at Localities 3 and 4 of this excursion.

Vehicles should now be collected and driven north-west on the A4069. Turn right in Llangadog, continuing on the A4069 to Llandovery. Here keep straight on along the eastbound A40 for 10 km to a point, about 2 km beyond Halfway, where a minor road leaves the A40 on its north side. A steep unmade track runs north-westwards from this same junction, past the ruined chapel of Capel Horeb.

2. Capel Horeb Quarry (SN 8445 3234) has its entrance about 100 m up the track. There is **parking for several vehicles** here and near the junction of the track with the main road. **Permission for access** to the quarry should be obtained in advance from the Nature Conservancy Council. The quarry is long disused, and the **faces are in a dangerous state**. **Safety helmets should be worn** and no attempts made to climb up the faces. The section can be examined from a distance and lithologies and sedimentary structures seen in fallen blocks.

The **north-west face of the quarry** is a steeply dipping bedding plane at the top of the Upper Roman Camp Beds (upper Ludfordian). The unconformity with the overlying Tilestones Formation is exposed at the north-east end of this face. The **north-eastern face** then exposes a continuous section through the Tilestones Formation and into the overlying Gwynfe Formation. The quarry is particularly famous for its yield of the early vascular land plants *Steganotheca*, *Cooksonia*, and *Hostinella* (Edwards 1970a; Edwards & Rogerson 1979; Edwards & Richardson 1978) preserved as coalified impressions in the Roman Camp Beds and the Capel Horeb Member of the Tilestones Formation.

The lowest 12 m of the Tilestones Formation forms the Capel Horeb Member (Fig. 6). Upper bed surfaces commonly display a small-scale wave or current ripple cross-lamination, the latter being well exposed on extensive bedding surfaces high on the main face. Planar cross-bedded sandstones are locally developed towards the top of the member. Individual sands are separated by thin lenticular bedded siltstones and mudstones. Bioturbation is locally intense, including *Planolites* and *Skolithos*. The sandstones are interpreted as high-energy lower to middle shoreface deposits (Almond 1983).

The conformably overlying Long Quarry Member is 23 m in thickness (Fig. 6) and comprises a stacked sequence of fine to medium grained, highly micaceous and quartzitic, grey to green or purple sandstones. It contains a diverse assemblage of decalcified brachiopods, bivalves and gastropods (Straw 1930; Potter & Price 1965) including *Orthoceras* sp., *Loxonema* sp., *Platischisma helicites*, *Modiolopsis complanata*, *Salopina* sp., *Bucanopsis* sp. and *Cyclonema* sp. Straw (1930) reported finding a solitary fish spine from an unspecified horizon.

The Long Quarry Member sandstones are medium to thick bedded, dominantly low-angle laminated and hummocky cross-bedded. Cross-bedded and ripple cross-laminated sandstones are locally developed. Decalcified shell fragments are common along laminae within beds. The sandstones are interpreted as high energy lower to middle shoreface deposits (Almond 1983).

The uppermost 11 m of the member consists of purple and green, medium to fine grained, highly micaceous sandstones. Stacked sequences of parallel laminated and trough cross-bedded sandstones occur interbedded with ripple cross-laminated, very fine-grained sandstones and siltstones. Skolithoform burrows are common in the upper parts of beds. Deposition within a zone of subtidal to intertidal sandy shoals is envisaged, such as that at the mouth of tidal inlets and estuaries.

The Pont-ar-llechau Member of the Gwynfe Formation conformably overlies the Tilestones in the **north-east corner of the quarry** (Fig. 6). The lowest 2 m consists of fine to medium grained quartzitic sandstones that are generally massive bedded and separated by impersistent mudstones. Concentrations of heavy minerals (zircon, rutile, tourmaline) occur. The remaining 5 m of the member are dull green mudstones or laminated siltstones, which are transitional to the dull green and red mudstones of the 'Middle division' of the Gwynfe Formation which comprises the last 10 m of exposed section. Incipient calcrete nodules are developed high in the quarry face. The sequence is interpreted as a vertical transition from a beach/barrier to muddy intertidal flats.

From Capel Horeb Quarry head eastward on the A40 for 10 km to Sennybridge. Take the A4215 southward and after 2.5 km take the right turn for a further 3 km to Heol Senni. Turn right at the road junction before the village (SN 9284 2350) and head west for 2 km, with

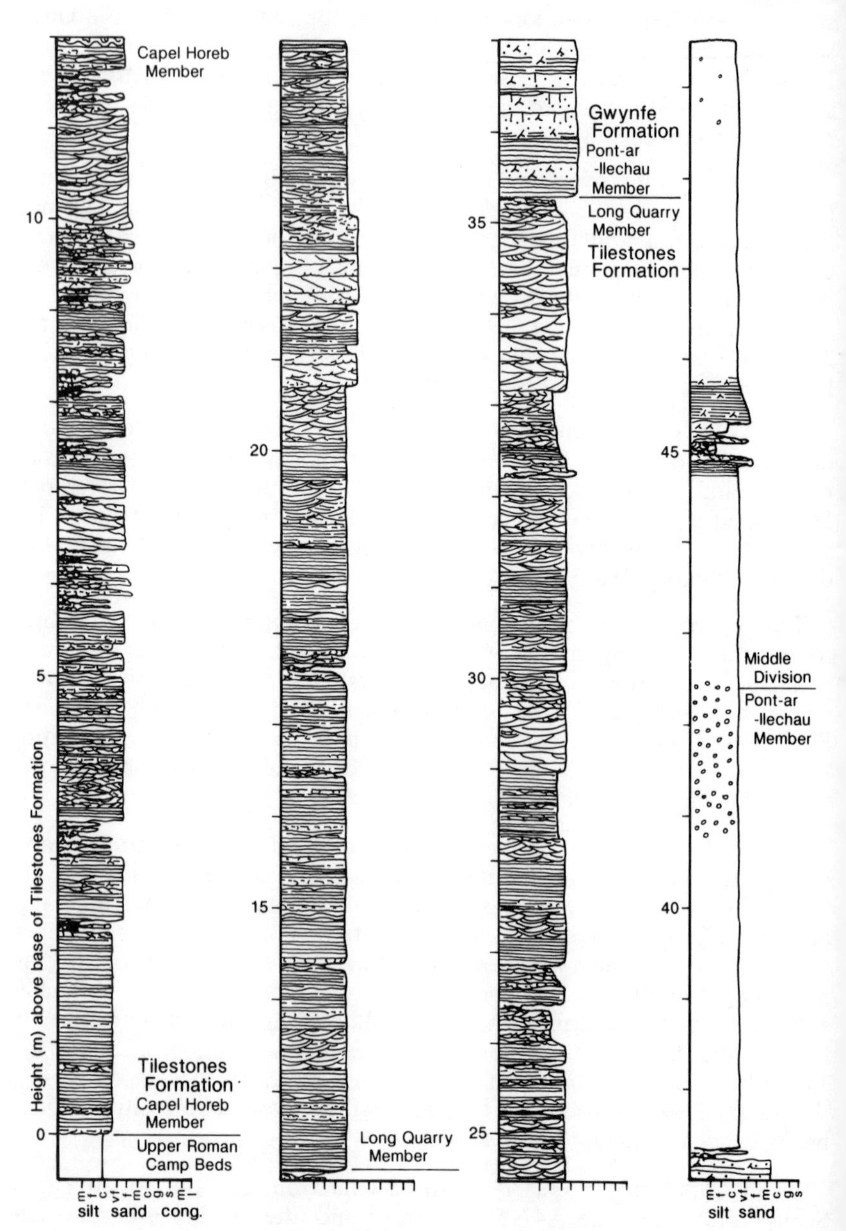

FIG. 6. Lithological log of the Tilestones Formation in Capel Horeb Quarry (Locality 2, SN 8445 3234). For key see Fig. 5.

Heol Senni Quarry visible on the north-eastern slope of Fan Bwlch Chwŷth to the south.

3. Heol Senni Quarry (SN 9145 2210) is approached by a drivable track that leaves the road at SN 9081 2288 and heads south-westward. There is **ample parking for small vehicles** near the disused quarry plant. **Care should be taken near the unstable faces**.

The quarry exposes a near-horizontal sequence within the upper-middle part of the Senni Beds. The whole formation is between 250 m and 400 m thick and is considered to be middle Middle to lower Upper Pragian in age (Thomas 1978). The sequence is dominated by sandstones and conglomerates, with subordinate mudstones, arranged with no recognizable vertical cyclicity.

The sandstones are grey-green, fine to medium grained litharenites. Sandstone beds overlie basal erosion surfaces, sometimes defining channels. The sandstones are often massive but may have laminae defined by grain size, mica flakes and carbonaceous debris. Flat to low-angle lamination is evident, along with occasional large-scale planar and trough cross-bedding. Ripple cross-lamination occurs in some of the finer units and linguoid ripple marks can be seen on some bedding surfaces. There is some record of soft-sediment deformation, particularly a persistent horizon of ball-and-pillow structures.

The conglomerates contain grey-green silt clasts up to 450 mm long, together with rolled calcrete nodules. There is no exotic debris. Many of these intraformational conglomerates show secondary cementation by carbonate. The mudstones are grey-green in the lower part of the sequence, but red and nodular near the top. In places they show sandstone-filled mudcracks.

The Senni Beds are particularly rich in plant fossils, over fifty varieties of spores being recorded from Heol Senni (Thomas 1978). Macroplant debris up to several centimetres long has been found in loose blocks of blue/grey-green very fine sandstones and siltstones on the quarry floor (Edwards *et al.* 1978).

The Senni Beds are interpreted as a medial alluvial facies, deposited from rivers closer to their source than those responsible for the underlying Llanddeusant Formation (Locality **1F**). The sandstones and conglomerates represent stacked bar complexes in low sinuosity sand-bed rivers, and the mudstones the rare remnants of the originally intervening floodflats.

Return from the quarry to Heol Senni village and at the road junction at SN 9284 2350 keep straight on in the direction of Brecon. After 3 km join the A4215 and continue eastward for 4 km to its junction with the A470. Turn right here and head southwards for 5 km.

4. Brecon Beacons Quarry (SN 9715 2084) is on the **west side of the A40** where the road bends sharply to the east beneath the crags of Craig y Fro. About 14 m of Senni Beds are exposed, the top being about 20–40 m below the top of the formation. The lithologies and inferred environment are similar to those of Heol Senni Quarry, though massive sandstones are more conspicuously channelled into finer sandstones and siltstones. The quarry has long been famous for its diverse macroflora (e.g. Heard 1927; Croft & Lang 1942; Edwards 1969, 1970*a*, *b*) including *Gosslingia breconensis*, *Psilophyton princeps*, *Cooksonia*, *Drepanophycus* and *Zosterophyllum*. There is also a rich spore assemblage confirming a Pragian age.

Continue a further 1 km along the A470 to Storey Arms (SN 9825 2032), which provides one of several possible access points for the mountain of Corn Du.

5. Corn Du (SO 008 214) is chosen as a representative outcrop of the Brownstones, which form the scarp faces of the Brecon Beacons but which are nowhere well exposed close to roads. The ascent of Corn Du is a **strenuous climb** from any direction, and **in bad weather becomes a serious expedition to be avoided**. From Storey Arms the more gentle ascent is north-north-eastward to the top of Y Gyrn, then eastward around the headwaters of the Afon Taf to the obelisk on the SE-trending ridge leading to Corn Du summit. Satisfactory views of the Brownstones sequence can be had from the ridge without making the full ascent.

The sequence on the **north scarp face** of Corn Du (Fig. 7) comprises laterally extensive sheets of interbedded sandstone and siltstone, which have been fully described and interpreted by Tunbridge (1981). There is an upward increase in the proportion of sandstones, interpreted as progressively more proximal depositional environments on an extensive alluvial plain. The lower 115 m shows alternations of relatively distal and medial sequences of fine to medium grained red sandstones interbedded with floodflat siltstones. The distal sandstone beds are 40–150 cm thick, parallel-laminated sometimes with a thin basal intraformational conglomerate, and are isolated within the floodflat siltstones in a ratio of about 2:3. They are interpreted as sandy sheetfloods. By contrast the sandstones of the medial facies mostly occur in multistorey sandbodies up to about 5 m thick, interbedded with siltstones in a ratio of about 3:1. Individual sandstone beds are 50–80 cm thick, have lenticular and channelled forms, and are often floored by thin intraformational conglomerates. They are interpreted as the deposits of low-sinuosity channels that merged downslope into the sheetfloods of the distal facies.

In the upper 35 m of the Brownstones, multistorey sandstones dominate interbedded siltstones in a ratio of about 12:1. Individual sandstone beds range up to 180 cm thick and form multistorey bodies up to 11 m thick.

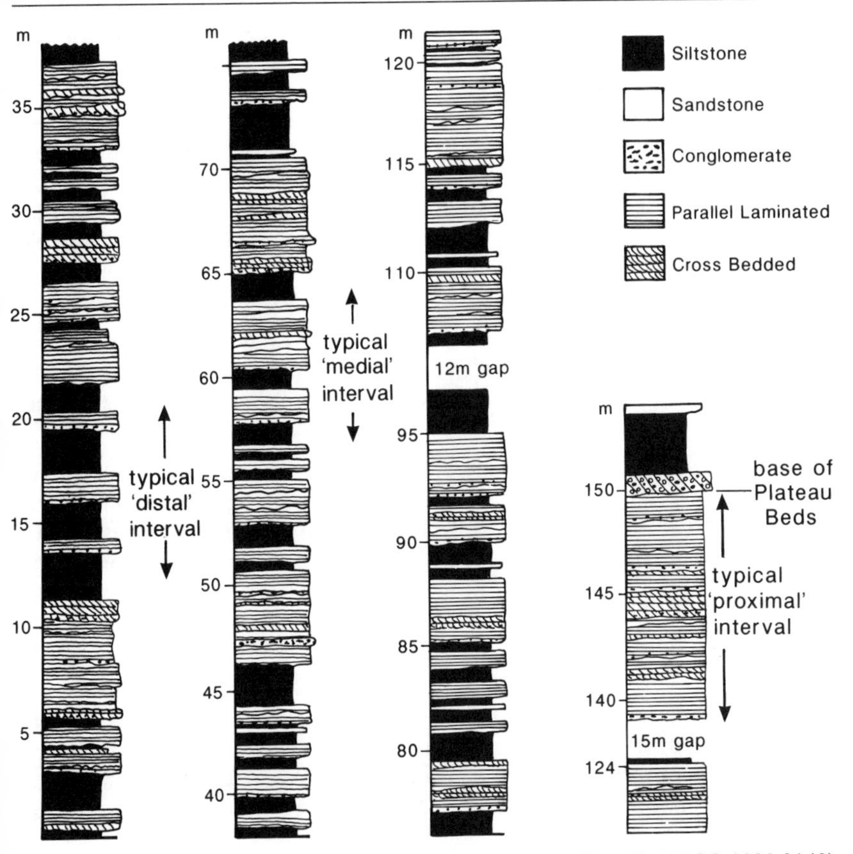

FIG. 7. Lithological log of the Brownstones on Corn Du (Locality 5, SO 0080 2140). After Tunbridge (1981).

They are interpreted as the deposits of a more proximal region of the same alluvial system that deposited the medial facies.

The top of the Brownstones on Corn Du is marked by an unconformity with a few metres of the Plateau Beds Formation. There is no angular unconformity here, but a few degrees of angular discordance develops 20 km further west on the Black Mountain, recording tilting during the mid Devonian deformation of the late Caledonian (Acadian) event. The Plateau Beds can be examined in detail at Locality 7.

Returning to Storey Arms, drive south on the A470 in the direction of Merthyr Tydfil. After about 14 km, at the junction with the A465, turn left onto the minor road for Ponsticill (5.5 km). At the far end of Ponsticill village fork right and descend to the reservoir. **Drive along the dam wall and park near its eastern end** (SO 0624 1186).

6. Abercriban Quarry (SO 0636 1273) is **east of the Pontsticill reservoir. Permission for access** must be obtained from the Taf Fechan Water Authority. The quarry is reached by walking about 500 m northwards along the road skirting the east side of the reservoir, passing under the old railway line and heading northwards obliquely up the hillside for a further 300 m.

The **main face of the disused quarry** exposes the Grey Grits of the Upper ORS, underlain by the top of the Plateau Beds and overlain by a few metres of the Carboniferous Lower Limestone Shales (Fig. 8a). The beds are almost horizontal, but are cut by a steep E–W-striking fault at the north edge of the quarry.

The Grey Grits are grey-white, fine to medium grained quartzose sandstones, with scattered lenses of pebbles, interbedded with subordinate siltstones and mudstones (Lovell 1978). The sandstone beds show abundant planar and trough cross-bedding and some parallel lamination. They are mostly tabular with little scouring at their bases. The Grey Grits lack body fossils here, but lingulides have been found in the underlying Plateau Beds (Hall *et al.* 1973). *Skolithos*-type burrows occur. Deposition from sandy braided streams has been proposed (Lovell 1978).

A lag conglomerate marks the erosive base of the Lower Limestone Shales. The persistence of this lagged horizon suggests some disconformity at what is a structurally conformable contact over South Wales. The Carboniferous sequence in this area is described by Dickson & Wright in this volume (Excursion 15).

The last locality can be approached on foot or by road. To walk from Abercriban, head eastward up the hillside above the quarry to reach a track along the foot of the limestone escarpment (SN 0700 1272). Walk north-eastward for about a further 3 km to the Duffryn Crawnon quarries. To drive to the last locality, head southward from Pontsticill dam for 4 km to regain the A465. Drive eastward for 9 km and turn left for Trefil. Drive northwards for about 5 km, past the disused Trefil quarries to the entrance to the Cwar Yr Ystrad limestone quarries near Pistyll Crawnon.

7. Duffryn Crawnon (SO 0945 1500) is reached from the quarry entrance by descending to the footpath running round the head of the valley and following this northward over a small waterfall. The top of the ORS is exposed **at the side of the footpath** at SO 0930 1518 where a small brook cuts across the path. The base of the Upper ORS can be seen above the path at SO 0908 1547. The sequence extends from the uppermost Brownstones, through

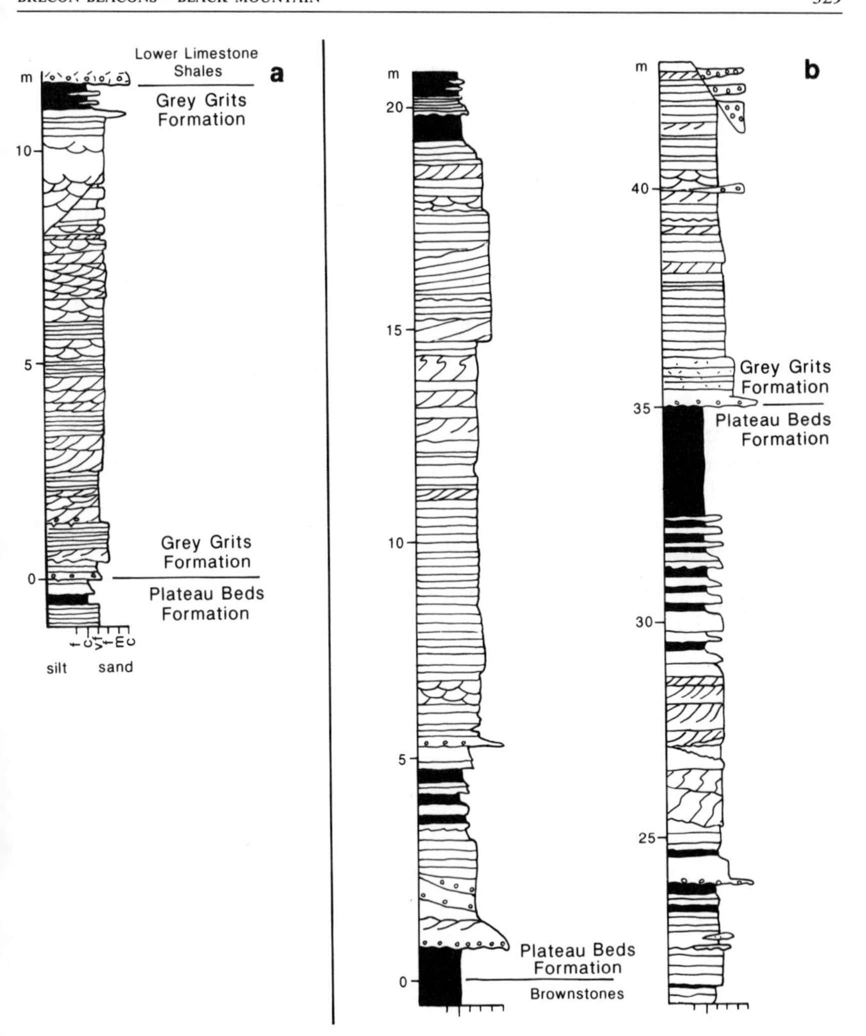

FIG. 8. Lithological logs of *a*) the Grey Grits Formation at Abercriban Quarry (Locality 6, SO 0636 1273) and *b*) the Upper ORS at Duffryn Crawnon (Locality 7, SO 0945 1500). After Lovell (1978). For key see Fig. 5.

the Plateau Beds and Grey Grits Formations to the lowermost Lower Limestone Shales (Fig. 8b).

The Brownstones here are interbedded red-brown sandstones and mudstones, the deposits of low-sinuosity channels on an alluvial plain. The Plateau Beds rest on the Brownstones with major disconformity, though here with no discernible angular unconformity. The lowest 18 m are mostly thin to medium bedded red-brown sandstones with pebble conglomerates near the base, interpreted as aeolian sands with intercalated waterlain sediments (Lovell 1978). The upper 18 m of the Plateau Beds are interbedded sandstones and mudstones probably deposited in marginal marine supra- and sub-tidal environments (Lovell 1978). SE-directed palaeoflow is suggested by cross-bedding in channel-fill sandstones in this upper division. Very fine-grained sandstone horizons have yielded the brachiopod cf. *Ptychmalotoechia omaluisi*, lingulides, and fragments of fish and plants. Fish fragments have also been found in conglomerates at the base and 12 m below the top of the formation (Taylor 1973; Hall *et al.* 1973).

The Grey-Grits are green-grey fine to medium grained sandstones with thin lenticular conglomerates. The sandstones are cross-bedded and parallel laminated. Fish fragments and the bivalve *Sanguinolites* sp. have been found in the conglomerate that marks the base of the formation (Hall *et al.* 1973; Taylor & Thomas 1974). The Grey Grits are interpreted as the deposits of braided streams (Lovell 1978). The base of the Lower Limestone Shales is marked by a thin lagged horizon.

A return from Duffryn Crawnon can be made on foot to Abercriban and Pontsticill dam or to the quarry entrance, and then by car southward to join the A465.

15. CARBONIFEROUS LIMESTONE OF THE NORTH CROP OF THE SOUTH WALES COALFIELD

by J. A. D. DICKSON *and* V. P. WRIGHT

Maps *Topographical:* 1:50 000 Sheets 160 Brecon Beacons and 161 Abergavenny

1:25 000 Outdoor Leisure maps.
Brecon Beacons Central Area 11
Brecon Beacons Eastern Area 13

Geological: 1:50 000 Merthyr Tydfil 231
Abergavenny 232

THE Carboniferous Limestone cropping out along the southern boundary of Powys represents deposition on the northern edge of a major carbonate platform over what is now southern Britain. Four localities in the Carboniferous Limestone of Powys are described in this itinerary (Fig. 1).

Regionally the Lower Carboniferous is a thick limestone and dolomite sequence sandwiched between the alluvial Old Red Sandstone (Devonian) beneath, and the deltaic–paralic Upper Carboniferous above. The carbonate unit constitutes a 100 km wide wedge, over 1000 m thick in southern outcrops (south Dyfed) but thinning to only 150 m along the northern outcrops in south Powys (Fig. 2). This thinning is due to lower rates of subsidence on the northern edge of the basin and to phases of uplift and erosion during the early Carboniferous. To the south of the limestone 'platform' was a deep water basin, the Culm or Cornubian Basin, while to the north was a small land mass (St George's Land), covering what is now mid Wales. The nature of the transition from the carbonate platform into the Cornubian Basin is obscured by later tectonic deformation.

The exact tectonic setting of the platform is rather unclear but the early Carboniferous seems to have been a transition phase from a late Caledonide minor strike-slip regime to a late Carboniferous Variscan foreland basin. During early Carboniferous times Britain was in an extensional regime with graben and half-graben structures dominating deposition over central and northern England. The carbonate wedge in

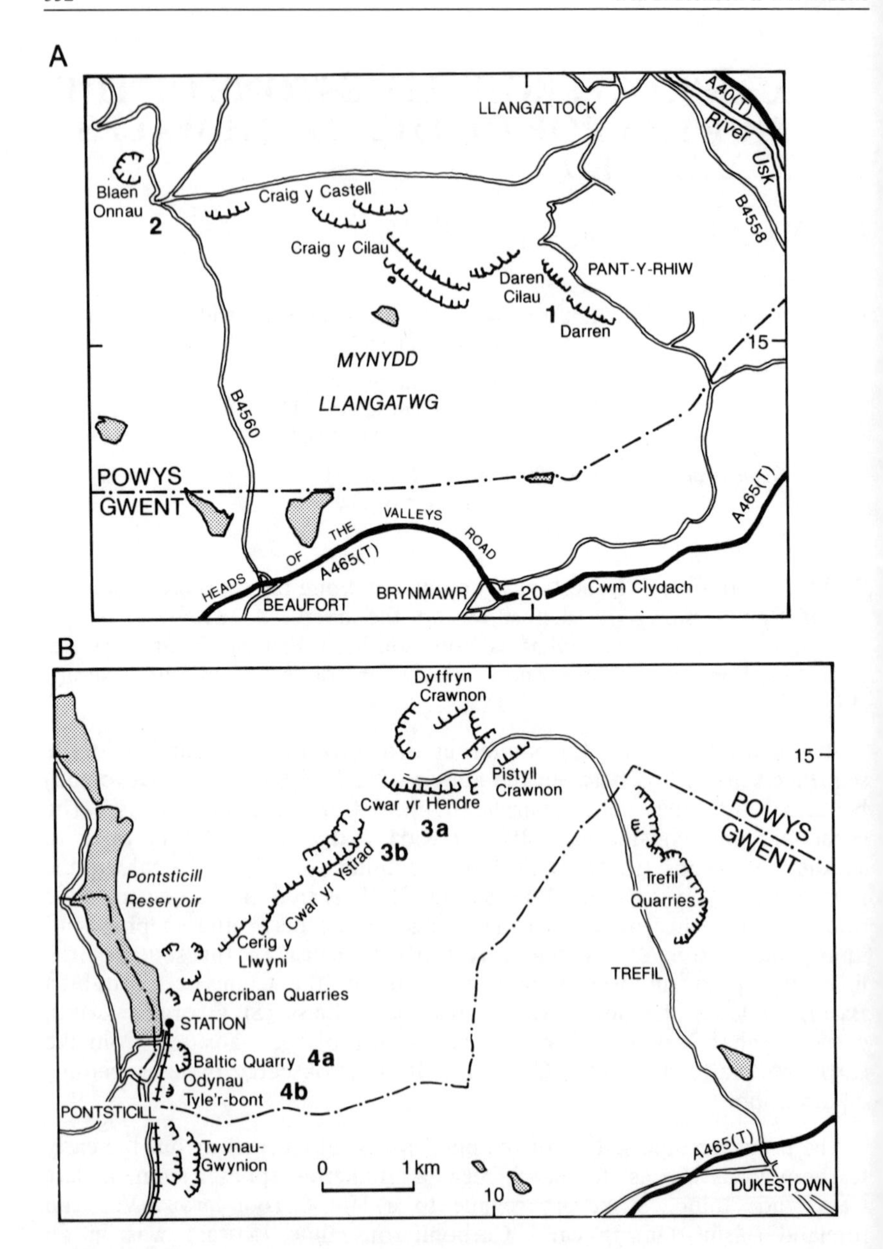

FIG. 1. *a*) Location map for Localities 1 and 2. Major road network shown, with 5 km National Grid lines around the margins. *b*) Location map for Localities 3 and 4.

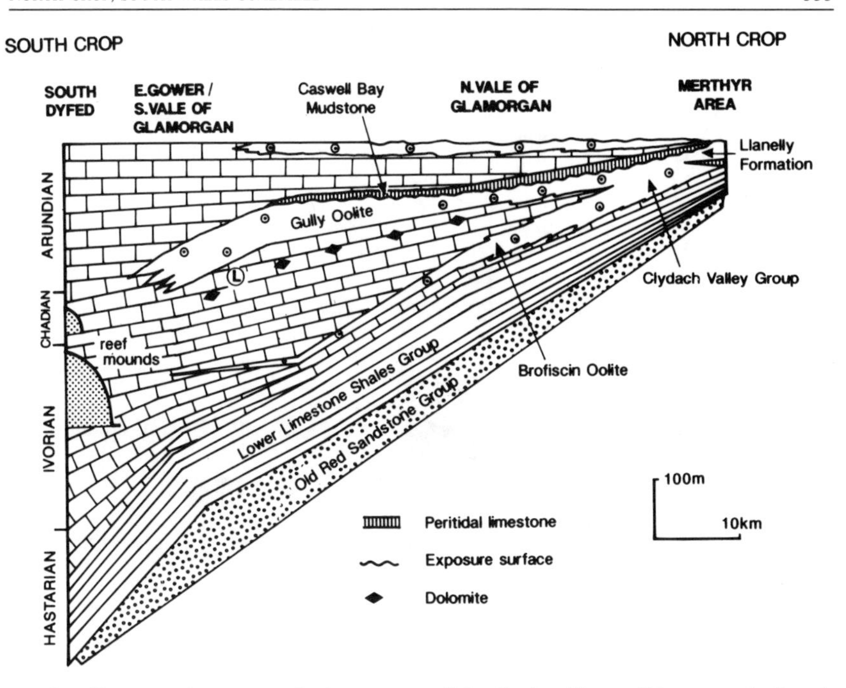

FIG. 2. Cross-section across the lower part of the Carboniferous Limestone in South Wales (from Wright 1986a) showing the possible relationship of the North Crop sequences to those in the South. The correlations between the Vale of Glamorgan and the Merthyr area are speculative.

south west Britain could be regarded as deposits on large, southerly dipping fault blocks (Wright 1987; Gayer & Jones 1989).

The platform evolved from a southerly dipping 'ramp' in the pre-Holkerian (Wright 1986a) to a more 'flat-topped' shelf in the Asbian. This change resulted both from a marked decrease in the rate of flexural subsidence over the region, and as a natural consequence that carbonate build-ups evolve into 'shelves' (Wright 1987).

After a period of major sea-level rise at the beginning of the Carboniferous, the 'Old Red Sandstone' alluvial plains were drowned, progressively, resulting in the deposition of the Lower Limestone Shales Group (represented by the Cwmyniscoy Mudstone; Fig. 3). This complex sequence, culminating in black shales representing a possible major period of regional bottom water anoxia, is generally poorly exposed but can be seen in the Pontsticill area.

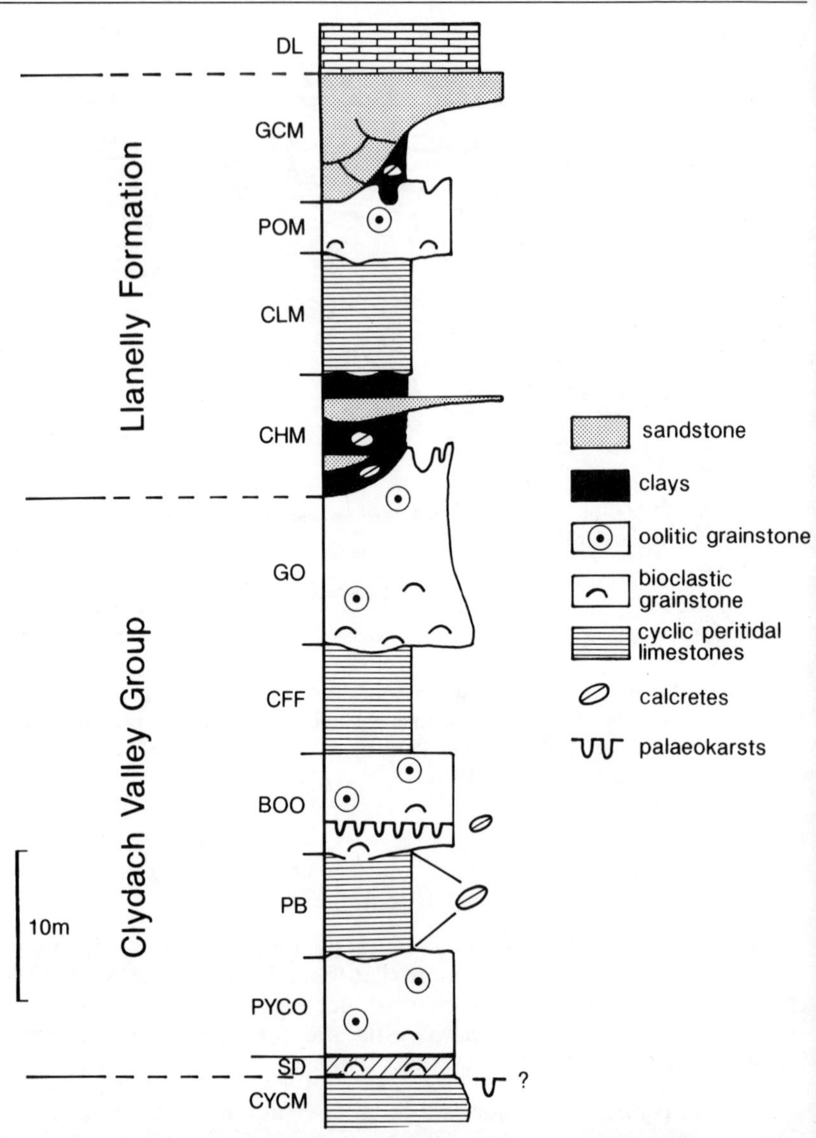

FIG. 3. Carboniferous Limestone sequence in the northern area. CYCM = Cwmyniscoy Mudstone; SD = Sychnant Dolomite; PYCO = Pwll-y-Cwm Oolite; PB = Pantydarren Beds; BOO = Blaen Onnen Oolite; CFF = Coed Ffyddlwn Formation; GO = Gilwern Oolite; CHM = Clydach Halt Mbr; CLM =Cheltenham Limestone Mbr; POM = Penllwyn Oolite Mbr; GCM = Gilwern Clay Mbr; DL = Dowlais Limestone (Holkerian).

The overlying Oolite Group (Clydach Valley Group) contains oolitic limestones, restricted peritidal deposits (formed in lagoonal and tidal flat settings) and unusual fossil soils (paleosols). This unit is well exposed in many quarry sections.

The overlying Llanelly Formation is a highly complex sequence of alluvial deposits, paleosols, peritidal limestones and oolites. It is also well exposed in the area.

The overlying Dowlais Limestone and higher units will not be discussed in this itinerary.

Disconformities of various types are very numerous in the sequence. The two most important ones separate the Clydach Valley Group from the Llanelly Formation and the Llanelly Formation from the Dowlais Limestone. The lower break was, in part, caused by local tectonic activity. As a result the Llanelly Formation oversteps the various formations of the Clydach Valley Group as it is traced northwards towards the Neath Disturbance (George 1954). To the east of the area in the Clydach Gorge near Bryn-mawr, the Llanelly Formation lies on the Gilwern Oolite whilst at Blaen Onneu it lies on the Blaen Onnen Oolite. [The customary mis-spelling (Blaen Onnen, see Barclay 1989 for history) is retained here for the stratigraphical unit, but the current Ordnance Survey spelling (Blaen Onneu) is used for the locality].

The second major disconformity results in the Dowlais Formation overstepping the Llanelly Formation and the Clydach Valley Group in a westerly direction. The Llanelly Formation is absent west of Penderyn and further west the Dowlais Formation lies on Lower Limestone Shales.

Clydach Valley/Group

Clydach Valley/Abercriban Oolite Groups

The most comprehensive lithostratigraphical terminology for the Dinantian rocks of the North Crop is that proposed by the British Geological Survey in the Merthyr Tydfil and Abergavenny memoirs (Barclay 1988, 1989).

Groups	*Formations*	*Members*
	Dowlais Limestone	
	Garn Caws Sandstone	
	Llanelly Formation	Gilwern Clay Penllwyn Oolite Member Cheltenham Limestone Member Clydach Halt Member

Abercriban Oolite Group (in west) Clydach Valley Group (in east)	Gilwern Oolite Coed Ffyddlwyn Formation Blaen Onnen Oolite Pantydarren Beds Pwll-y-Cwm Oolite Sychnant Dolomite
Lower Limestone Shale Group	Cwmyniscoy Mudstone Castell Coch Limestone

Within south-east Powys the first major carbonate unit above the Lower Limestone Shale Group is the Clydach Valley Group, which is overlain by the Llanelly Formation. West of Abercriban the Clydach Valley Group passes into the Abercriban Oolite Group. These groups are predominantly Courceyan in age but the uppermost Gilwern Oolite is difficult to date. The macro and microfauna of the Gilwern Oolite have been used to support ages which range from Courceyan through Chadian to Arundian (Barclay 1989). The Clydach Valley Group consists of three oolite formations separated by dolomitic/peritidal beds. The latter thin and die out and the upper units of the group are overstepped westwards; the Abercriban Oolite Group is predominantly oolitic limestone.

Synchnant Dolomite, Pantydarren Beds and Coed Ffyddlwn Formations

The Sychnant Dolomite consists of a few metres of fine crystalline dolomite intercalated with shales at its base. The Pantydarren Beds and Coed Ffyddlwn formation are principally stratiform dolomites intercalated between undolomitized grainstone units. Some ghost, primary laminations are preserved within these three dolomites: the younger two are associated with spherulitic carbonate, calcretes and thin coals. Subspherical vugs in these dolomites are filled with calcite interpreted to be calcium sulphate replacements by Searl (1988a).

Late stage coarsely crystalline dolomites associated with faulting (Raven 1983) are clearly distinguishable from the early stratiform dolomites. The origin of these dolomites is discussed by Searl (1988a) and Hird *et al.* (1987) and the association of the stratiform dolomites with pedogenic columnar calcites is discussed by Searl (1988a).

Pwll-y-Cwm, Blaen Onnen and Gilwern Oolites

These consist predominantly of mixed oolitic, peloidal and skeletal limestones; the best sorted, purest oolite occurs at the top of the Gilwern Oolite. Searl (1986) interprets these units as back-shoal deposits; graded bioclastic lenses (30 cm thick by 10 m lateral extent) she compared to storm blowouts as found in modern-day Florida Bay. The coarse bioclastic oolitic sands containing intraclasts and keystone vugs are beach sands which became emergent at the top of each unit. Emergence led to pedogenesis, mostly in an arid climate and widespread penetration of meteoric waters. In a general way the Clydach Valley Group can be thought of as a three-phase transgressive/regressive unit with subtidal grainstones being interspersed with calcretes and dolomite. The alternation of marine and freshwater fluids in these sediments has led to their very complex diagenesis (Raven, 1983; Searl, 1986). The most important break in this sequence is that at the top of the Clydach Valley Group. Not only is this clearly recognizable in the field as a karstic surface with solution pots and karstic brecciation developed several metres into underlying sediments but the unconformable overstepping relationships of the Llanelly formation was recognized by George (1954). The lithostratigraphy of the Clydach Valley Group emphasizes three major regressive episodes but as can be seen from the following itinerary, the three grainstone units contain many impersistent calcretes indicating repeated emergence during deposition.

Llanelly Formation

The Llanelly Formation consists of four members (Fig. 6). The lowest (Clydach Halt) and highest (Gilwern Clay) are siliciclastic alluvial deposits with paleosols, stream-flood, and high sinuosity channel deposits. The lower limestone member, the Cheltenham Limestone, is a crudely cyclic peritidal package, whilst the upper limestone unit, the Penllwyn Oolite, is an oolitic packstone-grainstone with stromatolites.

The contact between the Oolite Group and the Llanelly Formation is a prominent palaeokarstic horizon (Wright 1982a) (Localities **1**, **2**, **3a**). This zone, up to 6 m thick, appears as a rubbly horizon consisting of interconnected centimetre-size pipes and fissures filled with green clay. The degree of piping and fissuring is so intense that blocks of fluted limestone can be pulled out of the cliff faces making the sections locally hazardous.

Clydach Halt Member. The palaeokarst is capped by the first unit of the Llanelly Formation, the Clydach Halt Member. This is a highly variable unit consisting of up to 6 m of conglomerates and sandstone with calcrete paleosols and silty clays. The conglomerates were all locally derived and contain limestone and calcrete clasts but the sandstones (Localities **1**, **4a**) may represent floodplain or distal alluvial-fan deposits, possibly wave reworked during shallow lake phases (Wright 1981a). The calcretes are typical hardpan or petrocalcic horizons representing prolonged periods of pedogenesis (Wright 1982b) (Locality **4b**). Locally, this unit is missing because of erosion during the marine transgression, represented by the overlying Cheltenham Limestone Member, and the unit is missing at Blaen Onneu (Fig. 8). This member represents two phases of deposition on a complex, dissected landscape (Fig. 7). The first phase consists of calcretes (Locality **4b**) representing a semi-arid phase, overlying a palaeokarst formed under a sub-humid or humid climate (Wright 1988, 1990). The maturities on the calcretes, and their number (probably six at Locality **4b**) indicate a period of prolonged landscape stability, allowing soil formation. An estimate greater than 250,000 years for this phase of pedogenesis is not unreasonable, and perhaps as much as one million years. This phase was followed by erosion, dissection, and reworking to form the sandstones at Localities **1** and **4a**. These sandstones were largely externally derived, whereas the silts and clays in the earlier unit were probably wind-blown in origin, with some gravel-grade horizons consisting of locally-derived oolite and calcrete clasts.

Cheltenham Limestone Member. This unit, which is up to 8.5 m thick, onlaps the subaerial facies below and thins towards Blaen Onneu. The unit is composed of a series of weakly cyclic peritidal lithofacies (Fig. 4). Both non-skeletal (micritic) and skeletal (porostromate-bearing) oncoids occur, the latter containing mainly *Ortonella* and stromatolites built by *Ortonella* and vermiform gastropods (Wright 1983b; Wright & Wright 1981), but they are often difficult to see on the highly weathered rock faces. True skeletal algae are confined to Lithofacies A where the aberrant ?dasyclad *Koninckopora* is common (Locality **1**).

The prominent feature of this unit in outcrop is the regular bedding with numerous clay partings. Many, and perhaps all, represent thin paleosols. Two prominent paleosols occur near the top of the member, the

LITHOFACIES	STRUCTURES	TEXTURE	GRAIN TYPES	BIOTA	INTERPRETATION
A	Erosive based, cross-lamination	Grainstone	Intraclasts quartz sand	Oncoids *Koninckopora* crinoids, forams brachiopods cephalopods	Transgressive, shallow, open, marine lag
B	Bioturbated	Grainstone to mudstone	Peloids, superficial ooids	Ostracodes vermiform gastropods, oncoids	Restricted, shallow marine, "lagoonal"
C	Fenestral (irregular and laminoid) cryptalgal laminates	Grainstones to mudstones	Peloids	Ostracodes	Periodically exposed, "intertidal"
D	Replaced gypsum pseudomorphs	Mainly mudstones	Peloids	–	"Supratidal"
E	Carbonate nodules and thin beds, calcrete crusts, rhizocretions	Green, silty clay, carbonate layers with soil textures	Calcrete nodules, coated grains	–	Soil

FIG. 4. Lithofacies in the Cheltenham Limestone Member.

Darrenfelen and Cwm Dyar geosols. The former, which occurs about 1 m below the top of the member, is a typical, thin (under 0.2 m) calcrete crust', and actually represents a thin soil with abundant calcified soil aecal pellets, burrows and rhizocretions (Wright 1983a). It is very poorly developed in the itinerary area. The Cwm Dyar Geosol, capping the unit, s a prominent calcrete.

Penllwyn Oolite Member. This unit is only some 3–5 m thick, generally consists of a bioturbated peloidal, oolitic, grain aggregate grainstone to packstone. At some localities a stromatolitic level, or rarely a minor exposure urfaces, occurs within the unit. At Blaen Onneu (Locality **2**) the unit is a ross-stratified limestone capped by interference ripples upon which stromatolites grew.

The base of the member is erosive and the Cwm Dyar Geosol is locally absent. The basal horizon is a coarse bioclastic, locally oncolitic horizon, named the Uraloporella Bed because of the occurrence of the problematic ubiform organism *Uraloporella korde* (Wright 1982c). The oncoids in this bed (or series of beds) are spectacular and reach diameters of 0.12 m. They are generally well rounded and spherical with laminae of fascicular optic alcite, porostromates, or micrite (Wright 1981b). They exhibit a complex growth history and many have sparry-calcite filled nuclei which appear to have been open voids for at least some of their history (Locality **2**).

Gilwern Clay Member. This unit, some 8–12 m thick, exhibits two main lithoacies types: channel sandstones, and green, paleosol-bearing clays. The soft clays weather easily and exposures are not common, but at Blaen Onneu exposures are very good and the clay unit contains a thick sandstone unit, he Garn Caws Sandstone, which is at least 12 m thick and represents a fluvial channel (Locality **2**). The complex relationships in this unit have been discussed by Wright & Robinson (1988).

TINERARY

The aim of this itinerary is to illustrate the unusual and complex features in he Clydach Valley Group and Llanelly Formation, and in particular the various subaerial exposure features.

Cwar yr Ystrad is part of a working quarry (Gryphonn Quarries Ltd) and **permission to view the sections** must be sought. Blaen Onneu is worked periodically by Amey Roadstone Corporation and **permission should also be sought** to enter the quarry. The other sections are mainly disused quarries on privately owned land. All the sites, xcluding Blaen Onneu, have been recommended as Sites of Special Scientific Interest, under the supervision of the Nature Conservancy Council. The minor road giving access to Locality **1** is narrow, unfenced with steep hairpin bends; **it cannot**

be reached by coach but access from Bryn Mawr can be made by minibus. Localities **2** **3** and **4** can be reached by coach.

1. Daren Cilau (SO 194 156). Daren Cilau is approached by road from Llangattock or Bryn-mawr (Fig. 1). The minor road from Bryn-mawr affords spectacular views of the Carboniferous Limestone escarpment of the Clydach region. On reaching Pant-y-rhiw (**limited roadside parking**) a prominent path should be followed which climbs up to the base of the escarpment at Daren Cilau. At Daren Cilau a large cone of scree has built up above all the Clydach Valley Group and the first few metres of the Clydach Halt Member of the Llanelly Formation. It is **inadvisable to climb these scree slopes** as rocks frequently fall from the cliff. In any case it is preferable to examine large fallen blocks as these are clean and delicately weathered unlike the cliff face which is obscured by a lichen cover. The fallen blocks can be matched with the *in situ* sequence with a little practice.

All formations of the Clydach Valley Group except the Sychnant Dolomite can be seen stacked in the cliff (Fig. 5). The brown weathering stratiform dolomites and nodular calcite units (Pantydarren Beds and Coed Ffyddlwn Formation) are recessed in the cliff and can be used to locate the succession. Another feature readily seen from the base of the cliff is the rubbly nature of the top Gilwern Oolite which marks its karstic emergence. The general layout of the Clydach Valley Group is well seen at Daren Cilau but detailed examination of lithologies is best done elsewhere.

One of the most interesting features at Daren Cilau are the superb sedimentary structures displayed in recently fallen blocks of the Llanelly Formation. Planar laminated sandstones with wave-rippled and mudcracked tops from the Clydach Halt Member are common. These probably represent distal alluvial-fan or floodplain deposits, reworked by wave action in shallow ponds or lakes. Large blocks of oncolitic sandstones, representing the base of the Cheltenham Limestone Member are present, and large, stems up to 6 cm long, of the aberrant dasycladacean alga *Koninckopora* can be found, coated by oncolitic growths of *Ortonella*. The oncoid bands at the base of the Cheltenham Limestone Member are cut through by small channel-like features. Fenestral limestones, representing intertidal to supratidal deposits, from the Cheltenham Limestone Member, can also be seen. Large blocks of the Penllywn Oolite Member also occur showing *Chondrites* burrows and their stromatolitic layers.

2. Blaen Onneu (SO 155 169). The shortest route from Daren Cilau to Blaen Onneu is to follow the minor road from Pant-y-Rhiw towards Llangattock and double back in Llangattock past Craig-y-Castell to Blaen Onneu

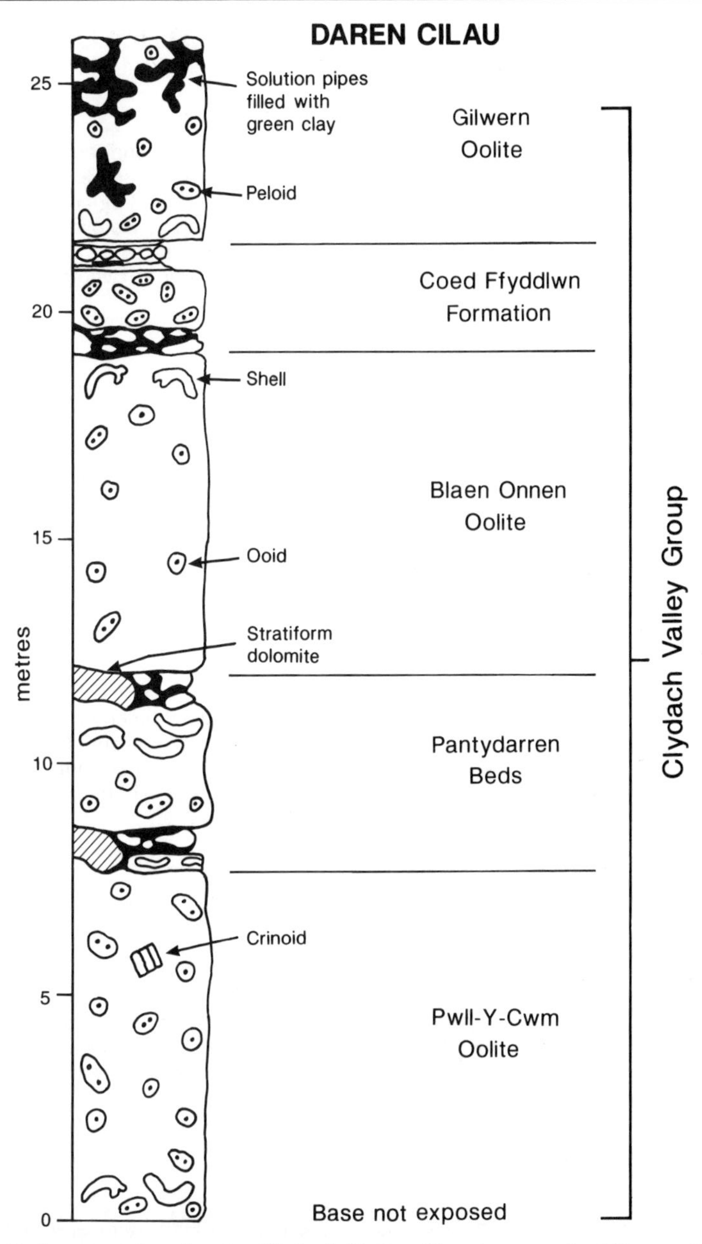

ɪɢ. 5. Sedimentary log for the Clydach Valley Group exposed at Daren Cilau, ,ocality 1.

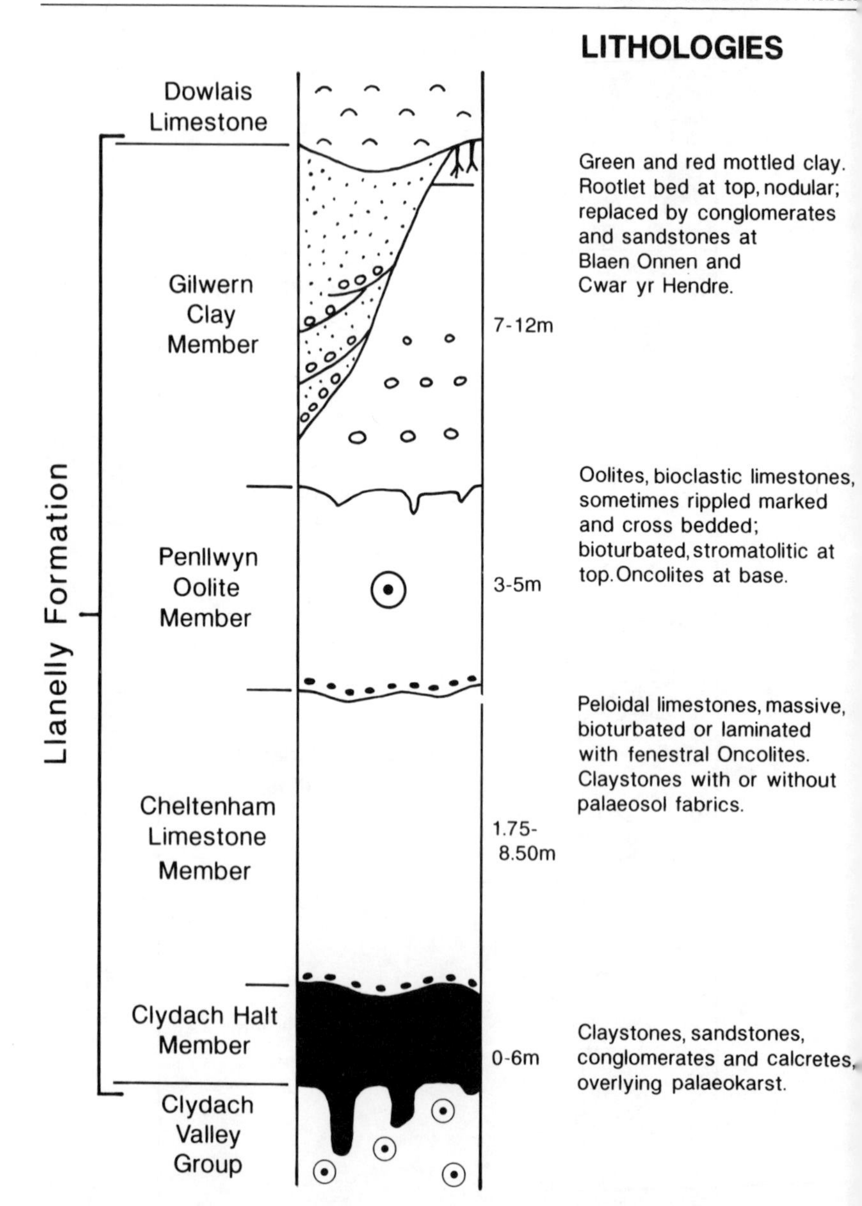
LITHOLOGIES
Dowlais Limestone
Gilwern Clay Member
7-12m
Green and red mottled clay. Rootlet bed at top, nodular; replaced by conglomerates and sandstones at Blaen Onnen and Cwar yr Hendre.
Llanelly Formation
Penllwyn Oolite Member
3-5m
Oolites, bioclastic limestones, sometimes rippled marked and cross bedded; bioturbated, stromatolitic at top. Oncolites at base.
Cheltenham Limestone Member
1.75-8.50m
Peloidal limestones, massive, bioturbated or laminated with fenestral Oncolites. Claystones with or without palaeosol fabrics.
Clydach Halt Member
0-6m
Claystones, sandstones, conglomerates and calcretes, overlying palaeokarst.
Clydach Valley Group

FIG. 6. Llanelly Formation subdivisions. The Clydach Halt Member is very thin o absent in places due to erosion prior to deposition of the Cheltenham Limestone. Th Cheltenham Limestone thins northwards (towards Blaen Onneu).

(Fig. 1). Access may also be made via the Heads of the Valleys Road (Fig. 1); exit the A465(T) to the south and turn right immediately on the B4560 over the A465(T) towards Llangynidr.

The **quarry at Blaen Onneu** is situated on open moorland. The lower faces of the quarry are principally in the Blaen Onnen Oolite but in the deepest, northern part of the Quarry the Pwll-y-Cwm Oolite is exposed. The Pantydarren Beds display a spectacular development, up to 1 metre thick, of large calcite spherulites and fibrous calcite layers (the calcite crystals may be up to 100 mm long) set in a green to black shale. In places these spherulites are replaced by pods and sheets of brown-weathering stratiform dolomite. Field and petrographic evidence suggest that these dolomite-spherulite horizons are paleosols although modern analogues are lacking (Searl 1988a, 1989a).

The most striking feature at Blaen Onneu is the rubbly palaeokarst developed at the top of the Blaen Onnen Oolite although locally the Coed Ffyddlwn Formation is present beneath the Llanelly Formation. The Gilwern Oolite is absent at Blaen Onneu. Recent quarrying has provided access to the palaeokarst on top of the first bench which displays clay-filled fissures and pipes. The intense level of solution piping and pocking has resulted in a very rubbly appearance. The grey-green clay which fills

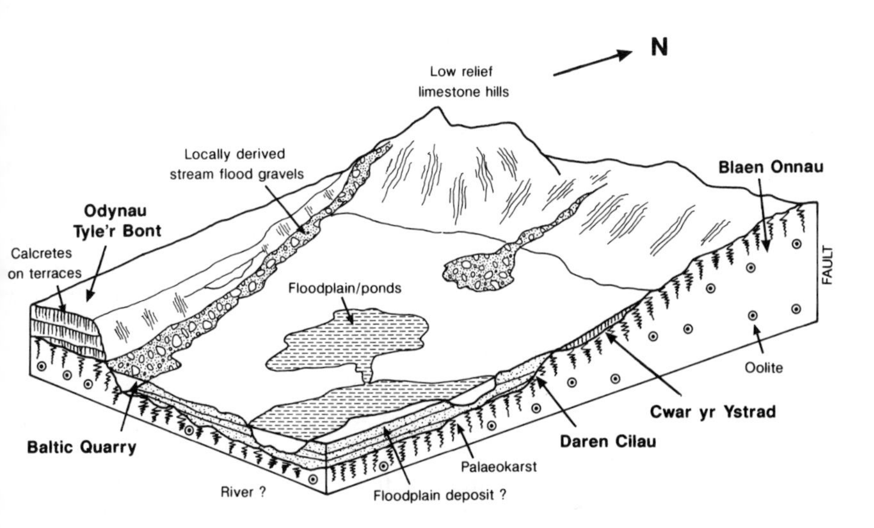

FIG. 7. Schematic reconstruction of the area during Clydach Halt Member times. The landscape was dissected both by locally-sourced small streams and externally-sourced, distal alluvial-fan or floodplain deposits. Movements on the nearby Neath Disturbance fault line resulted in erosion of the Clydach Valley Group, local topographic variations, and incision of drainage systems.

the pipes and fissures connects with a thin or locally missing Clydach Halt Member of the Llanelly Formation (Fig. 8). Another striking feature is the presence of large blocks (up to 5 m long and 1.5 m high) of nonporous oolite, lacking fissures, set within the intensely piped zone.

The basal unit of the Llanelly Formation, the Cheltenham Limestone Member, (Figs. 6, 8) is a coarse, quartz-rich bioclastic, intraclast grainstone of highly variable thickness (from a few centimetres to over one metre). Locally oncoids are abundant with numerous strongly-radial *Garwoodia* growths. (The 'oncoids' often lack a nucleus and do not show any obvious concentric coating laminae and so are arguably not oncoids *sensu stricto*).

The Cheltenham Limestone Member is very thin at this outcrop, only just onlapping on to the local palaeo-topographic high. This basal bioclastic horizon is immediately overlain by a prominent paleosol calcrete, the Cwm Dyar Geosol, up to 1.2 m thick (Fig. 8). Its base is irregular with stringers of calcrete a few centimetres wide extending into the underlying grainstone. The calcrete is highly brecciated and contains, at least locally along the section, numerous replaced evaporite rosettes up to 2 mm in diameter (Wright 1982b). This calcrete horizon, throughout the outcrop, has a high strontium and barium content compared to the other calcretes in the formation. The presence of evaporites concentrated near the top of the horizon suggests strong evaporation from saline groundwater, as also indicated by the high Sr and Ba content. The evaporite rosettes cannot normally be seen in the field but in polished samples comprise up to 10 per cent of the upper part of the horizon.

The overlying unit, the *Uraloporella* Bed (Fig. 8), has a strongly erosive base and contains numerous calcrete clasts. It is a bioclastic, quartz-rich intraclast, oncolitic grainstone which contains several thin beds, and is locally over one metre thick. The oncoids are prominent and occur in several layers. They are up to 120 mm in diameter with black laminae composed of fascicular optic calcite (Wright 1981b) and many oncoids have these layers exposed on their outer surface as black wart-like bumps. Discrete oncoids can be removed from the outcrop's face or collected in talus.

The rest of the Penllwyn Oolite Member consists mainly of a cross-stratified, quartz-rich, oolitic, peloidal bioclastic grainstone (Fig. 8). Locally overturned cross-sets and reactivation surfaces are present. The larger bedforms are low-angle sets probably indicating swash processes on a bar or beach. Smaller-scale sets occur within these larger sets with an opposed current direction.

Near the top of the unit is a 0.2 m stromatolitic horizon in which thin sets of stromatolitic laminae have coated interference ripples. These stromatolites

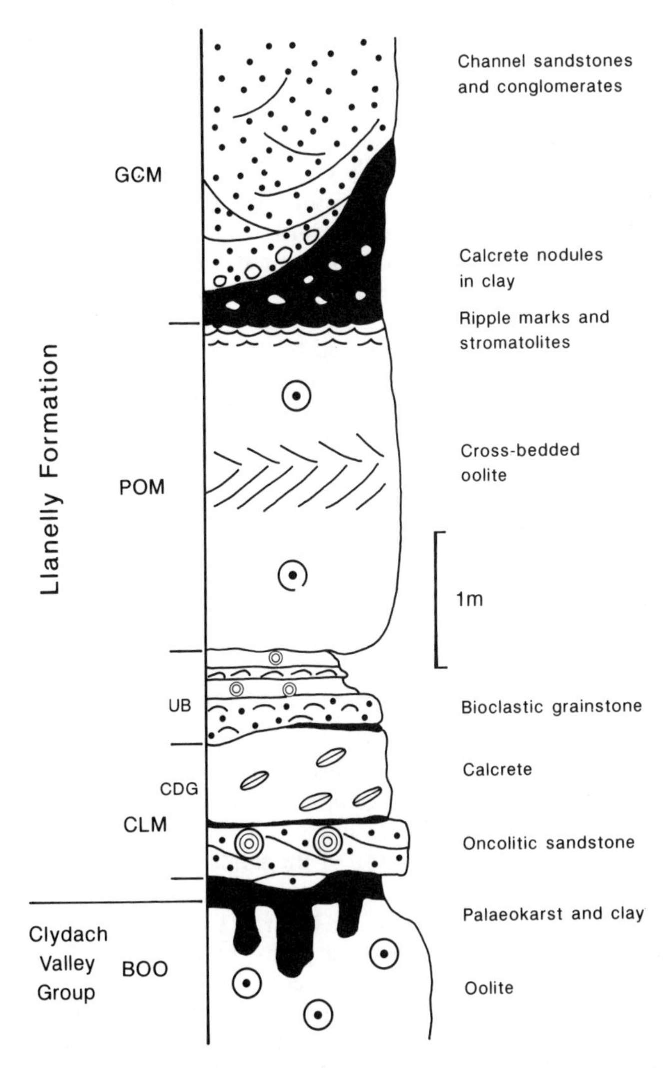

FIG. 8. Log of the section at Blaen Onneu. The oncolitic sandstone is very variable in thickness and only locally contains abundant oncoids. The calcrete corresponds to the Cwm Dyar Geosol (CDG) and numerous micritic stringers occur at its base. The overlying bioclastic grainstone (*Uraloporella* Bed) contains large oncoids. Locally a thin bioclastic limestone occurs above the stromatolitic horizon in the Oolite Penllwyn Member. BOO = Blaen Onnen Oolite; CLM = Cheltenham Limestone Member; UB = *Uraloporella* Bed; POM = Penllwyn Oolite Member; GCM = Gilwern Clay Member (with Garn Caws Sandstone).

(see above) show 'diurnal' laminations (Wright & Wright 1985) but seem to have only grown for a few cycles before bedform migration buried the mats. Relatively few samples of this horizon are now available and collecting should be kept to a minimum.

The **upper quarry bench** has been cut along the top of the oolite and a 6–8 m high, 200 m long strike-section is present in the overlying Gilwern Clay Member. The more resistant sandstone facies is exposed on the **eastern part of the face**. The unit fines upward from a very coarse, cobble and pebble, trough-cross stratified conglomerate with calcrete nodules, into medium sands. Earlier exposures revealed possible epsilon cross-sets indicating channel migration to the west. The channel has wing structures on its western margin and is replaced by a soft green-purple and buff clay, with calcrete nodules at the base, pseudo-anticlines and sandstone-filled desiccation cracks up to 2 m deep and 10 cm side. However, slumping of this clay and of the overlying blanket peat has obscured most of these features. At the **western end of the quarry** a series of planar and cross-stratified medium to coarse grade sandstones occurs in what appears to be a large depression, possibly a crevasse channel. These sands connect with the filled desiccation cracks. Presumably, after a long period of drought (seasonal) on the flood plains a major flood caused the breaching of the main channel to form a crevasse deposit which infilled the deep cracks. Soils with carbonate accumulations, pseudo-anticlines and deep dessication cracks (vertic soils) are common in seasonally arid climate zones in parent alluvium rich in smectite, as were these Carboniferous soils (Wright & Robinson 1988).

The top of the member is not exposed but outcrops of Dowlais Limestone occur some metres above. As a geological peculiarity, the present peat soils lie directly on the Carboniferous soils, with a 330 my time gap. Pleistocene glacial striae also occur on the top of the sandstone facies.

3. Cwar yr Hendre (SO 099 149) **and Cwar yr Ystrad** (SO 081 142). Access to this **working quarry complex** is made from the Heads of the Valleys Road A465(T) at the roundabout near Dukestown. The minor road is followed through Trefil village past the overgrown Trefil quarries to Pistyll Crawnon. Stop at Pistyll Crawnon before entering the quarry gates to observe excellent examples of normal faults which give rise to a graben-like structure. The palaeokarstic surface so well developed at Locality **2** should be recognizable and can be used to demonstrate the vertical throw of the faults at Pistyll Crawnon.

The Clydach Valley Group and Llanelly Formation are exposed entirely in the Cwar yr Hendre/Cwar yr Ystrad quarries but the **precipitous vertical faces** in both quarries means access in places is not possible. The lowest unit, the

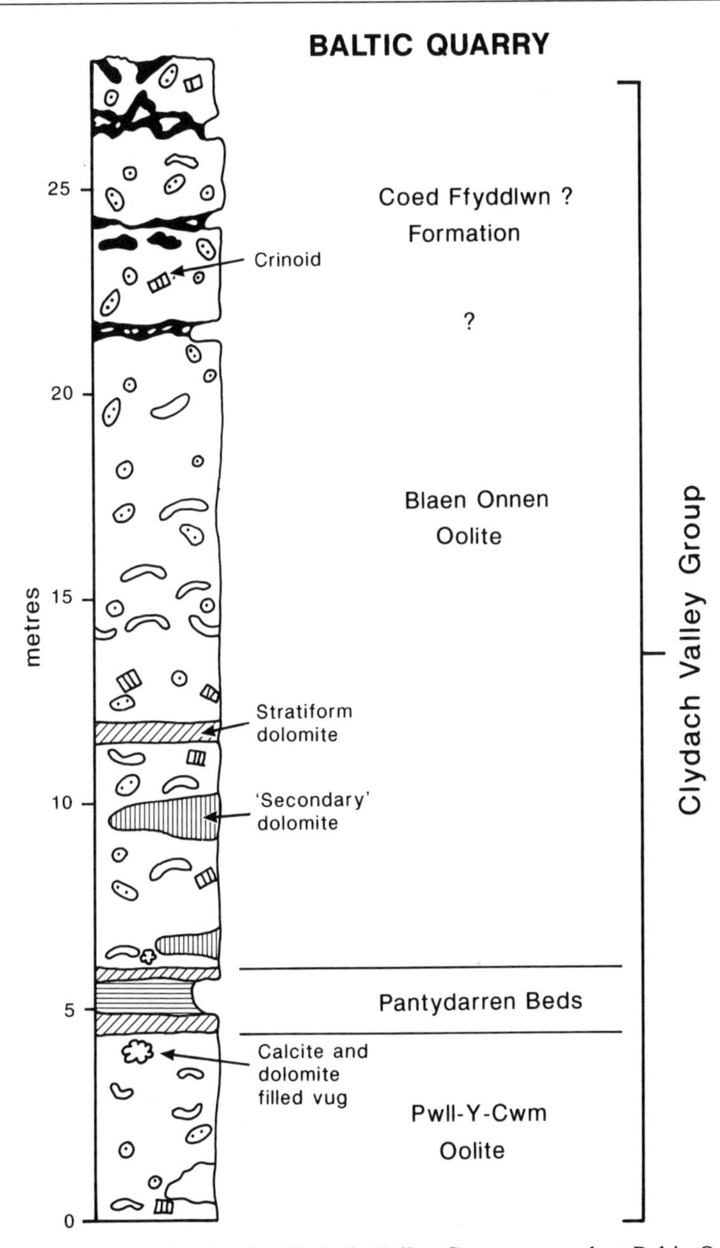

FIG. 9. Sedimentary log for the Clydach Valley Group exposed at Baltic Quarry, Locality 4a.

Sychnant Dolomite, is well exposed in a **small road-cutting** into Cwar yr Ystrad quarry and the **floor of the quarry** exposes the top of this formation. The dolomite in the road cutting shows replaced calcitic skeletons (particularly crinoids and bryozoa) and signs of primary sedimentary bedding. In the road cutting and on the quarry floor spar-filled vugs up to 80 mm diameter can be seen, some filled by well-formed saddle crystals. Searl (1988a) interprets these vugs as replaced evaporite (gypsum or anhydrite) nodules and the Sychnant Dolomite as a stratiform pedogenic dolomite.

The Pantydarren Beds show a similar development of spherulitic calcite nodules and dolomite pods to that noted from Locality **2**. These occur at about head height along the northern cliff of Cwar yr Ystrad but great care is needed and no hammering should be carried out beneath the vertical quarry wall. The karstic development on top of the Blaen Onnen Oolite and its contact with the overlying Llanelly Formation occurs in the vertical quarry wall. Some tracks rising up the quarry walls give access to the higher parts of the succession but often the long abandoned limestone faces become covered with a growth of black lichen which obscures all details. Part of the Llanelly Formation can be examined on the top of the quarries.

At the southern end of Cwar yr Ystrad the Cwm Dyar geosol of the Llanelly Formation occurs **high in the southern quarry wall**. This geosol is similar to the Pantydarren beds in that it contains calcite spherulites which are replaced laterally by lenses of ferroan stratiform dolomite. The impersistent brown dolomite can be seen readily from the floor of the quarry. Within the quarry faces large solution depressions filled with rotted Millstone Grit are clearly visible. These may represent an early Tertiary saprolite (deep chemical weathering zone; Battiau-Queney 1987).

4. Baltic (SO 0659 1184) **and Odynau Tyle'r Bont** (SO 0635 1125). To proceed by vehicle from Locality 3 to Locality 4 return to the Heads of the Valleys Road and travel west; exit to the north onto a minor road at Galon Uchaf; pass the Mountain Railway Station and proceed to the Pontsticill Reservoir. The Baltic and Odynau Tyle'r Bont quarries, both disused, are on the **hillside above the dam**. It is also possible to walk from Locality **3** to Locality **4**; proceed from the southern end of Cwar yr Ystrad along the limestone escarpment to Baltic Quarry, a distance of about one and a half miles across rough moorland.

The Clydach Valley or Abercriban Oolite Group is exposed in the lower, **main face of Baltic Quarry** (Fig. 9) whilst the base of the Llanelly Formation is exposed on an **upper step of the quarry**. A prominent recess is formed in the quarry wall of dark grey mudstone just less than one metre in thickness. This mudstone has no trace or body fossils and separates two stratiform dolomite layers, the triplet forming the Pantydarren Beds. In addition to these stratiform dolomites a distinctly different type of dolomite occurs in Baltic Quarry. This is coarsely crystalline and replaces 'normal' skeletal oolitic grainstones with preservation of much of the original limestone texture (labelled secondary dolomites on Fig. 9).

The **surface of the bench towards the top** of Baltic Quarry is a karstic surface developed on the Abercriban Oolite Group. The overlying Clydach Halt Member of the Llanelly Formation contains a coarse conglomerate at its base which is overlain by a series of interbedded clays and laminated sandstones, similar to those at Locality **1**. The upper part of this Member contains several fenestral calcretes. The Cheltenham Limestone Member lacks many clay beds, most units belonging to lithofacies A and B (Fig. 4).

South of Baltic lies the long disused quarry of Odynau Tyle'r Bont. The top of the Abercriban Oolite Group is poorly exposed with blackened surfaces in this old quarry. The main feature of note here is the Clydach Halt Member which is exposed at the **top of the lower quarry face**. It contains six calcretes, some of which exhibit striking prismatic structure. However, access to this level is difficult and requires a small cliff climb. The calcretes are best examined **behind the small tree** which grows out of the cliff face. The overlying Cheltenham Limestone Member is exposed only partially **above the bench** forming the top of the main quarry face.

REFERENCES

ALLEN, J. R. L. 1962. Petrology, origin and deposition of the highest Lower Old Red Sandstone of Shropshire, England. *Journal of Sedimentary Petrology*, **32**, 657–97.

—— 1974. The Devonian rocks of Wales and the Welsh Borderland. In OWEN, T. R. (ed.). *The Upper Palaeozoic and post-Palaeozoic rocks of Wales.* University of Wales Press, Cardiff, 47–84.

—— 1979. Old Red Sandstone facies in external basins, with particular reference to southern Britain. *Special Papers in Palaeontology*, **23**, 65–80.

—— 1984. *Sedimentary structures, their character and physical bases. Volume 2.* (Developments in Sedimentology, Vol. 30B) Elsevier, Amsterdam, xiii+663pp.

—— 1985. Marine to fresh water: the sedimentology of the interrupted environmental transition (Ludlow–Siegenian) in the Anglo-Welsh region. *Philosophical Transactions of the Royal Society of London*, **B309**, 85–104.

—— and WILLIAMS, B. P. J. 1981. Sedimentology and stratigraphy of the Townsend Tuff Bed (Lower Old Red Sandstone) in South Wales and the Welsh Borders. *Journal of the Geological Society, London*, **138**, 15–29.

ALLEN, P. M. 1982. Lower Palaeozoic volcanic rocks of Wales. In SUTHERLAND, D. (ed.). *Igneous rocks of the British Isles*, Wiley, London, 65–91.

ALLENDER, R. 1958. 'On the stratigraphy and structure of an area of Ludlovian and Lower Devonian rocks near Bishop's Castle, Shropshire.' Unpublished Ph.D. thesis, University of Wales.

ALMOND, J. 1983. 'The sedimentology of the earliest Lower Old Red Sandstone, South Central Wales.' Unpublished Ph.D. thesis, University of Bristol.

ANKETELL, J. M. 1987. On the geological succession and structure of south-central Wales. *Geological Journal*, **22**, 155–65.

AWAN, M. A. and WOODCOCK, N. H. 1991. A white mica crystallinity study of the Berwyn Hills, North Wales. *Journal of Metamorphic Geology*, **9**, 765–73

BAILEY, R. J. 1964. A Ludlovian facies boundary in south Central Wales. *Geological Journal*, **4**, 1–20.

—— 1969. Ludlovian sedimentation in south central Wales. In WOOD, A. (ed.). *The Pre-Cambrian and Lower Palaeozoic rocks of Wales.* University of Wales Press, Cardiff, 283–304.

—— and WOODCOCK, N. H. 1976. Field Meeting: the Ludlow Series slumps of east central Wales, 25–27 April 1975. *Proceedings of the Geologists' Association*, **82**, 183–9.

BAKER, J. W. and HUGHES, C. P. 1979. Summer (1973) Field Meeting in Central Wales 31 August to 7 September 1973. Report by the Organizing Directors. *Proceedings of the Geologists' Association*, **90**, 65–79.

BANCROFT, B. B. 1928. On the unconformity at the base of the Ashgillian in the Bala district. *Geological Magazine*, **65,** 484–93.

—— 1933. *Correlation tables of the Stages Costanian-Onnian in England and Wales.* Blakeney, Gloucestershire, 4pp.

BARCLAY, W. J., TAYLOR, K. and THOMAS, L. P. 1988. Geology of the South Wales Coalfield. Part V. The country around Merthyr Tydfil. *Memoirs of the British Geological Survey*, Sheet 231 (England and Wales), x+52pp.

—— —— —— 1989. Geology of the South Wales Coalfield. Part II. The Country around Abergavenny. *Memoirs of the British Geological Survey*, Sheet 232 (England and Wales), x+147pp.

BASSETT, D. A. 1955. The Silurian rocks of the Talerddig district, Montgomeryshire. *Quarterly Journal of the Geological Society, London*, **111**, 239–64.

—— 1969. Some of the major structures of early Palaeozoic age in Wales and the Welsh Borderland: An historical essay. In WOOD, A. (ed.). *The Pre-Cambrian and Lower Palaeozoic rocks of Wales*, University of Wales Press, Cardiff, 67–116.

—— WHITTINGTON, H. B. and WILLIAMS, A. 1966. The stratigraphy of the Bala district, Merionethshire. *Quarterly Journal of the Geological Society, London*, **122**, 219–71.

BASSETT, M. G. 1974. Review of the stratigraphy of the Wenlock Series in the Welsh Borderland and South Wales. *Palaeontology*, **17**, 745–77.

—— (ed.) 1982a. *Geological Excursions in Dyfed, South-West Wales*, National Museum of Wales Cardiff, 327pp.

—— 1982b. Ordovician and Silurian sections in the Llangadog–Llandilo area. In BASSETT, M. G. (ed.). *Geological Excursions in Dyfed, South-West Wales*, National Museum of Wales, Cardiff, 271–87.

—— LAWSON, J. D. and WHITE, D. E. 1982. The Downton Series as the fourth Series of the Silurian System. *Lethaia*, **15**, 1–24.

BATTIAU-QUENEY, Y. 1984. The pre-glacial evolution of Wales. *Earth Surface Processes and Landforms*, **10**, 137–42.

—— 1987. Buried palaeokarstic features in South Wales. Examples from Vaynor and Cwar-yr-Ystrad Quarries near Merthyr Tydfil. In PATERSON, K. and SWEETING, M. M. (eds). *New directions in karst*. GeoAbstracts, Norwich, 551–67.

BEVINS, R.E. 1985. Pumpellyite-dominated metadomain alteration at Builth Wells, Wales – evidence for a fossil hydrothermal system? *Mineralogical Magazine*, **49**, 451–6.

—— and HORÁK, J. M. 1985. A first occurrence of laumontite in Wales. *Journal of the Russell Society*, **1**, 78–9.

—— KOKELAAR, B. P. and DUNKLEY, P. N. 1984. Petrology and geochemistry of Lower to Middle Ordovician igneous rocks in Wales: a volcanic arc to marginal basin transition. *Proceedings of the Geologists' Association*, **95**, 337–47.

—— and ROBINSON, D. 1988. Low grade metamorphism of the Welsh Basin Lower Palaeozoic succession: an example of diastathermal metamorphism? *Journal of the Geological Society, London*, **145,** 363-6.

—— and ROWBOTHAM, G. 1983. Low grade metamorphism within the Welsh sector of the paratectonic Caledonides. *Geological Journal*, **18**, 141–67.

BICK, D. E. 1978. *The old metal mines of Mid-Wales. Part 5*. Pound House, Newent, 51pp.

BOSWELL, P. G. H. and DOUBLE, I. S. 1940. The geology of an area of Salopian rocks west of the Conway Valley, in the neighbourhood of Llanrwst, Denbighshire. *Proceedings of the Geologists' Association*, **51**, 151–87.

BOUMA, A. H. 1962. *Sedimentology of some flysch deposits: a graphic approach to facies interpretation*. Elsevier, Amsterdam, 168pp.

BRENCHLEY, P. J. 1969. The relationship between Caradocian volcanicity and sedimentation in North Wales. In WOOD, A. (ed.). *The Pre-Cambrian and Lower Palaeozoic rocks of Wales*. University of Wales Press, Cardiff, 181–202.

—— 1978. The Caradocian rocks of the north and west Berwyn Hills. *Geological Journal*, **13**, 137–64.

—— 1985. Storm influenced sandstone beds. *Modern Geology*, **9**, 369–96.

—— 1988. Environmental changes close to the Ordovician–Silurian boundary. In COCKS, L. R. M. and RICKARDS, R. B. (eds). A global analysis of the Ordovician-Silurian boundary. *Bulletin of the British Museum (Natural History)*, Geology, **43**, 377–85.

—— and CULLEN, B. 1984. The environmental distribution of associations belonging to the Hirnantia fauna – evidence from North Wales and Norway. In BRUTON, D. L. (ed.). *Aspects of the Ordovician System*, Palaeontological Contributions from the University of Oslo, No. 295, Universitetsforlaget, 113–25.

—— and NEWALL, G. 1980. A facies analysis of Upper Ordovician regressive sequences in the Oslo Region, Norway – a record of glacio-eustatic changes. *Palaeogeography, Palaeoclimatology, Palaeoecology*, **31,** 1–38.

—— —— 1984. Late Ordovician environmental changes and their effect on faunas. In BRUTON, D. L. (ed.). *Aspects of the Ordovician System*, Palaeontological Contributions from the University of Oslo, No. 295, Universitetsforlaget, 65–79.

BRITISH GEOLOGICAL SURVEY (1993). Rhayader. 1 : 50,000 Geology map series. Sheet 179 (Solid and drift editions).

BROWN, E. H. 1960. *The relief and drainage of Wales*, University of Wales Press, Cardiff, 1–186.

CAMPBELL, S. and BOWEN, D. Q. 1989. *Geological Conservation Review: Quaternary of Wales*, Nature Conservancy Council, Peterborough, 1–240.

CAMPBELL, S. D. G. 1984. Aspects of dynamic stratigraphic (Caradoc-Ashgill) in the northern part of the Welsh marginal basin. In Abstracts of current research in Wales. *Proceedings of the Geologists' Association*, **95,** 390–1.

CAVE, R. 1955. *The stratigraphy of the Welshpool area* (*Montgomeryshire*). Unpublished Ph.D. thesis, University of Cambridge.

—— 1957. *Salterolithus caractaci* (Murchison) from Caradoc strata near Welshpool, Montgomeryshire. *Geological Magazine*, **94,** 281–90.

—— 1965. The Nod Glas sediments of Caradoc age in North Wales. *Geological Journal*, **4**(2), 279–98.

—— 1979. Sedimentary environments of the basinal Llandovery of mid-Wales. In HARRIS, A. L., HOLLAND, C. H. and LEAKE, B. E. (eds). Caledonides of the British Isles, reviewed. *Special Publication of the Geological Society, London*, **8**, 517–26.

—— 1992. The Llandovery Series. In COPE, J. C. W., INGHAM, J. K. and RAWSON, P. K. (eds). *Atlas of palaeogeography and lithofacies*. Geological Society of London, Memoir, **13**, 37–40.

—— DEAN, W. T. and HAINS, B. A. 1988. Age of the Ordovician andesite conglomerate of Castle Hill, Montgomery, Powys, Wales. *Geological Journal*, **23**, 205–10.

—— and HAINS, B. 1986. Geology of the country between Aberystwyth and Machynlleth. *Memoirs of the British Geological Survey*, Sheet 163 (England and Wales), x+148pp.

—— and PRICE, D. 1978. The Ashgill Series near Welshpool, North Wales. *Geological Magazine*, **115**, 183–94.

—— and WHITE, D. E. 1971. The exposures of Ludlow rocks and associated beds at Tites Point near Newnham, Gloucestershire. *Geological Journal*, **7**, 239–54.

—— —— 1978. Stratigraphy of the Brookend (Vine Farm) Borehole. *Ludlow Research Group Bulletin*, **25**, 44–5.

CHERNS, L. 1979. The environmental significance of *Lingula* in the Ludlow Series of the Welsh Borderland and Wales. *Lethaia*, **12**, 35–46.

—— 1980. Hardgrounds in the Lower Leintwardine Beds (Silurian) of the Welsh Borderland. *Geological Magazine*, **117**, 311–25,

—— 1988. Faunal and facies dynamics in the upper Silurian of the Anglo-Welsh Basin. *Palaeontology*, **31**, 451–502.

—— In press. Faunal associations of the Lower Leintwardine Formation of the Welsh Borderland and Wales. In BOUCOT, A. J. and LAWSON, J. D. (eds). Project Ecostratigraphy.

COCKS, L. R. M. and McKERROW, W. S. 1978. In MCKERROW, W. S. (ed.). *The ecology of fossils, an illustrated guide*. Duckworth, London.

—— and RICKARDS, R. B. 1969. Five boreholes in Shropshire and the relationships of shelly and graptolitic facies in the Lower Silurian. *Quarterly Journal of the Geological Society, London* **124**, 213–38.

COPE, J. C. W. 1979. Early history of the southern margin of the Tywi Anticline in the Carmarthen area, South Wales. In HARRIS, A. L., HOLLAND, C. H. and LEAKE, B. E. (eds). Caledonides of the British Isles: reviewed. *Special Publication of the Geological Society, London*, **8**, 527–32.

—— 1984. The Mesozoic history of Wales. *Proceedings of the Geologists' Association*, **95**, 373–85.

COWARD, M. P. and SIDDANS, A. W. B. 1979. The tectonic evolution of the Welsh Caledonides. In HARRIS, A. L., HOLLAND, C. H. and LEAKE, B. E. (eds). Caledonides of the British Isles: reviewed. *Special Publication of the Geological Society, London*, **8**, 187–98.

CROFT, W. N. and LANG, W. H. 1942. The Lower Devonian flora of the Senni Beds of Monmouthshire and Breconshire. *Philosophical Transactions of the Royal Society of London*, **B231**, 131–63.

CUMMINS, W. A. 1957. The Denbigh Grits; Wenlock greywackes in Wales. *Geological Magazine*, **94**, 433–51.

—— 1959a. The Lower Ludlow Grits in Wales. *Liverpool and Manchester Geological Journal*, **2**, 168–79.

—— 1959b. The Nantglyn Flags; Mid-Salopian basin facies in Wales. *Liverpool and Manchester Geological Journal*, **2**, 159–67.

—— 1963. 'The geology of the northern part of the Llangadfan Syncline, Montgomeryshire.' Unpublished Ph.D. thesis, University of Liverpool.

DAVIES, D. C. 1875. The phosphorite deposits of North Wales. *Quarterly Journal of the Geological Society, London*, **31**, 351–67.

DAVIES, J. H. 1981. 'The structure and sedimentology of Lower Palaeozoic rocks in Brecknock, Wales.' Unpublished Ph.D. thesis, University of London.

—— HOLROYD, J., LUMLEY, R. G. and OWEN-ROBERTS, D. 1978. *Geology of Powys in outcrop*. Llandrindod, Powys County Council, 164pp.

DAVIES, K. A. 1928. The geology of the country between Rhayader (Radnorshire) and Abergwesyn (Breconshire). *Proceedings of the Geologists' Association*, **39**, 157–68.

—— 1933. The geology of the area between Abergwesyn (Breconshire) and Pumpsaint (Carmarthenshire). *Quarterly Journal of the Geological Society, London*, **89**, 172–200.

—— and PLATT, J. I. 1933. The conglomerates and grits of the Bala and Valentian rocks of the district between Rhayader (Radnorshire) and Llansawel (Breconshire). *Quarterly Journal of the Geological Society, London*, **89**, 202–18.

DEWEY, J. F. 1969. Structure and sequence in paratectonic British Caledonides, *Memoir of the American Association of Petroleum Geologists*, **12**, 309–35.

DIMBERLINE, A. J. 1987. 'The sedimentology and diagenesis of the Wenlock turbidite system.' Unpublished Ph.D. thesis, University of Cambridge.

—— BELL, A. and WOODCOCK, N. H. 1990. A laminated hemipelagic facies from the Wenlock and Ludlow of the Welsh Basin. *Journal of the Geological Society, London*, **147**, 693–701.

—— and WOODCOCK, N. H. 1987. The south-east margin of the Wenlock turbidite system, Mid-Wales. *Geological Journal*, **22**, 61–71.

DIXON, R. J. 1988. The Ordovician (Caradoc) volcanic rocks of Montgomery, Powys, N. Wales. *Geological Journal*, **23**, 149–56.

—— 1990. The Moel-y-Golfa Andesite: an Ordovician (Caradoc) intrusion into unconsolidated conglomeratic sediments, Breidden Hills Inlier, Welsh Borderland. *Geological Journal*, **25**, 35–46.

DUKE, W. L. 1985. Hummocky cross-stratification, tropical hurricanes, and intense winter storms. *Sedimentology*, **32**, 167–94.

DUNCAN, A. C. 1984. Transected folds: a re-evaluation with examples from the 'type area' at Sulphur Creek, Tasmania. *Journal of Structural Geology*, **7**, 409–19.

EARP, J. R. 1938. The higher Silurian rocks of the Kerry district, Montgomeryshire. *Quarterly Journal of the Geological Society, London*, **94**, 125–60.

—— 1940. The geology of the south-western part of Clun Forest. *Quarterly Journal of the Geological Society, London*, **96**, 1–11.

EDWARDS, D. 1969. *Zosterophyllum* from the Lower Old Red Sandstone of South Wales. *New Phytologist*, **68**, 923–31.

—— 1970a. Fertile Rhyniophytina from the Lower Devonian of Britain. *Palaeontology* **13**, 451–61.

—— 1970b. Further observations on the Lower Devonian plant *Gosslingia breconensis* Heard. *Philosophical Transactions of the Royal Society of London*, **B258**, 225–43.

—— 1977. A new non-calcified alga from the upper Silurian of mid Wales. *Palaeontology*, **20**, 823–32.

—— BASSETT, M. G. and ROGERSON, E. C. W. 1979. The earliest vascular land plants: continuing the search for proof. *Lethaia*, **12**, 313–24.

—— and RICHARDSON, J. B. 1978. Capel Horeb Quarry, Powys. In FRIEND, P. F. and WILLIAMS, B. P. J. (eds). *A field guide to selected outcrops of the Devonian of Scotland, the Welsh Borderland and South Wales*, Palaeontological Association, London, International Symposium on the Devonian System (PADS 78), 78–9.

—— —— and THOMAS, R. G. 1978. Heol Senni Quarry, Powys. In FRIEND, P. F. and WILLIAMS, B. P. J. (eds). *A field guide to selected outcrops of the Devonian of Scotland, the Welsh Borderland and South Wales*, Palaeontological Association, London, International Symposium on the Devonian System (PADS 78), 77–8.

—— and ROGERSON, E. C. W. 1979. New records of fertile Rhyniophytina from the Late Silurian of Wales. *Geological Magazine*, **116**, 93–8.

ELLES, G. L. 1900. The zonal classification of the Wenlock Shales of the Welsh Borderland. *Quarterly Journal of the Geological Society, London*, **56**, 370–413.

—— 1922. The Bala country: its structure and rock succession. *Quarterly Journal of the Geological Society, London*, **78**, 132–75.

FITCHES, W. R., CAVE, R., CRAIG, J. and MALTMAN, A. J. 1986. Early veins as evidence of detachment in the Lower Palaeozoic rocks of the Welsh Basin. *Journal of Structural Geology*, **8**(6), 607–15.

FORTEY, R. A. and COCKS, L. R. M. 1988. Arenig to Llandovery faunal distributions in the Caledonides. In HARRIS, A. L. and FETTES, D. J. (eds). The Caledonian-Appalachian Orogen. *Special Publication of the Geological Society, London*, **38**, 233–46.

FURNES, H. 1978. 'A comparative study of Caledonian volcanics in Wales and West Norway.' Unpublished Ph.D. thesis, University of Oxford.

GARWOOD, E. J. and GOODYEAR, E. 1918. On the geology of the Old Radnor district, with special reference to an algal development in the Woolhope Limestone. *Quarterly Journal of the Geological Society, London*, **74**, 1–30.

GAYER, R. and JONES, J. 1989. The Variscan foreland in South Wales. *Proceedings of the Ussher Society*, **7**, 177–9.

GEORGE, T. N. 1954. Pre-Seminulan Main Limestone of the Avonian Series of Breconshire. *Quarterly Journal of the Geological Society, London*, **110**, 283–322.

—— 1963. Palaeozoic growth of the British Caledonides. In JOHNSON, M. R. W. and STEWART, F. H. (eds). *The British Caledonides*. Oliver and Boyd, Edinburgh, 1–33.

HALL, I. H. S., TAYLOR, K. and THOMAS, L. R. 1973. The stratigraphy of the Upper Old Red Sandstone in South Breconshire. *Bulletin of the Geological Survey of Great Britain*, **44**, 45–62.

HEARD, A. 1927. On Old Red Sandstone plants showing structure from Brecon (South Wales). *Quarterly Journal of the Geological Society, London*, **83**, 195–207.

HIRD, K., TUCKER, M. E. and WATERS, R. A. 1987. Petrography, geochemistry and origin of Dinantian dolomites from south-east Wales. In MILLER, J., ADAMS, A. E. and WRIGHT, V. P. (eds). *European Dinantian environments*. Geological Journal Special Issue No. 12, Wiley, Chichester etc., 359–78.

HOLGATE, N. and HALLOWES, K. A. K. 1941. The igneous rocks of the Stanner-Hanter District, Radnorshire. *Geological Magazine*, **78**, 241–67.

HOLLAND, C. H. 1959. The Ludlovian and Downtonian rocks of the Knighton district, Radnorshire. *Quarterly Journal of the Geological Society, London*, **114**, 449–82.

—— and LAWSON, J. D. 1963. Facies patterns in the Ludlovian of Wales and the Welsh Borderlands. *Geological Journal*, **3**, 269–88.

—— —— and WALMSLEY, V. G. 1963. The Silurian rocks of the Ludlow district. *Bulletin of the British Museum (Natural History), Geology*, **8**, 93–171.

—— and PALMER, D. C. 1974. *Bohemograptus*, the youngest graptoloid known from the British Silurian sequence. *Special Papers in Palaeontology*, **13**, 215–36.

HOUSE, M. R., RICHARDSON, J. B., CHALONER, W. G., ALLEN, J. R. L., HOLLAND, C. H. and WESTOLL, T. S. 1977. A correlation of Devonian rocks of the British Isles. *Special Report of the Geological Society, London*, **7**, 1–110.

HOWELLS, M. F., LEVERIDGE, B. E. and REEDMAN, A. J. 1981. *Snowdonia*. Unwin Paperbacks, London, 119pp.

HURST, J. M., HANCOCK, N. J. and McKERROW, W. S. 1978. Wenlock stratigraphy and palaeogeography of Wales and the Welsh Borderland. *Proceedings of the Geologists' Association*, **89**, 197–226.

INSTITUTE OF GEOLOGICAL SCIENCES. 1972. Newtown. 1 : 10,560 geological map.
JAMES, D. M. D. 1983. Observations and speculations on the north-east Towy 'axis', mid-Wales *Geological Journal*, **18**, 283–96.
—— 1986. The Rhiwnant Inlier, Powys, Mid-Wales. *Geological Magazine*, **123**, 585–7.
—— 1987. Tectonics and sedimentation in the Lower Palaeozoic back-arc basin of S. Wales, UK: some quantitative aspects of basin development. *Norsk Geologisk Tidsskrift*, **67**, 419–28.
—— 1990. Late Ordovician stratigraphy north-west of the Sugarloaf, mid-Wales – a discussion. *Geological Journal*, **25**, 199–204.
JONES, O. T. 1909. The Hartfell-Valentian succession in the district around Plynlimon and Pont Erwyd (North Cardiganshire). *Quarterly Journal of the Geological Society, London*, **65**, 463–537.
—— 1912. The geological structure of central Wales and the adjoining regions. *Quarterly Journal of the Geological Society, London*, **68**, 328–44.
—— 1922. Lead and zinc. The mining district of north Cardiganshire and west Montgomeryshire. *Memoirs of the Geology Survey. Special Reports on the Mineral Resources of Great Britain*, **20**, 207pp.
—— 1938. On the evolution of a geosyncline. *Quarterly Journal of the Geological Society, London*, **94**, lx–cx.
—— 1947. The geology of the Silurian rocks west and south of the Carneddau Range, Radnorshire. *Quarterly Journal of the Geological Society, London*, **103**, 1–36.
—— 1956. The geological evolution of Wales and the adjacent regions. *Quarterly Journal of the Geological Society, London*, **111**, 323–51.
—— and PUGH, W. J. 1916. The geology of the district around Machynlleth and the Llyfnant Valley. *Quarterly Journal of the Geological Society, London*, **71**, 343–85.
—— —— 1941. The Ordovician rocks of the Builth District: a preliminary account. *Geological Magazine*, **78**, 185–91.
—— —— 1946. The complex intrusion of Welfield near Builth Wells, Radnorshire. *Quarterly Journal of the Geological Society, London*, **102**, 157–88.
—— —— 1948a. A multi-layered dolerite complex of laccolithic form near Llandrindod Wells, Radnorshire. *Quarterly Journal of the Geological Society, London*, **104**, 43–71.
—— —— 1948b. The form and distribution of dolerite masses in the Builth-Llandrindod Inlier, Radnorshire. *Quarterly Journal of the Geological Society, London*, **104**, 71–98.
—— —— 1949. An early Ordovician shoreline in Radnorshire near Builth Wells. *Quarterly Journal of the Geological Society, London*, **105**, 65–99.
JONES, W. D. V. 1946. The Valentian succession around Llanidloes, Montgomeryshire. *Quarterly Journal of the Geological Society, London*, **100**, 309–32.
KELLING, G. and WOOLLANDS, M. A. 1969. The stratigraphy and sedimentation of the Llandoverian rocks of the Rhayader district. In WOOD, A. (ed.). *The Pre-Cambrian and Lower Palaeozoic rocks of Wales*, University of Wales Press, Cardiff, 255–82.
KING, W. B. R. 1923. The Upper Ordovician rocks of the south-western Berwyn Hills. *Quarterly Journal of the Geological Society, London*, **79**, 487–507.
—— 1928. The geology of the district around Meifod, Montgomeryshire. *Quarterly Journal of the Geological Society, London* **84**, 671–702.

KING, W. W. 1934. The Downtonian and Dittonian strata of Great Britain and north-western Europe. *Quarterly Journal of the Geological Society, London*, **90**, 89–121.

KIRK, N. H. 1947. 'The geology of the anticlinal disturbance of Breconshire and Radnorshire: Pont Faen to Presteigne.' Unpublished Ph.D. thesis, University of Cambridge.

—— 1951a. The upper Llandovery and lower Wenlock rocks of the area between Dolyhir and Presteigne, Radnorshire. *Proceedings of the Geological Society, London*, **1471**, 56–8.

—— 1951b. The Silurian and Downtonian rocks of the anticlinal disturbance of Breconshire and Radnorshire: Pont Faen to Presteigne. *Abstracts of the Proceedings of the Geological Society, London*, **1474**, 72–4.

KOKELAAR, B. P. 1982. Fluidization of wet sediments during the emplacement and cooling of various igneous bodies. *Journal of the Geological Society, London*, **139**, 21–33.

—— 1988. Tectonic controls of Ordovician arc and marginal basin volcanism in Wales. *Journal of the Geological Society, London*, **145**, 759–76.

—— BEVINS, R. E. and ROACH, R. A. 1985. Submarine silicic volcanism and associated sedimentary and tectonic processes, Ramsey Island, S.W. Wales. *Journal of the Geological Society, London*, **142**, 589–91.

—— HOWELLS, M. F., BEVINS, R. E., ROACH, R. A. and DUNKLEY, P. N. 1984, The ordovician marginal basin of Wales. In KOKELAAR, B. P. and HOWELLS, M. F. (eds). *Marginal Basin Geology*. Geological Society of London Special Publication, **16**, 245–69.

LAPWORTH, H. 1900. The Silurian sequence of Rhayader. *Quarterly Journal of the Geological Society, London*, **56**, 67–137.

LAWSON, J. D. 1971. Some problems and principles in the classification of the Silurian System. *Mémoires du Bureau de Recherches Géologiques et Minières, Paris*, **73**, 301–8.

LENG, M. J. 1990. 'Late Ordovician/early Silurian palaeo-environmental analysis in the Tywyn-Corris area of mid-Wales.' Unpublished Ph.D. thesis, University College of Wales, Aberystwyth.

LOWE, D. R. 1982. Sediment gravity flows II. Depositional models with specific reference to the deposits of high-density turbidity currents. *Journal of Sedimentary Petrology*, **52**, 279–97.

LOYDELL, D. K. 1991. The biostratigraphy and formational relationships of the upper Aeronian and lower Telychian (Llandovery, Silurian) formations of western mid-Wales. *Geological Journal*, **26**, 209–44.

LYNAS, B. D. T. 1987. Geological notes and local details for 1 : 10 000 sheet: SO 28 NE (Mainstone). *BGS Open-file Report*.

—— 1988. Evidence for dextral oblique-slip faulting in the Shelve Ordovician Inlier, Welsh Borderland: implications for the South British Caledonides. *Geological Journal*, **23**, 39–57.

MACKIE, A. H. 1987. 'The geology of the Llyn Brianne area, central Wales.' Unpublished Ph.D. thesis, University of Cambridge.

—— and SMALLWOOD, S. 1987. A revised stratigraphy and sedimentology of the Abergwesyn-Pumpsaint area, Mid-Wales. *Geological Journal*, **22** (thematic issue), 45–60.

METCALFE, R. 1989. Metadomains as indicators of fluid/rock interaction in the Welsh Marginal Basin. In MILES, D. L. (ed.). *Proceedings of the Sixth International Symposium on Water/Rock Interaction*. A. A. Balkema, Rotterdam, 479–81.

—— 1990. 'Fluid/rock interaction and metadomain formation during low grade metamorphism in the Welsh Marginal Basin.' Unpublished Ph.D. thesis, University of Bristol.

MIDDLETON, G. V. and HAMPTON, M. A. 1973. Sediment gravity flows: mechanics of flow and deposition. In *Turbidites and deep-water sedimentation*. Society of Economic Paleontologists and Mineralogists. Short course lecture notes, 1–38.

MURCHISON, R. I. 1839. *The Silurian System founded on geological researches in the counties of Salop, Hereford, Radnor, Montgomery, Caermarthen, Brecon, Pembroke, Monmouth, Gloucester, Worcester, and Stafford with the descriptions of the coal-fields and overlying formations*. Murray, London, xxxii+768pp (in 2 vols.)

MUTTI, E. and RICCI-LUCCI, F. 1978. Turbidites of the northern Appennines: introduction to facies analysis. *International Geology Review*, **20**, 127–66.

NATLAND, M. L. 1976. Classification of clastic sediments. *Bulletin of the American Association of Petroleum Geologists*, **60**, p.702.

OWEN, T. R. and WEAVER, J. D. 1983. The structure of the main South Wales coalfield and its margins. In HANCOCK, P. L. (ed.). *The Variscan fold belt in the British Isles*. Adam Hilger, Bristol, 47–73.

PALMER, D. 1970. A stratigraphical synopsis of the Long Mountain, Montgomeryshire and Shropshire. *Proceedings of the Geological Society, London*, **1660**, 341–6.

—— 1972. 'The geology of the Long Mountain, Montgomeryshire and Shropshire.' Unpublished Ph.D. thesis, Trinity College, Dublin.

PATCHETT, P. J., GALE, N. H., GOODWIN, R. and HUMM, M. 1980. Rb–Sr whole-rock isochron ages of late Precambrian to Cambrian igneous rocks from southern Britain. *Journal of the Geological Society, London*, **137**, 649–56.

PAULEY, J. C. 1990. Sedimentology, structural evolution and tectonic setting of the late Precambrian Longmyndian Supergroup of the Welsh Borderland, UK. *Special Publication of the Geological Society, London*, **51**, 341–51.

PICKERILL, R. K. 1977. Trace fossils from the Upper Ordovician (Caradoc) of the Berwyn Hill, central Wales. *Geological Journal*, **12**, 1–16.

—— and BRENCHLEY, P. J. 1979. Caradoc benthic communities of the South Berwyn Hills, North Wales. *Palaeontology*, **22**, 229–64.

PICKERING, K. T., BASSETT, M. G. and SIVETER, D. J. 1988. Late Ordovician–early Silurian destruction of the Iapetus Ocean: Newfoundland, British Isles and Scandinavia – a discussion. *Transactions of the Royal Society of Edinburgh*, **79**, 361–82.

—— HISCOTT, R. N. and HEIN, F. J. 1989. *Deep marine environments*. London, Unwin Hyman, 416pp.

—— STOW, D. A. V., WATSON, M. P. and HISCOTT, R. N. 1986. Deep-water facies, processes and models: a review and classification scheme for modern and ancient sediments. *Earth Science reviews*, **23**, 75–174.

PIPER, J. D. A. and BRIDEN, J. C. 1973. Palaeomagnetic studies in the British Caledonides – I. Igneous rocks of the Builth Wells Ordovician Inlier, Radnorshire, Wales. *Geophysical Journal of the Royal Astronomical Society*, **34**, 1–12.

POTTER, J. F. and PRICE, J. H. 1965. Comparative sections through rocks of Ludlovian–Downtonian age in the Llandovery and Llandeilo districts. *Proceedings of the Geologists' Association*, **76**, 379–401.

PRICE, D. 1984. The Pusgillian Stage in Wales. *Geological Magazine*, **21**, 99–105.

PUGH, W. J. 1923. The geology of the district around Corris and Aberllefenni (Merionethshire). *Quarterly Journal of the Geological Society, London* **79**, 508–45.

RAD, U. von and WISSMAN, G. 1982. Cretaceous-Cenozoic history of the West Saharan continental margin (NW Africa): development, destruction and gravitational sedimentation. In RAD, U. von, HINZ, K., SARNTHEIN, M. and SEIBOLD, E. (eds). *Geology of the Northwest African continental margin*, Springer-Verlag, Berlin, 106–31.

RAVEN, M. R. 1983. 'The diagenesis of the Oolite Group between Blaen Onneu and Pwll Du, Lower Carboniferous, South Wales.' Unpublished Ph.D. thesis, University of Nottingham.

ROBERTS, B. and MERRIMAN, R. J. 1985. The distinction between Caledonian burial and regional metamorphism in metapelites from North Wales: an analysis of isocryst patterns. *Journal of the Geological Society, London*, **142**, 615–24.

ROBERTS, R. O. 1929. The geology of the district around Abbey-Cwmhir (Radnorshire). *Quarterly Journal of the Geological Society, London*, **85**, 651–76.

ROBINSON, D. and BEVINS, R. E. 1986. Incipient metamorphism in the Lower Palaeozoic marginal basin of Wales. *Journal of Metamorphic Geology* **4**, 101–13.

SANDERSON, R. W. 1975. Andesitic agglomerate in the Caradocian succession at Montgomery Castle, Wales. *Bulletin of the Geological Survey of Great Britain*, **52**, 43–9.

—— and CAVE, R. 1980. Silurian volcanism in the Central Welsh Borderland. *Geological Magazine*, **117**, 455–62.

SAXOV, S. and NIEUWENHUIS, J. K. 1982 (eds). *Marine slides and other mass movements.* Plenum Press, New York, 353 pp.

SCOTT, M. M. and CAVE, R. 1990a. Geological notes and local details for 1 : 10,000 sheet SH70SW (Machynlleth). *British Geological Survey Technical Report*, **WA/90/46**.

—— —— 1990b. Geological notes and local details for 1 : 10,000 sheet SH70SE (Llanwrin). *British Geological Survey Technical Report*, **WA/90/47**.

SCHIENER, E. J. 1970. Sedimentology and petrography of three tuff horizons in the Caradocian sequence of the Bala area (North Wales). *Geological Journal*, **7**, 25–46.

SEARL, A. 1986. 'The diagenesis of some Lower Carboniferous oolitic limestones from South Wales.' Unpublished Ph.D. thesis, University of Cambridge.

—— 1988a. Pedogenic dolomites from the Oolite Group (Lower Carboniferous), South Wales. *Geological Journal*, **23**, 147–56.

—— 1988b. The limitation of 'cement stratigraphy' as revealed in some Lower Carboniferous oolites from South Wales. *Sedimentary Geology*, **57**, 171–83.

—— 1989. Pedogenic columnar calcites from the Oolite Group (Lower Carboniferous), South Wales. *Sedimentary Geology*, **58**, 157-69.

SIVETER, D. J., OWENS, R. M. and THOMAS, A. T. 1989. *Silurian field excursions: a geotraverse across Wales and the Welsh Borderland.* National Museum of Wales, Geological Series No. 10, Cardiff, 133pp.

SMALLWOOD, S. D. 1986a. Sedimentation across the Tywi Lineament, mid Wales. *Philosophical Transactions of the Royal Society of London*, **A317,** 279–88.

—— 1986b. 'The Lower Palaeozoic history of the Tywi Lineament, mid-Wales.' Unpublished Ph.D. thesis, University of Cambridge.

SMITH, R. D. A. 1987a. The *griestoniensis* Zone turbidite system, Welsh Basin. In LEGGETT, J. K. and ZUFFA, G. G. (eds). *Marine clastic sedimentology: models and case histories*. Graham and Trotman, 89–107.

—— 1987b. Structure and deformation history of the Central Wales Synclinorium, north-east Dyfed: evidence for a long-lived basement structure. *Geological Journal*, **22** (Thematic issue), 183–98.

SOPER, N. J. and WOODCOCK, N. H. 1990. Silurian collision and sediment dispersal patterns in southern Britain. *Geological Magazine*, **127**, 527–42.
STAMP, L. D. 1923. The base of the Devonian with special reference to the Welsh Borderland. *Geological Magazine*, **60**, 276–410.
—— and WOOLDRIDGE, S. W. 1923. The igneous and associated rocks of Llanwrtyd (Brecon). *Quarterly Journal of the Geological Society, London*, **79**, 16–46.
STOW, D. A. V. and PIPER, D. J. W. 1984. Deep water fine-grained sediments: facies models. In STOW, D. A. V. and PIPER, D. J. W. (eds). Fine-grained sediments: deep-water processes and facies. *Special Publication of the Geological Society, London*, **15**, 611–45.
STRAW, S. H. 1930. The Siluro-Devonian boundary in South Central Wales. *Journal of the Manchester Geological Association*, **1**, 79-102.
—— 1937. The higher Ludlovian rocks of the Builth district. *Quarterly Journal of the Geological Society, London*, **93**, 406–56.
—— 1953. The Silurian succession at Cwm Craig Ddu (Breconshire). *Liverpool and Manchester Geological Journal*, **1**, 208–19.
TAYLOR, G. K. and STRACHAN, R. A. 1990. Palaeomagnetic and tectonic constraints on the development of Avalonian-Cadomian terranes in the North Atlantic region. In STRACHAN, R. A. and TAYLOR, G. K. (eds). *Avalonian and Cadomian geology of the North Atlantic*. Blackie, Glasgow, London, 237–48.
TAYLOR, K. 1973. New fossiliferous localities in the Upper Old Red Sandstone of the Ystradfellte–Cwm Taf district of Breconshire. *Bulletin of the Geological Survey of Great Britain*. **38**, 11–14.
—— and THOMAS, L. P. 1974. Field meeting in South Breconshire. *Proceedings of the Geologists' Association*. **85**, 423–32.
TEMPLE, J. T. 1988. Ordovician–Silurian boundary strata in Wales. In COCKS, L. R. M. and RICKARDS, R. B. (eds). A global analysis of the Ordovician–Silurian boundary. *Bulletin of the British Museum (Natural History)*, Geology, **43**, 65–71.
THOMAS, R. G. 1978. 'The stratigraphy, palynology and sedimentology of the Lower Old Red Sandstone Cosheston Group, S.W. Dyfed, Wales.' Unpublished Ph.D. thesis, University of Bristol, 522pp.
THORPE, R. S., BECKINSALE, R. D., PATCHETT, P. J., PIPER, J. D. A., DAVIES, G. R. and EVANS, J. A. 1984. Crustal growth and late Precambrian–early Palaeozoic plate tectonic growth of England and Wales. *Journal of the Geological Society, London*, **141**, 521–36.
TRAYNOR, J. J. 1988. The Arenig in South Wales: sedimentary and volcanic processes during the initiation of a marginal basin. *Geological Journal*, **23**, 275–92.
—— 1990. Arenig sedimentation and basin tectonics in the Harlech Dome area (Dolgellau Basin), North Wales. *Geological Magazine*, **127**, 13–30.
TRENCH, A., TORSVIK, T. H. and McKERROW, W. S. 1991. The palaeogeographic evolution of southern Britain during Early Palaeozoic times: a reconciliation of palaeomagnetic and biogeographic evidence. *Tectonophysics*.
TUNBRIDGE, I. P. 1981. Old Red Sandstone sedimentation – an example from the Brownstones (highest Lower Old Red Sandstone) of south central Wales. *Geological Journal*, **16**, 111–24.
—— 1986. Mid-Devonian tectonics and sedimentation in the Bristol Channel area. *Journal of the Geological Society, London*, **143**, 107–16.
TUNNICLIFFE, S. P. 1987. Caradocian bivalve molluscs from Wales. *Palaeontology*, **30**, 677–90.

TYLER, J. E. 1987. 'Clastic marine facies in the Ludlow of the Central Welsh region.' Unpublished Ph.D. thesis, University of Cambridge.

—— and WOODCOCK, N. H. 1987. Bailey Hill Formation: Ludlow Series turbidites in the Welsh Borderland reinterpreted as distal storm deposits. *Geological Journal*, **22**, 73–86.

VANNIER, J. M. C., SIVETER, D. J. and SCHALLREUTER, R. E. L. 1989. The composition and palaeogeographical significance of the Ordovician ostracode faunas of southern Britain, Baltoscandia and Ibero-Armorica. *Palaeontology*, **12**, 163–222.

WARREN, P. T., PRICE, D., NUTT, M. J. C. and SMITH, E. G. 1984. Geology of the country around Rhyl and Denbigh. *Memoirs of the British Geological Survey*, Sheets 95 & 107 (England and Wales), x+128pp.

WEDD, C. B., SMITH, B., KING, W. B. R. and WRAY, D. A. 1929. The country around Oswestry. *Memoirs of the Geological Survey*, Sheet 137 (England and Wales), xix+234pp.

WHITHAM, A. G. and SPARKS, R. S. J. 1986. Pumice. *Bulletin of Volcanology*, **48**, 209–33.

WHITTINGTON, H. B. 1938. The geology of the district around Llansantffraid ym Mechain, Montgomeryshire. *Quarterly Journal of the Geological Society, London*, **94**, 423–57.

—— 1962–1968. The Ordovician trilobites of the Bala area, Merioneth. *Monograph of the Palaeontographical Society, London*, Parts 1–4, 138pp. (Publication numbers 497, 504, 512, 520).

WILLIAMS, A. 1963. The Caradocian brachiopod faunas of the Bala district, Merionethshire. *Bulletin of the British Museum (Natural History)*, Geology, **8**, 327–471.

—— STRACHAN, I., BASSETT, D. A., DEAN, W. T., INGHAM, J. K., WRIGHT, A. D. and WHITTINGTON, H. B. 1972. A correlation of Ordovician rocks in the British Isles. *Special Report of the Geological Society, London*, **3**, 1–74.

WILLIAMS, R. A. 1985. *The old mines of the Llangynog district, North Powys, Mid-Wales.* Monograph of the Northern Mine Research, British Mining No. 26, Sheffield, 128pp.

WOOD, E. M. R. 1900. The Lower Ludlow Formation and its graptolite fauna. *Quarterly Journal of the Geological Society of London*, **56**, 415–92, pls 25, 26.

—— 1906. The Tarannon Series of Tarannon. *Quarterly Journal of the Geological Society, London*, **62**, 644–701.

WOODCOCK, N. H. 1973. 'The structure of the slump sheets in the Ludlow Series of east Central Wales.' Unpublished Ph.D. thesis, University of London.

—— 1976a. Structural style in slump sheets: Ludlow Series, Powys, Wales. *Journal of the Geological Society, London*, **132**, 399–415.

—— 1976b. Ludlow Series slumps and turbidites and the form of the Montgomery Trough, Powys, Wales. *Proceedings of the Geologists' Association, London*, **87**, 169–82.

—— 1984a. The Pontesford Lineament, Welsh Borderland. *Journal of the Geological Society, London*, **141**, 1001–14.

—— 1984b. Early Palaeozoic sedimentation and tectonics in Wales. *Proceedings of the Geologists' Association*, **95**, 323–35.

—— 1987a. Structural geology of the Llandovery Series in the type area, Dyfed, Wales. *Geological Journal*, **22**, 199–209.

—— 1987b. Kinematics of strike-slip faulting, Builth Inlier, Mid-Wales. *Journal of Structural Geology*, **9**, 353–63.

—— 1988. Strike-slip faulting along the Church Stretton Lineament, Old Radnor, Welsh Borderland. *Journal of the Geological Society, London*, **145**, 925–33.

—— 1990. Sequence stratigraphy of the Palaeozoic Welsh Basin. *Journal of the Geological Society, London*, **147**, 537–48.

—— AWAN, M. A., JOHNSON, T. E., MACKIE, A. H. and SMITH, R. D. A. 1988. Acadian tectonics of Wales during Avalonia/Laurentia convergence. *Tectonics*, **7**, 483–95.

—— and GIBBONS, W. 1988. Is the Welsh Borderland Fault System a terrane boundary? *Journal of the Geological Society, London*, **145**, 915–23.

—— and PAULEY, J. C. 1989. The Longmyndian rocks of the Old Radnor Inlier, Welsh Borderland. *Geological Journal*, 24, 113–20.

—— and SMALLWOOD, S. 1987. Late Ordovician shallow marine environments due to glacio-eustatic regression: Scrach Formation, Mid-Wales. *Journal of the Geological Society, London*, **144**, 393–400.

WREN, W. J. 1968. *The Tanat Valley: its railways and industrial archaeology*. David and Charles, Newton Abbott, 192pp.

WRIGHT, V. P. 1981a. 'The stratigraphy and sedimentology of the Llanelly Formation between Blorenge and Penderyn, South Wales.' Unpublished Ph.D. thesis, University of Wales.

—— 1981b. Algal aragonite-encrusted pisoids from a Lower Carboniferous schizohaline lagoon. *Journal of Sedimentary Petrology*, **51**, 479–89.

—— 1982a. The recognition and interpretation of palaeokarsts: two examples from the Lower Carboniferous of South Wales. *Journal of Sedimentary Petrology*, **52**, 83–94.

—— 1982b. Calcrete bearing palaesols from the Lower Carboniferous Llanelly Formation, South Wales. *Sedimentary Geology*, **33**, 1–33.

—— 1982c. *Uraloporella* Korde from the Lower Carboniferous of South Wales. *Bulletin of the British Museum (Natural History)*, Geology, **36**, 151–5.

—— 1983a. A rendzina from the Lower Carboniferous of South Wales. *Sedimentology*, **30**, 159–79.

—— 1983b. Morphogenesis of oncoids in the Lower Carboniferous Llanelly Formation, South Wales. In PERYT, T. M. (ed.). *Coated grains*. Springer, Berlin, 424–34.

—— 1986. Facies sequences on a carbonate ramp: the Carboniferous Limestone of South Wales. *Sedimentology*, **33**, 221–41.

—— 1987. The evolution of the early Carboniferous Limestone province in southwest Britain. *Geological Magazine*, **124**, 477–80.

—— 1987. Equatorial aridity and climatic oscillations during the early Carboniferous, southern Britain. *Journal of the Geological Society, London*, **147**, 359–63.

—— 1988. Paleokarsts and paleosols as indicators of paleo-climate and porosity evolution. In JAMES, N. P. and CHOQUETTE, P. W. (eds). *Paleokarst*. Springer, New York, 149–63.

—— 1990. Equatorial aridity and climatic oscillation during the early Carboniferous. *Journal of the Geological Society, London*, **147**, 359–63.

—— and ROBINSON, D. 1988. Early Carboniferous floodplain deposits from South Wales: a case study of the controls on paleosol development. *Journal of the Geological Society, London*, **145**, 847–857.

—— and WRIGHT, E. V. G. 1981. The palaeoecology of some algal-gastropod bioherms in the Lower Carboniferous of South Wales. *Neues Jahrbuch für Geologie und Paläontologie, Monatshefte*, **9**, 546–58.

—— and WRIGHT, J. M. 1985. A stromatolite built by a Phormidium-like alga from the Lower Carboniferous of South Wales. In NITECKI, M. and TOOMEY, D. F. (eds). *Paleoalgology*. Springer, Berlin, 40–54.

ZIEGLER, A. M., COCKS, L. R. M. and McKERROW, W. S. 1968. The Llandovery transgression of the Welsh Borderland. *Palaeontology*, **11**, 736–82.

—— and McKERROW, W. S. 1975. Silurian marine red beds. *American Journal of Science*, **275**, 31–56.

LIST OF CONTRIBUTORS

Dr J. ALMOND, Shell U.K. Exploration and Production, Shell-Mex House, The Strand, London WC2R 0DX

Professor M. G. BASSETT, Department of Geology, National Museum of Wales, Cathays Park, Cardiff CF1 3NP

Dr R. E. BEVINS, Department of Geology, National Museum of Wales, Cathays Park, Cardiff CF1 3NP

Dr P. J. BRENCHLEY, Department of Earth Sciences, University of Liverpool, Brownlow Street, PO Box 147, Liverpool L69 3BX

Dr R. CAVE, Institute of Earth Studies, University College of Wales, Llandinam Building, Aberystwyth, Dyfed SY23 3DB

Dr L. CHERNS, Department of Geology, University of Wales College of Cardiff, PO Box 914, Cardiff CF1 3YE

Dr J. R. DAVIES, British Geological Survey, Bryn Eithyn Hall, Llanfarian, Aberystwyth, Dyfed SY23 4BY

Dr J. A. D. DICKSON, Department of Earth Sciences, University of Cambridge, Downing Street, Cambridge CB2 3EQ

Dr A. J. DIMBERLINE, Department of Earth Sciences, University of Cambridge, Downing Street, Cambridge CB2 3EQ

Dr R. J. DIXON, Department of Geology, University of Wales College of Cardiff, PO Box 914, Cardiff CF1 3YE *Present Address* BP Exploration, 301 St. Vincent Street, Glasgow G2 5DD

Dr C. J. N. FLETCHER, British Geological Survey, Bryn Eithyn Hall, Llanfarian, Aberystwyth, Dyfed SY23 4BY

Dr B. A. HAINS, British Geological Survey, Bryn Eithyn Hall, Llanfarian, Aberystwyth, Dyfed SY23 4BY

Dr M. J. LENG, Institute of Earth Studies, University College of Wales, Llandinam Building, Aberystwyth, Dyfed SY23 3DB *Present address* NIGL, c/o British Geological Survey, Keyworth, Nottingham NG12 5GG

Dr A. H. MACKIE, Department of Earth Sciences, University of Cambridge, Downing Street, Cambridge CB2 3EQ

Dr R. METCALFE, Department of Geology, Wills Memorial Building, Queens Road, Bristol BS8 1RJ *Present address* British Geological Survey, Keyworth, Nottingham NG12 5GG

Dr J. E. TYLER, Department of Earth Sciences, University of Cambridge, Downing Street, Cambridge CB2 3EQ

Dr R. A. WATERS, British Geological Survey, Bryn Eithyn Hall, Llanfarian, Aberystwyth, Dyfed SY23 4BY

Dr D. E. WHITE, British Geological Survey, Keyworth, Nottingham NG12 5GG *Present address* 115 Vale Road, Chesham, Bucks HP5 3HP

Professor B. P. J. WILLIAMS, Department of Geology and Petroleum Geology, University of Aberdeen, Meston Building, King's College, Aberdeen AB9 2UE

Dr D. WILSON, British Geological Survey, Bryn Eithyn Hall, Llanfarian, Aberystwyth, Dyfed SY23 4BY

Dr N. H. WOODCOCK, Department of Earth Sciences, University of Cambridge, Downing Street, Cambridge CB2 3EQ

Dr V. P. WRIGHT, Postgraduate Research Institute for Sedimentology, University of Reading, PO Box 227, Whiteknights, Reading, RG6 2AB

Dr J. A. ZALASIEWICZ, British Geological Survey, Keyworth, Nottingham NG12 5GG